Mechanics of Flight

Mechanics of Flight

11th EDITION

A. C. KERMODE CBE, MA, CEng, FRAeS

Revised by
R. H. BARNARD PhD, CEng, FRAeS
and D. R. PHILPOTT PhD, CEng, MRAes, MAIAA

PEARSON
Education

ISBN 978-81-317-1657-1

First Impression, 2007
Second Impression, 2008
Third Impression, 2009

Published by Dorling Kindersley (India) Pvt. Ltd., licensees of Pearson Education in South Asia.

Head Office: 482, F.I.E., Patparganj, Delhi 110 092, India.
Registered Office: 14 Local Shopping Centre, Panchsheel Park,
New Delhi 110 017, India.

Printed in India by India Binding House.

Contents

Preface to Eleventh Edition		vi
Acknowledgements		vii
Chapter 1	Mechanics	1
Chapter 2	Air and airflow – subsonic speeds	28
Chapter 3	Aerofoils – subsonic speeds	67
Chapter 4	Thrust	120
Chapter 5	Level flight	147
Chapter 6	Gliding and landing	180
Chapter 7	Performance	215
Chapter 8	Manoeuvres	230
Chapter 9	Stability and control	257
Chapter 10	A trial flight	287
Chapter 11	Flight at transonic speeds	297
Chapter 12	Flight at supersonic speeds	342
Chapter 13	Space flight	377

Appendixes

1 Aerofoil data	412
2 Scale effect and Reynolds Number	431
3 Numerical questions	435
4 Answers to numerical questions	471
5 Answers to non-numerical questions	479

Index 491

Preface to eleventh edition

The lasting popularity of this classic book is aptly demonstrated by the fact that this is the eleventh edition. This is also the third time that the current reviewers have undertaken the task of updating it, and we hope that the changes will be as well received this time as previously.

It would be unreasonable to try to include details of all recent developments, and furthermore, we wanted to retain as much as possible of the practical detail that Kermode supplied. This detail nowadays relates mostly to light general aviation and initial training aircraft, of the type that will be encountered by anyone who wishes to learn to fly. However, transonic, supersonic and even space flight are given their place.

The late A. C. Kermode was a high-ranking Royal Air Force officer responsible for training. He also had a vast accumulation of practical aeronautical experience, both in the air and on the ground. It is this direct knowledge that provided the strength and authority of his book.

Most chapters have some simple non-numerical questions that are intended to test students' undertstanding, and our answers to these are provided. There are also numerical questions and solutions for each chapter. For engineering and basic scientific questions we have used the SI unit system, but aircraft operations are an international subject, and anyone involved in the practical business will need to be familiar with the fact that heights are always given in feet, and speeds in knots. We have therefore retained several appropriate qestions where these units are involved.

R. H. Barnard
D. R. Philpott

Acknowledgements

We are grateful to the following for permission to reproduce copyright material:

Figures 1B, 2G, 2E, 8D courtesy of the Lockheed Aircraft Corporation, USA; Figures 2B, 2C, 3A, 3B, 6B, 11B, 12D courtesy of the former British Aircraft Corporation; Figures 3C, 6E, 9F, 13B courtesy of General Dynamics Corporation, USA; Figure 3D courtesy of Paul MacCready; Figures 3E, 5H courtesy of the Grumman Corporation, USA; Figure 3F courtesy of Fiat Aviazione, Torino, Italy; Figures 4D, 13D, courtesy of the Bell Aerospace Division of Textron Inc., USA; Figure 4G courtesy of Beech Aircraft Corporation, USA; Figures 4H, 5B courtesy of Cessna Aircraft Company, USA; Figure 4I courtesy of the former Fairey Aviation Co. Ltd; Figures 5C, 8C courtesy of *Flight*; Figure 6A courtesy of Slingsby Sailplanes Ltd; Figure 6C (bottom) courtesy of Terry Shwetz, de Havilland, Canada; Figure 6F courtesy of Bell Helicopter Textron; Figure 6G courtesy of Nigel Cogger; Figures 7C, 13C courtesy of the Boeing Company; Figure 9H courtesy of SAAB, Sweden; Figure 9I courtesy of Piaggio, Genoa, Italy; Figure 11A courtesy of the Shell Petroleum Co. Ltd; Figure 11E courtesy of McDonnell Douglas Corporation, USA; Figure 12A courtesy of the Lockheed-California Company, USA; Figure 12B courtesy of British Aerospace Defence Ltd, Military Aircraft Division; Figure 12C courtesy of Avions Marcel Dassault, France; Figures 13A, 13E courtesy of NASA.

Quotation from *The Stars in their Courses* on p.391 (Sir James Jeans) reprinted courtesy of Cambridge University Press.

CHAPTER 1

Mechanics

Flying and mechanics

The flight and manoeuvres of an aeroplane provide glorious examples of the principles of mechanics. However, this is not a book on mechanics. It is about flying, and is an attempt to explain the flight of an aeroplane in a simple and interesting way; the mechanics are only brought in as an aid to understanding. In the opening chapter I shall try to sum up some of the principles with which we are most concerned in flying.

Force, and the first law of motion

An important principle of mechanics is that any object that is at rest will stay at rest unless acted upon by some force, and any object that is moving will continue moving at a steady speed unless acted upon by a force. This statement is in effect a simple statement of what is known as Newton's First Law of Motion.

There are two types of forces that can act on a body. They are:

(1) externally applied mechanical forces such as a simple push or pull

(2) the so-called body forces such as those caused by the attraction of gravity and electromagnetic and electrostatic fields.

External forces relevant to the mechanics of flight include the **thrust** produced by a jet engine or a propeller, and the **drag** resistance produced by movement through the air. A less obvious external force is that of **reaction**. A simple example of a reactive force is that which occurs when an object is placed on a fixed surface. The table produces an upward reactive force that exactly balances the weight. The only body force that is of interest in the mechanics of flight is the force due to the attraction of gravity, which we know simply as the weight of the object.

Forces (of whatever type) are measured in the units of newtons (N) in the metric SI system or pounds force (lbf) in the Imperial or Federal systems. In this book, both sets of units are used in the examples and questions.

Mass

The mass of an object can be loosely described as the quantity of matter in it. The greater the mass of an object, the greater will be the force required to start it moving from rest or to change its speed if it is already moving.

Mass is measured in units of kilograms (kg) in the SI metric system or pounds (lb) in the Imperial and Federal systems. Unfortunately, the same names are commonly used for the units of weight (which is a force), and this causes a great deal of confusion, as will be explained a little later under the heading Units. In this book, we will always use kilograms for mass, and newtons for weight.

Momentum

The quantity that decides the difficulty in stopping a body is its momentum, which is the product of its mass and the velocity of movement.

A body having a 20 kg mass moving at 2 m/s has a momentum of 40 kg m/s, and so does a body having a 10 kg mass moving at 4 m/s. The first has the greater mass, the second the greater velocity, but both are equally difficult to stop. A car has a larger mass than a bullet, but a relatively low velocity. A bullet has a much lower mass, but a relatively high velocity. Both are difficult to stop, and both can do considerable damage to anything that tries to stop them quickly.

To change the momentum of a body or even a mass of air, it is necessary to apply a force. Force = **Rate of change of momentum.**

Forces in equilibrium

If two tug-of-war teams pulling on a rope are well matched, there may for a while be no movement, just a lot of shouting and puffing! Both teams are exerting the same amount of force on the two ends of the rope. The forces are therefore in equilibrium and there is no change of momentum. There are,

Fig 1.1 Forces in equilibrium

however, other more common occurrences of forces in equilibrium. If you push down on an object at rest on a table, the table will resist the force with an equal and opposite force of reaction, so the forces are in equilibrium. Of course, if you press too hard, the table might break, in which case the forces will no longer be in equilibrium, and a sudden and unwanted acceleration will occur.

As another example, consider a glider being towed behind a small aircraft as in Fig. 1.1. If the aircraft and glider are flying straight and level at constant speed, then the pulling force exerted by the aircraft on the tow-rope must be exactly balanced by an equal and opposite aerodynamic resistance or drag force acting on the glider. The forces are in equilibrium.

Some people find it hard to believe that these forces really are exactly equal. Surely, they say, the aircraft must be pulling forward just a bit harder than the glider is pulling backwards; otherwise, what makes them go forward? Well, what makes them go forward is the fact that they are going forward, and the law says that they will continue to do so unless there is something to alter that state of affairs. If the forces are balanced then there is nothing to alter that state of equilibrium, and the aircraft and glider will keep moving at a constant speed.

Forces not in equilibrium

In the case of the glider mentioned above, what would happen if the pilot of the towing aircraft suddenly opened the engine throttle? The pulling force on the tow-rope would increase, but at first the aerodynamic resistance on the glider would not change. The forces would therefore no longer be in equilibrium. The air resistance force is still there of course, so some of the pull on the tow-rope must go into overcoming it, but the remainder of the force will cause the glider to accelerate as shown in Fig. 1.2 (overleaf), which is called a free-body diagram.

This brings us to **Newton's second law,** which says in effect that if the forces are not in balance, then the acceleration will be proportional to force and inversely proportional to the mass of the object:

$$a = F/m$$

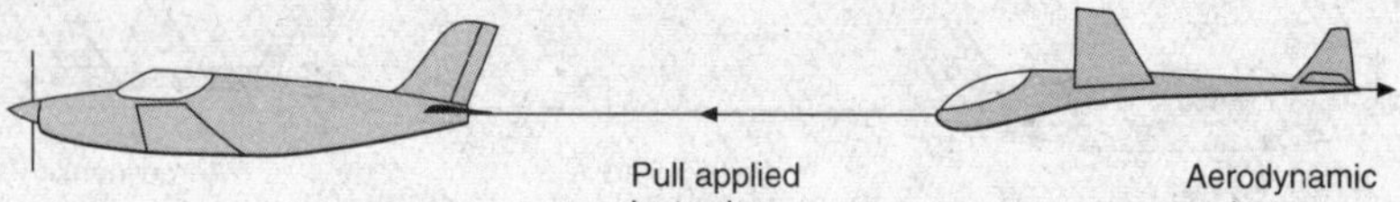

Fig 1.2 Forces not in equilibrium

where a is the acceleration, m is the mass of the body, and F is the force. This relationship is more familiarly written as:

$$F = m \times a$$

Inertia forces

In the above example, of the accelerating glider, the force applied to one end of the rope by the aircraft is greater than the air resistance acting on the glider at the other end. As far as the rope is concerned, however, the force it must apply to the glider tow-hook must be equal to the air resistance force plus the force required to accelerate the glider. In other words, the forces on the two ends of the rope are in equilibrium (as long as we ignore the mass of the rope). The extra force that the rope has to apply to produce the acceleration is called an **inertia force**.

As far as the rope is concerned, it does not matter whether the force at its far end is caused by tying it to a wall to create a reaction or by attaching it to a glider which it is causing to accelerate, the effect is the same – it feels an equal and opposite pull at the two ends. From the point of view of the glider, however, the situation is very different; if there were a force equal and opposite to the pull from the rope, no acceleration would take place. The forces on the glider are not in equilibrium.

Great care has to be taken in applying the concept of an inertia force. When considering the stresses in the tow-rope it is acceptable to apply the pulling force at one end, and an equal and opposite force at the other end due to the air resistance plus the inertia of the object that it is causing to accelerate. When considering the motion of the aircraft and glider, however, no balancing inertia force should be included, or there would be no acceleration. A free-body diagram should be drawn as in Fig. 1.2.

This brings us to the much misunderstood **third law of Newton**: to every action there is an equal and opposite reaction. If a book rests on a table then the table produces a reaction force that is equal and opposite to the weight force. However, be careful; the force which is accelerating the glider produces a reaction, but the reaction is not a force, but an acceleration of the glider.

Weight

There is one particular force that we are all familiar with; it is known as the force due to gravity. We all know that any object placed near the earth is attracted towards it. What is perhaps less well known is that this is a mutual attraction like magnetism. The earth is attracted towards the object with just as great a force as the object is attracted towards the earth.

All objects are mutually attracted towards each other. The force depends on the masses of the two bodies and the distance between them, and is given by the expression:

$$F = \frac{Gm_1m_2}{d^2}$$

where G is a constant which has the value 6.67×10^{-11} N m²/kg², m_1 and m_2 are the masses of the two objects, and d is the distance between them. Using the above formula you can easily calculate the force of attraction between two one kilogram masses placed one metre apart. You will see that it is very small. If one of the masses is the earth, however, the force of attraction becomes large, and it is this force that we call the force of gravity. In most practical problems in aeronautics, the objects that we consider will be on or relatively close to the surface of the earth, so the distance d is constant, and as the mass of the earth is also constant, we can reduce the formula above to a simpler one:

$$F = m \times g$$

Fig 1A Weight and thrust The massive Antonov An-255 Mriya, with a maximum take-off weight of 5886 kN (600 tonnes). The six Soloviev D-18T turbofans deliver a total maximum thrust of 1377 kN.

where m is the mass of the object and g is a constant called the gravity constant which takes account of the mass of the earth and its radius. It has the value 9.81 m/s^2 in the SI system, or 32 ft/s^2 in the Imperial or Federal systems.

The force in the above expression is what we know as **weight.** Weight is the force with which an object is attracted towards the centre of the earth. In fact g is not really a constant because the earth is not an exact sphere, and large chunks of very dense rock near the surface can cause the force of attraction to increase slightly locally. For most practical aeronautical calculations we can ignore such niceties. We cannot, however, use this simple formula once we start looking at spacecraft or high-altitude missiles.

Weight is an example of what is known as a **body force.** Body forces unlike mechanical forces have no visible direct means of application. Other examples of body forces are electrostatic and electromagnetic forces.

When an aircraft is in steady level flight, there are two vertical forces acting on it, as shown in Fig. 1.3. There is an externally applied force, the lift force provided by the air flowing over the wing, and a body force, the weight.

The acceleration due to gravity

All objects near the surface of the earth have the force of gravity acting on them. If there is no opposing force, then they will start to move, to accelerate. The rate at which they accelerate is independent of their mass.

The force due to gravity (weight) $F = m \times g$

but, from Newton's second law, $F = m \times acceleration$

By equating the two expressions above, we can see that the acceleration due to gravity will be numerically equal to the gravity constant g, and will be independent of the mass. Not surprisingly, many people confuse the two terms 'gravity constant' and 'acceleration due to gravity', and think that they are the same thing. The numerical value is the same, but they are different things. If a book rests on a table, then the weight is given by the product of the gravity

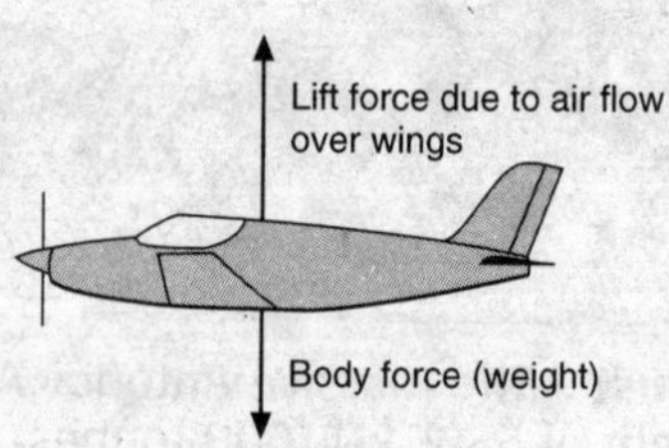

Fig 1.3 Aerodynamic and body forces

constant and the mass, but it is not accelerating. If it falls off the table, it will then accelerate at a rate equal to the value of the gravity constant.

This brings us to the old problem of the feather and the lump of lead; which will fall fastest? Well, the answer is that in the vacuum of space, they would both fall at the same rate. In the atmosphere, however, the feather would be subjected to a much larger aerodynamic resistance force in relation to the accelerating gravity force (the weight), and therefore the feather would fall more slowly.

For all objects falling through the atmosphere, there is a speed at which the aerodynamic resistance is equal to the weight, so they will then cease to accelerate. This speed is called the **terminal velocity** and will depend on the shape, the density and the orientation of the object. A man will fall faster head first than if he can fall flat. Free-fall sky-divers use this latter effect to control their rate of descent in free fall.

Mass weight and *g*

The mass of a body depends on the amount of matter in it, and it will not vary with its position on the earth, nor will it be any different if we place it on the moon. The weight (the force due to gravity) will change, however, because the so-called gravity constant will be different on the moon, due to the smaller mass of the moon, and will even vary slightly between different points on the earth. Also, therefore, the rate at which a falling object accelerates will be different. On the moon it will fall noticeably slower, as can be observed in the apparently slow-motion moon-walking antics of the Apollo astronauts.

Units

The system of units that we use to measure quantities, feet, metres, etc., can be a great source of confusion. In European educational establishments and most of its industry, a special form of the metric system known as the **Système International** or **SI** is now in general use. The basic units of this system are the kilogram for mass (not weight) (kg), the metre for distance (m) and the second for time (s).

Temperatures are in degrees Celsius (or Centigrade) (°C) when measured relative to the freezing point of water, or in Kelvin (K) when measured relative to absolute zero; 0°C is equivalent to 273 K. A temperature change of one degree Centigrade is exactly the same as a change of one degree Kelvin, it is just the starting or zero point that is different. Note that the degree symbol ° is not used when temperatures are written in degrees Kelvin, for example we write 273 K.

Forces and hence weights are in newtons (N) not kilograms. Beware of weights quoted in kilograms; in the old (pre-SI) metric system still commonly used in parts of Europe, the name kilogram was also used for weight or force. To convert weights given in kilograms to newtons, simply multiply by 9.81.

The SI system is known as a coherent system, which effectively means that you can put the values into formulae without having to worry about conversion factors. For example, in the expression relating force to mass and acceleration: $F = m \times a$, we find that a force of 1 newton acting on a mass of 1 kilogram produces an acceleration of 1 m/s^2. Contrast this with a version of the old British 'Imperial' system where a force of 1 pound acting on a mass of 1 pound produces an acceleration of 32.18 ft/sec^2. You can imagine the problems that the latter system produces. Notice how in this system, the same name, the pound, is used for two different things, force and mass.

Because aviation is dominated by American influence, American Federal units and the similar Imperial (British) units are still in widespread use. Apart from the problem of having no internationally agreed standard, the use of Federal or Imperial units can cause confusion, because there are several alternative units within the system. In particular, there are two alternative units for mass, the pound mass, and the slug (which is equivalent to 32.18 pounds mass). The slug may be unfamiliar to most readers, but it is commonly used in aeronautical engineering because, as with the SI units, it produces a coherent system. A force of 1 pound acting on a mass of one slug produces an acceleration of 1 ft/sec^2. The other two basic units in this system are, as you may have noticed, the foot and the second. Temperatures are measured in degrees Fahrenheit.

You may find all this rather confusing, but to make matters worse, in order to avoid dangerous mistakes, international navigation and aircraft operations conventions use the foot for altitude, and the knot for speed. The knot is a nautical mile per hour (0.5145 m/s). A nautical mile is longer than a land mile, being 6080 feet instead of 5280 feet. Just to add a final blow, baggage is normally weighed in kilograms (not even newtons)!

To help the reader, most of the problems and examples in this book are in SI units. If you are presented with unfamiliar units or mixtures of units, convert them to SI units first, and then work in SI units. One final tip is that when working out problems, it is always better to use basic units, so convert millimetres or kilometres to metres before applying any formulae. In the real world of aviation, you will have to get used to dealing with other units such as slugs and knots, but let us take one step at a time. Below, we give a simple example of a calculation using SI units (see Example 1.1).

EXAMPLE 1.1

The mass of an aeroplane is 2000 kg. What force, in addition to that required to overcome friction and air resistance, will be needed to give it an acceleration of 2 m/s^2 during take-off?

SOLUTION

Force = ma
= 2000 × 2
= 4000 newtons

This shows how easy is the solution of such problems if we use the **SI** units.

Many numerical examples on the relationship between forces and masses involve also the principles of simple kinematics, and the reader who is not familiar with these should read the next paragraph before he tackles the examples.

Kinematics

It will help us in working examples if we summarise the relations which apply in kinematics, that is, the study of the movement of bodies irrespective of the forces acting upon them.

We shall consider only the two simple cases, those of uniform velocity and uniform acceleration.

Symbols and units will be as follows –

Time = t (sec)
Distance = s (metres)
Velocity (initial) = u (metres per sec)
Velocity (final) = v (metres per sec)
Acceleration = a (metres per sec per sec)

Uniform velocity

If velocity is uniform at u metres per sec clearly

Distance travelled = Velocity × Time
or $\boldsymbol{s = ut}$

Uniform acceleration

Final velocity = Initial velocity + Increase of velocity

or $\boldsymbol{v = u + at}$

Distance travelled = Initial velocity × Time
+ $\frac{1}{2}$ Acceleration × Time squared

i.e. $s = ut + \frac{1}{2}at^2$

Final velocity squared = Initial velocity squared + 2 × Acceleration × Distance

or $v^2 = u^2 + 2as$

With the aid of these simple formulae – all of which are founded on first principles – it is easy to work out problems of uniform velocity or uniform acceleration (see Examples 1.2 to 1.4).

EXAMPLE 1.2

If, during a take-off run an aeroplane starting from rest attains a velocity of 90 km/h in 10 seconds, what is the average acceleration?

SOLUTION

Initial velocity $u = 0$

Final velocity $v = 90$ km/h = 25 m/s

Time $t = 10$ sec

$a = ?$

Since we are concerned with u, v, t and a, we use the formula

$v = u + at$

$25 = 0 + 10a$

$a = 25/10 = \mathbf{2.5\ m/s^2}$

EXAMPLE 1.3

How far will the aeroplane of the previous example have travelled during the take-off run?

SOLUTION

$u = 0$, $v = 25$ m/s, $t = 10$ sec, $a = 2.5$ m/s^2

To find s, we can either use the formula

Final velocity squared = Initial velocity squared + 2 × Acceleration × Distance

$$s = ut + \tfrac{1}{2}at^2$$
$$= 0 + \tfrac{1}{2} \times 2.5 \times 10^2$$
$$= \mathbf{125\ m}$$

$$\text{or } v^2 = u^2 + 2as$$
$$25 \times 25 = 0 + 2 \times 2.5 \times s$$
$$\therefore s = (25 \times 25)/(2 \times 2.5)$$
$$= \mathbf{125\ m}$$

EXAMPLE 1.4

A bomb is dropped from an aeroplane which is in level flight at 200 knots at a height of 3500 m. Neglecting the effect of air resistance, how long will it be before the bomb strikes the ground, and how far horizontally before the target must the bomb be released?

SOLUTION

To find the time of fall we are concerned only with the vertical velocity, which was zero at release.

$\therefore u = 0$

a = acceleration of gravity = 9.81 m/s^2

s = vertical distance from aeroplane to ground = 3500 m

t = ?

We need the formula connecting u, a, s and t, i.e.

$$s = ut + \tfrac{1}{2}at^2$$

$$\therefore 3500 = 0 + \tfrac{1}{2} \times 9.81 \times t^2$$

$$\therefore t^2 = (3500/9.81) \times 2 = 713$$

$$\therefore \boldsymbol{t = 27\ \text{sec}}\ \text{(approx)}$$

Since we are neglecting the effect of air resistance, the horizontal velocity of the bomb will, throughout the fall, remain the same as it was at the moment of release, i.e. the same as the velocity of the aeroplane, namely 200 knots or, converting into metres per second, (200 × 1852)/3600 = **103 m/s** (approx).

Therefore the distance that the bomb will travel forward during the falling time of 27 *s* will be 103 × 27 = **2781 m.**

This, of course, is the distance before the target that the bomb must be released.

Note that in Example 1.4 we have neglected air resistance. Since we are interested in flying this may seem rather a silly thing to do, because we are only able to fly by making use of the same principles that are responsible for air resistance. In fact, too, the effects of air resistance on bombs are of vital importance and are always taken into account when bombing. But it is better to learn things in their most simple form first, then gradually to add the complications. As these complications are added we get nearer and nearer to the truth, but if we are faced with them all at once the picture becomes blurred and the fundamental principles involved fail to stand out clearly.

Other examples on kinematics will be found in Appendix 3, and the reader who is not familiar with examples of this type is advised to work through them.

Motion on curved paths

It has already been emphasised that bodies tend to continue in the same state of motion, and that this involves direction as well as speed. It is clear, therefore, that **if we wish to make a body change its motion by turning a corner or travelling on a curved path, we must apply a force to it in order to make it do so,** and that this will apply even if the speed of the body does not change. This is a force exactly similar to the one that is required to accelerate an aircraft, that is to say: **the force must be proportional to the mass of the body and to the acceleration which it is desired to produce.** But what is the acceleration of a body that is going round a corner? Is there, in fact, any acceleration at all if the speed remains constant? And in what direction is the acceleration?

Let us deal with the last question first. There is another part of Newton's second law which has not so far been mentioned, namely that **the rate of change of momentum of the body will be in the direction of the applied force.** If the mass of the body does not change as it goes round the corner the acceleration must be in the direction of the force. But is there any acceleration if the speed does not change? Yes – because velocity is what we call a vector quantity, that is to say, it has both magnitude and direction, while speed has only magnitude. Thus if the direction of motion changes, the velocity changes even though the speed remains unaltered. But at what rate does the velocity change? – in other words, what is the acceleration? and in what direction is it?

Centripetal force and centripetal acceleration

We all know the direction of the force as a result of practical experience. Swing a stone round on the end of a piece of string. In what direction does the string pull on the stone to keep it on its circular path? Why, towards the centre of the circle, of course, and since force and acceleration are in the same direction, **the acceleration must also be towards the centre.**

We know too that the greater the velocity of the stone, and the smaller the radius of the circle on which it travels, the greater is the pull in the string, and therefore the greater the acceleration. The acceleration is actually given by the simple formula v^2/r, where v is the velocity of the body and r the radius of the circle.

The force towards the centre is called centripetal force (centre-seeking force), and will be equal to the mass of the body × the centripetal acceleration, i.e. to $m \times v^2/r$ (Fig. 1.4).

We have made no attempt to prove that the acceleration is v^2/r – the proof will be found in any textbook on mechanics – but since it is not easy to conceive of an acceleration towards the centre as so many metres per second per second when the body never gets any nearer to the centre, it may help if we

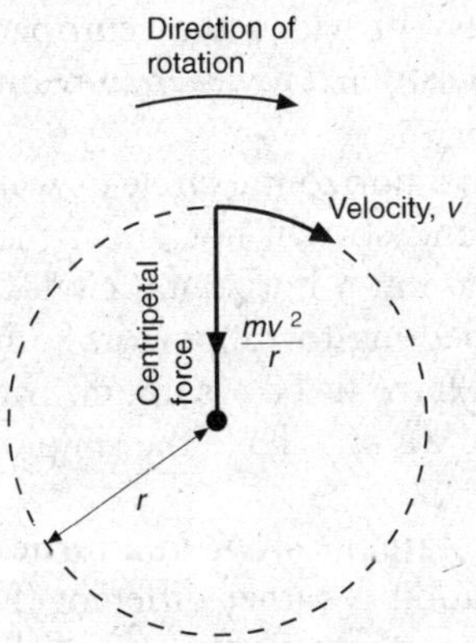

Fig 1.4 Centripetal force

translate the algebraic expression into some actual figures. Taking the simple example of a stone on the end of a piece of string, if the stone is whirled round so as to make one revolution per second, and the length of the string is 1 metre, the distance travelled by the stone per second will be $2\pi r$, i.e. $2\pi \times 1$ or 6.28 m. Therefore

$$v = 6.28 \text{ m/s}, \; r = 1 \text{ m}$$

$$\begin{aligned} \therefore \text{acceleration towards centre} &= v^2/r \\ &= (6.28 \times 6.28)/1 \\ &= \mathbf{39.5\ m/s^2}\ \text{(approx)} \end{aligned}$$

Notice that this is nearly four times the acceleration of gravity, or nearly $4g$. Since we are only using this example as an illustration of principles, let us simplify matters by assuming that the answer **is** $4g$, i.e. 39.24 m/s^2.

This means that the velocity of the stone towards the centre is changing at a rate 4 times as great as that of a falling body. Yet it never gets any nearer to the centre! No, but what would have happened to the stone if it had not been attached to the string? It would have obeyed the tendency to go straight on, and in so doing would have departed farther and farther from the centre.

What centripetal force will be required to produce this acceleration of $4g$? The mass of the stone $\times\ 4g$.

So, if the mass is 1/2 kg, the centripetal force will be $1/2 \times 4g = 2 \times 9.81 = 19.62$, say 20 newtons.

Therefore the pull in the string is 20 N in order to give the mass of 1/2 kg an acceleration of $4g$.

Notice that the **force is 20 newtons**, the **acceleration is 4 g**. There is a horrible tendency to talk about '*g*' as if it were a force; it is not, it is an acceleration.

Now this is all very easy provided the centripetal force is the only force acting upon the mass of the stone. However, in reality there must be a force of gravity acting upon it.

If the stone is rotating in a horizontal circle its weight will act at right angles to the pull in the string, and so will not affect the centripetal force. But of course a stone cannot rotate in a horizontal circle, with the string also horizontal, unless there is something to support it. So let us imagine the mass to be on a table – but it will have to be a smooth, frictionless table or we shall introduce yet more forces. We now have the simple state of affairs illustrated in Fig. 1.5.

Now suppose that we rotate the stone in a **vertical** circle, like an aeroplane looping the loop, the situation is rather different (Fig. 1.6). Even if the stone were not rotating, but just hanging on the end of the string, there would be a tension in the string, due to its weight, and this as near as matters would be very roughly 5 newtons, for a mass of 1/2 kg. If it must rotate with an acceleration of 4g the string must also provide a centripetal force of 20 newtons. So when the stone is at the bottom of the circle, D, the total pull in the string will be 25 N. When the stone is in the top position, C, its own weight will act towards the centre and this will provide 5 N, so the string need only pull with an additional 15 N to produce the total of 20 N for the acceleration of 4g. At the side positions, A and B, the weight of the stone acts at right angles to the string and the pull in the string will be 20 N.

To sum up: the pull in the string varies between 15 N and 25 N, but the acceleration is all the time 4g and, of course, the centripetal force is all the time −20 N. From the practical point of view, what matters most is the pull in the string, which is obviously most likely to break when the stone is in position D and the tension is at the maximum value of 25 N.

To complicate the issue somewhat, suppose the stone rotates in a horizontal circle, but relies on the pull of the string to hold it up (Fig. 1.7), and that the string has been lengthened so that the radius on which the stone is rotating is

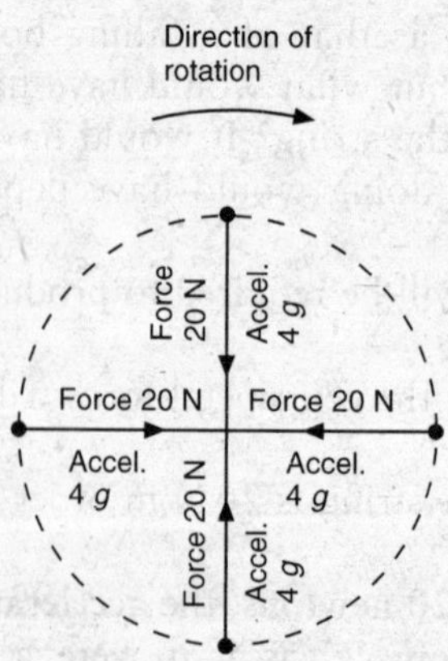

Fig 1.5 Stone rotating in a horizontal circle, supported on a table

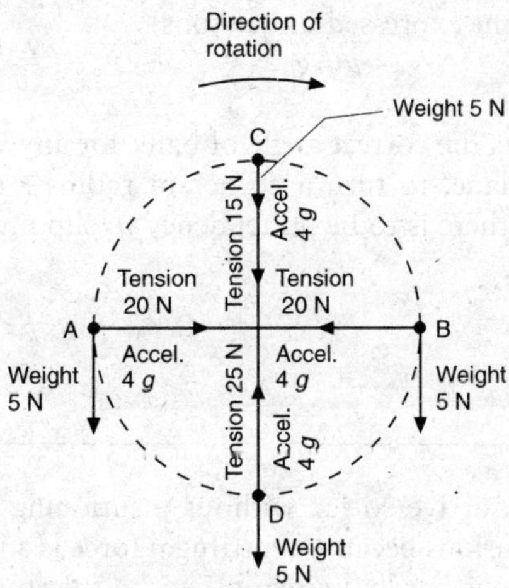

Fig 1.6 Stone rotating in a vertical circle

still 1 metre. The string cannot of course be horizontal since the pull in it must do two things – **support the weight of the stone** and **provide the centripetal force.**

Here we must introduce a new principle.

A force of 5 N, vertically, is required to support the weight.

A force of 20 N, horizontally, is required to provide the centripetal force.

Now five plus twenty does not always make twenty-five! It does not in this example, and for the simple reason that they are not pulling in the same direction. We must therefore represent them by vectors (Fig. 1.7), and the diagonal will represent the total force which, by Pythagoras' Theorem, will be

$$\sqrt{(20^2 + 5^2)} = \sqrt{425} = 20.6 \text{ N}$$

The tangent of the angle of the string to the vertical will be 20/5 = 4.0. So the angle will be approx 76°. Expressing the angle, θ, in symbols –

$$\tan\theta = \frac{\text{Centripetal force}}{\text{Weight}} = (m \times v^2/r)/W$$

$$= (m \times v^2/r)/mg$$

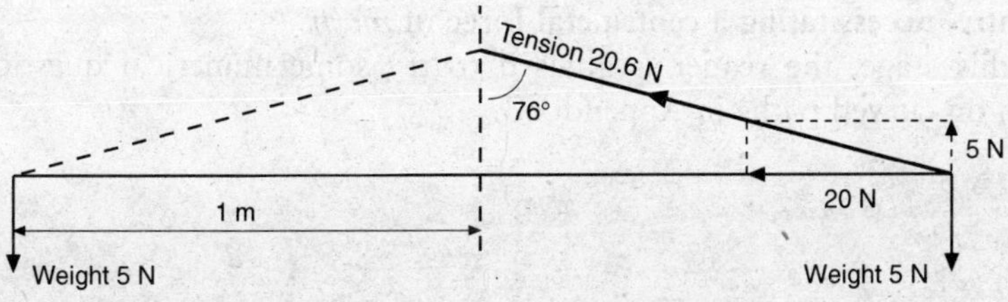

Fig 1.7 Stone rotating in a horizontal circle, with string support

(*mg* being the weight expressed in newtons)

$$= v^2/rg$$

This angle θ represents the correct angle of bank for any vehicle, whether it be bicycle, car or aeroplane, to turn a corner of radius *r* metres, at velocity *v* metres per second, if there is to be no tendency to slip inwards or to skid outwards.

Centrifugal force

We have managed to arrive so far without mentioning the term **centrifugal force.** This is rather curious because centrifugal force is a term in everyday use, while centripetal force is hardly known except to the student of mechanics.

Consider again the stone rotating, on a table, in a horizontal circle. We have established the fact that there is an **inward** force on the stone, **exerted by the string,** for the set purpose of providing the acceleration **towards the centre** – yes, centripetal force, however unknown it may be, is a real, practical, physical force. But is there also an outward force?

The situation is similar to that of the accelerating aircraft towing a glider that we described earlier. There is an outward reaction force on the outer end of the **string** caused by the fact that it is accelerating the stone inwards: an inertia force, and we could call this a centrifugal reaction force. This keeps the string in tension in a state of equilibrium just as if it were tied to a wall and pulled. Note, however, that **there is no outward force on the stone,** only an inward one applied by the string to produce the centripetal acceleration. As with the accelerating glider described previously and shown in Fig. 1.2, the forces on the two ends of the string are in balance, but the forces on the object, the stone or the glider are not, and hence acceleration occurs.

The concept of inertia forces is a difficult one. In a free-body diagram of the horizontally whirling stone, the only externally applied horizontal force is the inward force applied by the string. This force provides the necessary acceleration. There may be outward forces on the internal components of the system like the string, but not on the overall system. Note that if you let go of the string, the stone will not fly outwards, it will fly off at a tangent.

To sum up motion on curved paths. **There is an acceleration (v^2/r) towards the centre, necessitating a centripetal force of mv^2/r.**

At this stage, the reader is advised to try some numerical questions on motion on curved paths in Appendix 3.

The mechanics of flight

A knowledge of the principles of mechanics – and particularly of the significance and meaning of the force of gravity, of accelerated motion, of centripetal and centrifugal force and motion on curved paths – will help us to understand the movements and manoeuvres of an aeroplane. This knowledge will also help us to understand the movement of satellites and spacecraft.

Work, power and energy

These three terms are used frequently in mechanics, so we must understand their meaning. This is especially important because they are common words too in ordinary conversation, but with rather different shades of meaning.

A force is said to do work on a body when it moves the body in the direction in which it is acting, and the amount of work done is measured by the product of the force and the distance moved in the direction of the force. Thus if a force of 10 newtons moves a body 2 m (along its line of action), it does 20 newton metres (Nm) of work. **A newton metre**, the unit of work, is called a **joule (J)**. Notice that, according to mechanics, you do no work at all if you push something without succeeding in moving it – no matter how hard you push or for how long you push. Notice that you do no work if the body moves in the opposite direction, or even at right angles to the direction in which you push. Someone else must be doing some pushing – and some work!

Power is simply the rate of doing work. If the force of 10 N moves the body 2 m in 5 seconds, then the power is 20 Nm (20 joules) in 5 seconds, or 4 joules per second. A joule per second **(J/s)** is called a **watt (W), the unit of power.** So the power used in this example is 4 watts. Readers who have studied electricity will already be familiar with the watt as a unit of electrical power; this is just one example of the general trend towards the realisation that all branches of science are inter-related. Note the importance of the time taken, i.e. of the rate at which the work is done; the word power or powerful, is apt to give an impression of size and brute force. The unit of 1 watt is small for practical use, and kilowatts are more often used. The old unit of a horse-power was never very satisfactory but, as a matter of interest, it was the equivalent of 745.7 watts (Fig. 1B).

A body is said to have **energy** if it has the ability to do work, and the amount of energy is reckoned by the amount of work that it can do. The units of energy will therefore be the same as those of work. We know that petrol can do work by driving a car or an aeroplane, a man can do work by propelling a bicycle or even by walking, a chemical battery can drive an electric motor which can do work on a train, an explosive can drive a shell at high speed from the muzzle of a gun. All this means that energy can exist in many forms, heat, light, sound, electrical, chemical, magnetic, atomic – and, most useful of all, mechanical. A little thought will convince us how much of our time and energy is spent in converting, or trying to convert, other forms of energy into mechanical energy, the eventual form which enables us to get somewhere. The human body is simply a form of engine – not a simple form of engine – in which the energy contained in food is converted into useful, or useless, work. Unfortunately there is a tendency for energy to slip back again, we might almost say deteriorate, into other forms, and our efforts to produce mechanical energy are not always very efficient.

Even mechanical energy can exist in more than one form; a weight that is high up can do work in descending, and it is said to possess **potential energy** or **energy of position**; a mass that is moving rapidly can do work in coming to rest, and it is therefore said to have **kinetic energy** or **energy of motion**; a spring that is wound up, a gas that is compressed, even an elastic material that is stretched, all can do work in regaining their original state, and all possess energy which is in a sense potential but which is given various names according to its application.

In figures, a weight of 50 newtons raised to a height of 2 metres above its base has 100 joules of potential energy or, to be more correct, it has 100 joules more potential energy than it had when at its base. This was the work done to raise it to the new position, and it is the work that it should be able to do in returning to its base.

In symbols, **W newtons at height h metres has –**
Wh joules of energy.

What is the kinetic energy of mass of m kg moving at v m/s?

We don't know, of course, how it got its kinetic energy, but the actual process is unimportant, so let us suppose that it was accelerated uniformly at a metres per second per second from zero velocity to v metres per second by being pushed by a constant force of F newtons.

If the distance travelled during the acceleration was s metres, then the work done, i.e. its kinetic energy, will be Fs joules.

$$\text{But } v^2 = u^2 + 2as \text{ (and } u = 0)$$
$$\therefore v^2 = 2as$$
$$\therefore s = v^2/2a$$
$$\text{But } F = ma$$
$$\text{So K.E.} = Fs = ma \times v^2/2a = \tfrac{1}{2}mv^2 \text{ joules.}$$

Fig 1B Power (opposite) (By courtesy of the Lockheed Aircraft Corporation, USA)
The Lockheed C-5 Galaxy with four turbofan engines, each of 183 kN thrust represents a total power of 183 000 kW at the maximum level speed of about 900 km/h.

Thus the kinetic energy of 2 kg moving at 10 m/s

$$= \tfrac{1}{2}mv^2$$
$$= \tfrac{1}{2} \times 2 \times 10^2$$
$$= 100 \text{ joules}$$

Energy and momentum

Let us be sure that we understand the differences between energy and momentum, because we shall be concerned with this later on.

Energy is 1/2 mv^2. Momentum is mv.

So the mass of 2 kg, moving at 10 m/s, has 100 units of energy (joules), but 2×10, i.e. 20 units, of momentum (kg $\times$ m/s).

Yes, but there is more to it than that.

Consider two bodies colliding, e.g. billiard balls.

The total momentum after the collision is the same as the total momentum before; the momentum lost by one ball is exactly the same as the momentum gained by the other. This is the principle of the **conservation of momentum.** (In considering this it must be remembered that momentum has **direction,** because velocity has direction.) The law will apply whether the balls rebound, or whether they stick together, or whatever they do.

But the total mechanical energy after the collision will not be the same as before; energy will be dissipated, it will go into the air in the form of heat, sound, etc.; the total energy of the universe will not be changed by the collision – but that of the balls will be.

So momentum is a more permanent property than energy, the latter is often wasted and we shall sometimes find it unfortunate that in order to give a body momentum we must also give it energy.

Fluid pressure

In the mechanics of flight we shall be chiefly concerned with fluid pressure, that is, the pressure in a liquid or gas, and the force that it produces. The reason why a fluid exerts a force is because its molecules are in rapid motion and bombard any surface that is placed in the fluid; each molecule exerts only a tiny force on the surface but the combined effect of the bombardment of millions upon millions of molecules results in an evenly distributed force on the surface.

Pressure is a scalar quantity. That means that it has a magnitude, but unlike a vector, there is no direction involved. When a fluid at rest is in contact with a surface, the pressure produces a force which acts at right angles to the surface (Fig. 1.8). Note that the force *does* have a direction whereas the pressure that causes it does not. It is not surprising that people often confuse cause and effect when talking about pressure. You will find many old books that say that pressure acts equally in all directions. It does not. Pressure does not act in any direction; it is the force **due to pressure** that acts in a direction. The direction of the force is always at right angles to the surface that the pressure is exposed to. We measure pressure in terms of the force that it will produce on an area, so pressure has the units of newtons per square metre (N/m^2).

A pressure of 1 N/m^2 is also called a **pascal** (**Pa**). Another common metric unit of pressure is the **millibar** (**mb**), which is 1/1000 th of a **bar**: a bar being 10^5 N/m^2. This seemingly odd unit comes about, because 1 bar is very close to the standard atmospheric pressure at sea level. It had been adopted by meteorologists many years before the metric SI units were introduced, and the reader may often encounter atmospheric pressure given in millibars, particularly in flying manuals. In this book, we will use millibars for atmospheric pressure when appropriate, as in dealing with the effects of altitude. However, for most straightforward aerodynamic calculations we will use N/m^2 since this is normal practice in European educational institutions.

Density

Density is defined as the **mass per unit volume** of a substance, so it has the units of kg/m^3 in the SI metric system. Notice that it is mass and not weight that is used in this definition. The symbol commonly used for density is the Greek letter ρ.

As we will see later, the density of air changes with height and with the weather conditions. The density of water is conveniently 1000 kg/m^3, and hardly varies at all, even if subjected to a very high pressure.

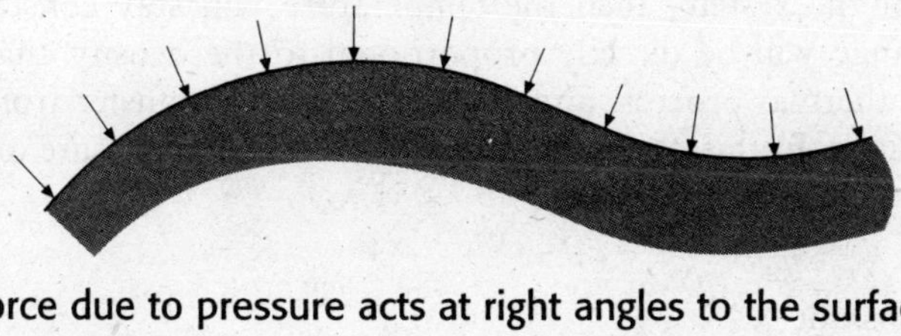

Fig 1.8 Force due to pressure acts at right angles to the surface

Pressure and density variation in a stationary fluid

The pressure in a stationary fluid increases with depth. This variation is rather complicated in the case of air because the density also increases with depth. These variations are very important for aircraft flight and in the next chapter the changes of pressure, density and temperature with height are described in some detail.

In a liquid, matters are much simpler; the density remains almost constant, and the pressure change is directly proportional to the change in depth. In a liquid, the pressure variation is given by the simple expression:

Change in pressure = density × gravity constant × change in depth

Or: $\Delta p = \rho \times g \times$ **(change in depth)**

This book is about the mechanics of aircraft flight, so you might wonder why we should have any interest in the way that pressure varies in a liquid. The answer is that it is because the easiest way to measure pressure is to use a U-shaped tube containing liquid. This is known as a **U-tube manometer,** and is described in the next chapter.

The behaviour of gases

In the study of the flight of aircraft, we are really only interested in the behaviour of one particular gas, air. The most important relationship that we need to know is called the gas law, which can be written as:

$$\frac{p}{\rho} = RT$$

where p is the pressure, ρ is the density, T is the temperature measured relative to absolute zero (i.e. in degrees Kelvin in the SI system), and R is a constant called **the gas constant.**

If the gas is compressed its density increases, so either or both the other quantities, temperature or pressure, must change. The way that they change depends on how the compression takes place. If the compression is very slow, and the gas is contained in a poorly insulated vessel so that heat is transferred out of the system, then the temperature will stay constant, and the pressure change will be directly proportional to the density change. This is called an isothermal process, and it involves a heat transfer from the gas to its surroundings. In this case the relationship between pressure and density is given by:

$$\frac{p}{\rho} = \text{a constant}$$

If the compression takes place with no transfer of heat, which commonly occurs when compression is very rapid, then the change is said to be adiabatic. If the change also takes place without any increase in turbulence, so there is no increase in the disorder (entropy) of the system, then the process is called isentropic, and the relationship between pressure and density is given by:

$$\frac{p}{\rho^{\gamma}} = \text{a constant}$$

where γ is the ratio of the specific heat at constant pressure to the specific heat at constant volume, and has a value of approximately 1.4 for air.

We cannot go much further down this path without becoming embroiled in the complexities of thermodynamics, however, and as the relationships above are the only ones that are relevant to the understanding of the contents of this book, we will not pursue the subject any further.

Composition and resolution of forces, velocities, etc.

A force is a vector quantity – it has magnitude and direction, and can be represented by a straight line, passing through the point at which the force is applied, its length representing the magnitude of the force, and its direction corresponding to that in which the force is acting. Forces can be added, or subtracted, to form a resultant force, or they can be resolved, that is to say, split into component parts, by drawing the vectors to represent them (Fig. 1.9).

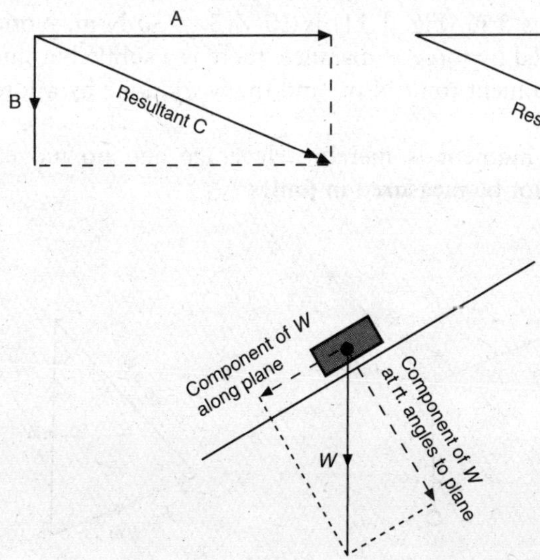

Fig 1.9 Composition and resolution of vector quantities

Note that velocity and momentum are also vector quantities and can be represented in the same way by straight lines. Mass, on the other hand, is not; a mass has no direction, and this is yet another distinction between a force and a mass.

The triangle, parallelogram, and polygon of forces

If three forces which act at a point are in equilibrium, they can be represented by the sides of a triangle taken in order (Fig. 1.10). This is called the principle of the **triangle of forces,** and the so-called **parallelogram of forces** is really the same thing, two sides and the diagonal of the parallelogram corresponding to the triangle.

If there are more than three forces, the principle of the **polygon of forces** is used – when any number of forces acting at a point are in equilibrium, the polygon formed by the vectors representing the forces and taken in order will form a closed figure, or, conversely, **if the polygon is a closed figure the forces are in equilibrium.**

Moments, couples and the principles of moments

The **moment** of a force about any point is the **product of the force and the perpendicular distance from the point to the line of action of the force.**

Thus the moment of a force of 10 N about a point whose shortest distance from the line of action is 3 m (Fig. 1.11) is $10 \times 3 = 30$ N-m. Notice that, though both are measured by force × distance, there is a subtle but important distinction between a moment (unit N-m) and the work done by a force (unit Nm, or joules).

The distance in the moment is merely a leverage and no movement is involved; **moments cannot be measured in joules.**

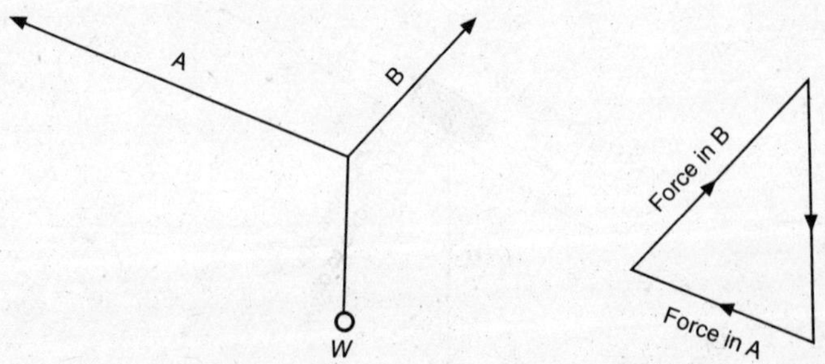

Fig 1.10 Triangle of forces

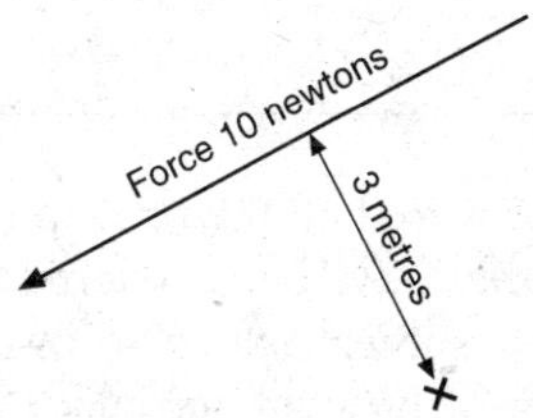

Fig 1.11 Moment of a force (Anti-clockwise)

A moment is normally taken to be positive if it is in a clockwise direction, and negative if it is in an anti-clockwise direction.

If a body is in equilibrium under the influence of several forces in the same plane, the sum of the clockwise moments about any point is equal to the sum of the anti-clockwise moments about that point, or, what amounts to the same thing and is much shorter to express, **the total moment is zero.** This is called the **principle of moments**, and applies whether the forces are parallel or not.

When considering the forces acting on a body, the weight of the body itself is often one of the most important forces to be considered. The weight may be taken as acting through the **centre of gravity**, which is defined as the point through which the resultant weight acts whatever position the body may be in.

Two equal and opposite parallel forces are called a couple. The moment of a couple is one of the forces multiplied by the distance between the two, i.e. by the arm of the couple. Notice that the moment is the same about any point (Fig. 1.12), and a couple has no resultant.

Moments about O. P $10 \times 1 = 10$ clock, Q $10 \times 1 = 10$ clock,
Total 20 clock.
Moments about A. P zero, Q $10 \times 2 = 20$ clock,
Total 20 clock.
Moments about B. P $10 \times 2 = 20$ clock, Q zero,
Total 20 clock.
Moments about C. P $10 \times 6 = 60$ clock, Q $10 \times 4 = 40$ anti,
Total 20 clock.

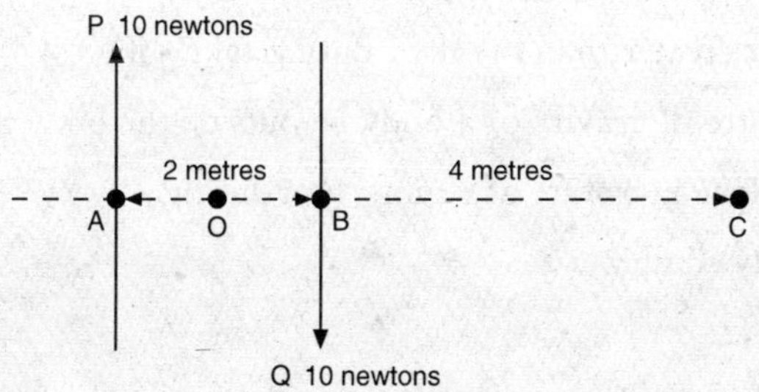

Fig 1.12 A couple

Mechanics of flight

We do not pretend to have covered all the principles of mechanics, nor even to have explained fully those that have been covered. All we have done has been to select some aspects of the subject which seem to form the chief stumbling blocks in the understanding of how an aeroplane flies; we have attempted to remove them as stumbling blocks, and perhaps even so to arrange them that, instead, they become stepping stones to the remainder of the subject. In the next chapter we will turn to our real subject – the *Mechanics of Flight*.

Before continuing, try to answer some of the questions below, and the numerical questions in Appendix 3.

Can you answer these?

These questions are tests not so much of mechanical knowledge as of mechanical sense. Try to puzzle them out. Some of them are easy, some difficult; the answers are given in Appendix 5.

1 A lift is descending, and is stopping at the ground floor. In what direction is the acceleration?

2 What is the difference between –

(*a*) Pressure and Force?

(*b*) Moment and Momentum?

(*c*) Energy and Work?

3 Why does it require less force to pull a body up an inclined plane than lift it vertically? Is the same work done in each case?

4 Distinguish between the mass and weight of a body.

5 If the drag of an aeroplane is equal to the thrust of the propeller in straight and level flight, what makes the aeroplane go forward?

6 Is the thrust greater than the drag during take-off?

7 Can the centre of gravity of a body be outside the body itself?

8 Is an aeroplane in a state of equilibrium during –

(*a*) A steady climb?

(*b*) Take-off?

9 Are the following the same, or less, or more, on the surface of the moon as on the surface of the earth –

(*a*) The weight of a given body as measured on a spring balance?

(*b*) The apparent weight of a given body as measured on a weigh-bridge (using standard set of weights)?

(*c*) The time of fall of a body from 100 m?

(*d*) The time of swing of the same pendulum?

(*e*) The thrust given by a rocket?

10 In a tug-o'-war does the winning team exert more force on the rope than the losing team?

11 Are the following in equilibrium –

(*a*) A book resting on a table?

(*b*) A train ascending an incline at a steady speed?

12 A flag is flying from a vertical flag pole mounted on the top of a large balloon. If the balloon is flying in a strong but steady east wind, in what direction will the flag point?

For solutions see Appendix 5.

For numerical examples on mechanics see Appendix 3.

CHAPTER 2

Air and airflow – subsonic speeds

Introduction – significance of the speed of sound

As was explained in Chapter 1, the remainder of the book will be concerned almost entirely with fluids in motion or, what comes to much the same thing, with motion through fluids. But it would be misleading even to start explaining the subject without a mention of the significance of the speed of sound.

The simple fact is that fluids behave quite differently when they move, or when bodies move through them, at speeds below and at speeds above the speed at which sound travels in that fluid. This virtually means that in order to understand modern flight we have to study two subjects – flight at speeds below that at which sound travels in air, and flight at speeds above that speed – in other words, flight at **subsonic** and flight at **supersonic** speeds.

To complicate things still further, the airflow at speeds near the speed of sound, **transonic speeds**, is complex enough to be yet a third subject in its own right. We shall cover these subjects as fully as we can, but we must not let our natural interest in supersonic flight tempt us to try to run before we can walk, and in the early chapters the emphasis will be on flight at subsonic speeds, though we shall point out from time to time where we may expect to find differences at supersonic speeds.

But first let us have a closer look at the fluid, air, with which we are most concerned.

Invisibility of the atmosphere

Air is invisible, and this fact in itself makes flight difficult to understand. When a ship passes through water we can see the 'bow wave', the 'wash' astern, and all the turbulence which is caused; when an aeroplane makes its way through air nothing appears to happen – yet in reality there has been even more commotion (Fig. 2.1).

If only we could see this commotion, many of the phenomena of flight would need much less explanation, and certainly if the turbulence formed in the atmosphere were visible no one could have doubted the improvement to be gained by such inventions as streamlining. After some experience it is possible to cultivate the habit of 'seeing the air' as it flows past bodies of different shapes, and the ability to do this is made easier by introducing smoke into the air or by watching the flow of water, which exhibits many characteristics similar to those of air.

Density of the air

Another property of air which is apt to give us misleading ideas when we first begin to study flight is its low density. The air feels thin, it is difficult for us to obtain any grip upon it, and if it has any mass at all we usually consider it as negligible for all practical purposes. Ask anyone who has not studied the question, 'What is the mass of air in any ordinary room?' – you will probably receive answers varying from 'almost nothing' up to 'about 5 kilograms.' Yet the real answer will be nearer 150 kilograms, and in a large hall may be over a metric tonne! Again, most of us who have tried to dive have experienced the sensation of coming down 'flat' onto the surface of the water; since then we

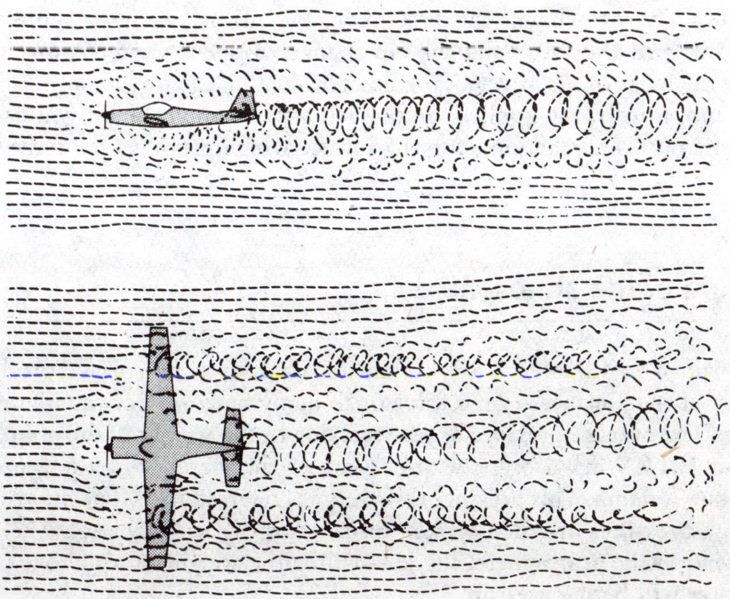

Fig 2.1 'Seeing the air'

have treated water with respect, realising that it has substance, that it can exert forces which have to be reckoned with. We have probably had no such experience with air, yet if we ever try we shall find that the opening of a parachute after a long drop will cause just such a jerk as when we encountered the surface of the water. It is, of course, true that the density of air – i.e. **the mass per unit volume** – is low compared with water (the mass of a cubic metre of air at ground level is roughly 1.226 kg – whereas the mass of a cubic metre of water is a metric tonne, 1000 kg, nearly 800 times as much); yet it is this very property of air – its density – which makes all flight possible, or perhaps we should say airborne flight possible, because this does not apply to rockets. The balloon, the kite, the parachute and the aeroplane – all of them are supported in the air by forces which are entirely dependent on its density; the less the density, the more difficult does flight become; and for all of them flight becomes impossible in a vacuum. So let us realise the fact that, however thin the air may seem to be, it possesses the property of density.

Inertia of the air

It will now be easy to understand that air must also possess, in common with other substances, the property of inertia and the tendency to obey the laws of mechanics. Thus air which is still will tend to remain still, while air which is moving will tend to remain moving and will resist any change of speed or direction (First Law); secondly, if we wish to alter the state of rest or uniform motion of air, or to change the direction of the airflow, we must apply a force to the air, and the more sudden the change of speed or direction and the greater the mass of air affected, the greater must be the force applied (Second Law); and, thirdly, the application of such a force upon the air will cause an equal and opposite reaction upon the surface which produces the force (Third Law).

Pressure of the atmosphere

As explained in Chapter 1, the weight of air above any surface produces a pressure at that surface – i.e. a force of so many newtons per square metre of surface. **The average pressure at sea-level due to the weight of the atmosphere is about 101 kN/m^2**, a pressure which causes the mercury in a barometer to rise about 760 mm. This pressure is sometimes referred to as 'one atmosphere', and high pressures are then spoken of in terms of 'atmospheres'. The higher we ascend in the atmosphere, the less will be the weight of air above us, and so the less will be the pressure.

Decrease of pressure and density with altitude

The rate at which the pressure decreases is much greater near the earth's surface than at altitude. This is easily seen by reference to Fig. 2.2 (overleaf); between sea-level and 10 000 ft (3480 m) the pressure has been reduced from 1013 mb to 697 mb, a drop of 316 mb; whereas for the corresponding increase of 10 000 ft between 20 000 ft (6096 m) and 30 000 ft (9144 m), the decrease of pressure is from 466 mb to 301 mb, a drop of only 165 mb; and between 70 000 ft (21 336 m) and 80 000 ft (24 384 m) the drop is only 17 mb.

This is because air is compressible; the air near the earth's surface is compressed by the air above it, and as we go higher the pressure becomes less, the air becomes less dense, so that if we could see a cross-section of the atmosphere it would not appear homogeneous – i.e. of uniform density – but it would become thinner from the earth's surface upwards, the final change from atmosphere to space being so gradual as to be indistinguishable. In this respect air differs from liquids such as water; in liquids there is a definite dividing line or surface at the top; and beneath the surface of a liquid the pressure increases in direct proportion to the depth because the liquid, being practically incompressible, remains of the same density at all depths.

Temperature changes in the atmosphere

Another change which takes place as we travel upwards through the lower layers of the atmosphere is the gradual drop in temperature, a fact which unhappily disposes of one of the oldest legends about flying – that of Daedalus and his son Icarus, whose wings were attached by wax which melted because he flew too near the sun. In most parts of the world, the atmospheric temperature falls off at a steady rate called the **lapse rate** of about −6.5°C for every 1000 metres increase in height up to about 11 000 metres. Above 11 000 metres, the temperature remains nearly constant until the outer regions of the atmosphere are reached. The portion of the atmosphere below the height at which the change occurs is called the **troposphere**, and the portion above, the **stratosphere**. The interface between the two is called the **tropopause**. The lapse rate and the height of the tropopause vary with latitude. In Arctic regions, the rate of temperature change is lower, and the stratosphere does not start until around 15 500 m. The temperature in the stratosphere varies between about −30°C at the equator to −95°C in the Arctic. In temperate regions such as Europe the temperature in the stratosphere is around −56.5°C.

For aircraft performance calculations, it is normal practice to use a standard set of conditions called the **International Standard Atmosphere** (ISA). This defines precise values of lapse rate, height of the tropopause, and sea-level values of temperature, pressure and density. For temperate regions the ISA

Speed of sound m/s	Density kg/m³	Press. kN/m²	Temp. °K	Height above sea-level km
299	0.040	2.5	222	25
298	0.047	3.0	221	24
298	0.055	3.5	220	23
297	0.065	4.0	219	22
296	0.076	4.7	218	21
295	0.089	5.5	217	20
295	0.104	6.5	217	19
295	0.122	7.6	217	18
295	0.142	8.8	217	17
295	0.166	10.4	217	16
295	0.195	12.1	217	15
295	0.228	14.2	217	14
295	0.267	16.6	217	13
295	0.312	19.4	217	12
295	0.365	22.7	217	11
299	0.414	26.5	223	10
304	0.467	30.8	230	9
308	0.526	35.7	236	8
313	0.590	41.1	243	7
317	0.660	47.2	249	6
321	0.736	54.0	256	5
325	0.819	61.7	262	4
329	0.909	70.1	269	3
333	1.007	79.5	275	2
337	1.112	89.9	282	1
340	1.225	101.3	288	Sea level

Height above sea-level Thousands of feet	Temp. °C	Press. mb	Relative density ρ/ρ_0	Relative pressure p/p_0
80	−52	28	0.036	0.028
75	−54	35	0.046	0.035
70	−55	45	0.059	0.044
65	−56	57	0.075	0.056
60	−56	72	0.095	0.071
55	−56	92	0.121	0.091
50	−56	117	0.153	0.115
45	−56	148	0.194	0.145
40	−56	188	0.247	0.186
35	−54	239	0.311	0.236
30	−44	301	0.375	0.298
25	−34	377	0.449	0.372
20	−25	466	0.533	0.460
15	−15	572	0.629	0.565
10	−5	697	0.739	0.688
5	5	843	0.862	0.832
Sea level	15	1013	1.000	1.000

Upper stratosphere
Lower stratosphere
Troposphere
Mt Everest
Mt Blanc
Cirrus clouds
High Cumulus clouds
Stratus and Nimbus clouds

Fig 2.2 The International Standard Atmosphere
Based on the US Standard Atmosphere, 1962, which was prepared under the sponsorship of NASA, the USAF and the US Weather Bureau.

value of the lapse rate is −6.5°C per 1000 m, the tropopause is at 11 km, and the sea-level values of pressure and temperature are 101.325 kN/m², and 15°C respectively.

Modern long- and medium-range airliners cruise in or very close to the stratosphere, and the supersonic airliner Concorde used to fly in the stratosphere well above the tropopause. When piston-engined aircraft first started to fly in the stratosphere, conditions were very uncomfortable for the crew. The low density and pressure meant that oxygen masks had to be worn, and at temperatures of −56°C, even the heavy fur-lined clothing was barely adequate. Nowadays, the cabins of high-flying airliners are pressurised, and the air is heated, so that the passengers are unaware of the external conditions. Nevertheless, above every seat there is an emergency oxygen mask to be used in the event of a sudden failure of the pressurisation system.

Despite the low external air temperature in the stratosphere, supersonic aircraft have the problem that surface friction heats the aircraft up during flight, so means have to be provided to keep the cabin cool enough.

Effect of temperature and pressure on density

Although air is not quite a 'perfect gas' it does obey the gas law within reasonable limits, so we can say that

$$\frac{p}{\rho} = RT$$

It is often convenient to express the density in terms of the ratio of the density at some height ρ to the value at standard sea-level conditions ρ_0. This ratio ρ/ρ_0 is usually denoted by the Greek letter σ, and is called the **relative density.**

EXAMPLE 2.1

If the temperature at sea-level in the temperate ISA is 15°C, and the lapse rate is 6.5°C per km, find the temperature and density at 6 km altitude where the pressure is 47 200 N/m². The gas constant R = 287 J/kg K.

SOLUTION

First we must convert all temperatures to absolute (Kelvin). Sea-level temperature is

$273 + 15 = 288$ K.

Temperature at 6000 m is

$288 + 6.5 \times 6 =$ **249 K**

The gas law states that

$$\frac{p}{\rho} = RT \text{ or } \rho = \frac{p}{RT}$$

so the density at 6 km is

$$\frac{47\,200}{287 \times 249} = 0.66\ \text{kg/m}^3$$

Viscosity

An important property of air in so far as it affects flight is its viscosity. This is a measure of the resistance of one layer of air to movement over the neighbouring layer; it is rather similar to the property of friction between solids. It is owing to viscosity that eddies are formed when the air is disturbed by a body passing through it, and these eddies are responsible for many of the phenomena of flight. Viscosity is possessed to a large degree by fluids such as treacle and certain oils, and although the property is much less noticeable in air, it is none the less of considerable importance.

Winds and up-and-down currents

The existence of separate regions of high and low pressure is the cause of wind, or bodily movement of large portions of the atmosphere. Winds vary from the extensive trade winds caused by belts of high and low pressure surrounding the earth's surface to the purely local gusts and 'bumps', caused by local differences of temperature and pressure. On the earth's surface we are usually only concerned with the horizontal velocity of winds, but when flying the rising convection currents and the corresponding downward movements of the air are also important. The study of winds, of up-and-down convection currents, of cyclones and anti-cyclones, and the weather changes produced by them – all these form the fascinating science of *meteorology*, and the reader who is interested is referred to books on that subject.

In the lower regions of the atmosphere conditions are apt to be erratic; this is especially so within the first few hundred feet. It often happens that as we begin to climb the temperature rises instead of falling – called an **inversion** of temperature. This in itself upsets the stability of the air, and further disturbances may be caused by the sun heating some parts of the earth's surface more than others, causing thermal up-currents, and by the wind blowing over uneven ground, hangars, hills, and so on. On the windward side of a large building, or of a hill, the wind is deflected upwards, and on the leeward side it is apt to leave the contour altogether, forming large eddies which may result in a flow of air near the ground back towards the building or up the far side of the hill, that is to say in the opposite direction to that of the main wind. Even when the surface of the ground is comparatively flat, as on the average airfield, the wind is retarded near the ground by the roughness of the surface, and successive layers are held back by the layers below them – due to viscosity

– and so the wind velocity gradually increases from the ground upwards. This phenomenon is called **wind gradient**. When the wind velocity is high it is very appreciable, and since most of the effect takes place within a few metres of the ground it has to be reckoned with when landing.

Quite apart from this wind gradient very close to the ground, there is often also a wind gradient on a larger scale. Generally, it can be said that on the average day the wind velocity increases with height for many thousands of feet, and it also tends to veer, i.e. to change in a clockwise direction (from north towards east, etc.); at the same time it becomes more steady and there are fewer bumps.

Air speed and ground speed

But our chief concern with the wind at the present moment is that we must understand that when we speak of the speed of an aeroplane we mean its speed relative to the air, or **air speed** as it is usually termed. Now the existence of a wind simply means that portions of the air are in motion relative to the earth, and although the wind will affect the speed of the aeroplane **relative to the earth** – i.e. its **ground speed** – it will not affect its speed relative to the air.

For instance, suppose that an aeroplane is flying from A to B (60 km apart), and that the normal speed of the aeroplane (i.e. its air speed) is 100 km/h (see Fig. 2.3). If there is a wind of 40 km/h blowing from B towards A, the ground speed of the aeroplane as it travels from A to B will be 60 km/h, and it will take one hour to reach B, but the air speed will be 100 km/h. If, when the aeroplane reaches B, it turns and flies back to A, the ground speed on the return journey will be 140 km/h (Fig. 2.4); the time to regain A will be less than half an hour, but the air speed will still remain 100 km/h – that is, the wind will strike the aeroplane at the same speed as on the outward journey. Similarly, if the wind had been blowing across the path, the pilot would have inclined his aeroplane several degrees towards the wind on both journeys so that it would have travelled crabwise, but again, on both outward and homeward journeys the air speed would have been 100 km/h and the wind would have been a headwind straight from the front as far as the aeroplane was concerned.

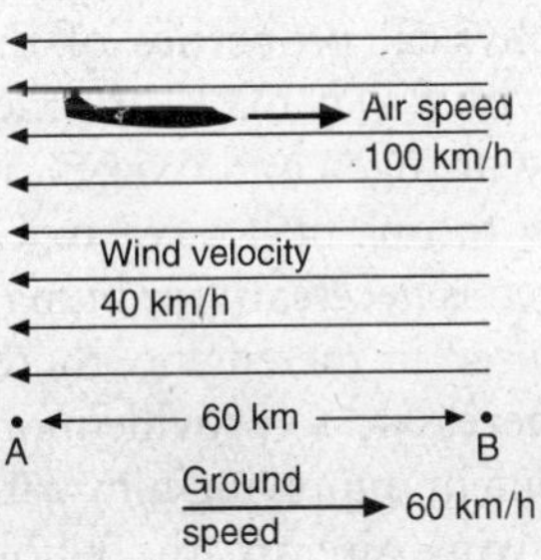

Fig 2.3 Outward flight

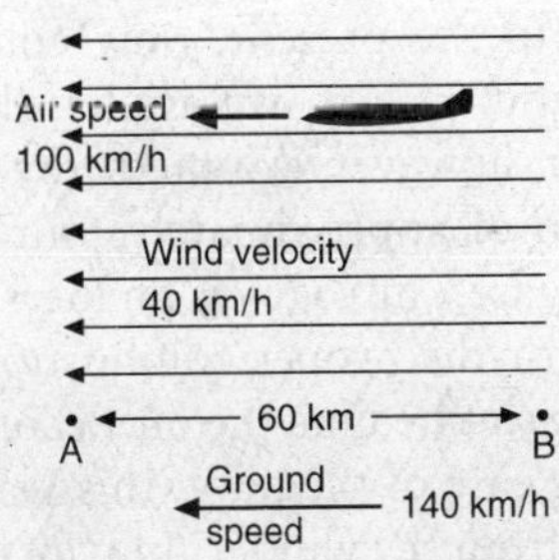

Fig 2.4 Return flight

An aeroplane which encounters a headwind equal to its own air speed will appear to an observer on the ground to stay still, yet its air speed will be high. A free balloon flying in a wind travels over the ground, yet it has no air speed – a flag on the balloon will hang vertically downwards.

All this may appear simple, and it is in fact simple, but it is surprising how long it sometimes takes a student of flight to grasp the full significance of air speed and all that it means. There are still pilots who say that their engine is overheating because they are flying 'down wind'! It is not only a question of speed, but of direction also; a glider may not lose height in a rising current of air (it may, in fact, gain height), yet it is all the time descending **relative to the air**. In short, the only true way to watch the motion of an aeroplane is to imagine that one is in a balloon floating freely with the wind and to make all observations relative to the balloon.

Ground speed is, of course, important when the aeroplane is changing from one medium to another, such as in taking-off and landing, and also in the time taken and the course to be steered when flying cross-country – this is the science of navigation, and once again the student who is interested must consult books on that subject.

The reader may have noticed that we have not been altogether consistent, nor true to the SI system, in the units that we have used for speed; these already include m/s, km/h and knots. There are good reasons for this inconsistency, the main one being that for a long time to come it is likely to be standard practice to use knots for navigational purposes both by sea and by air, km/h for speeds on land, e.g. of cars, while m/s is not only the proper SI unit but it must be used in certain formulae. We shall continue to use these different units throughout the book as and when each is most appropriate, and the important thing to remember is that it is only a matter of simple conversion from one to the other –

1 knot = 0.514 m/s = 1.85 km/h

Chemical composition of the atmosphere

We have, up to the present, only considered the physical properties of the atmosphere, and, in fact, we are hardly concerned with its chemical or other properties. Air, however, is a mixture of gases, chiefly nitrogen and oxygen, in the proportion of approximately four-fifths nitrogen to one-fifth oxygen. Of the two main gases, nitrogen is an inert gas, but oxygen is necessary for human life and also for the proper combustion of the fuel used in the engine, therefore when at great heights the air becomes thin it is necessary to provide more oxygen. In the case of the pilot, this was formerly done by supplying him with pure oxygen from a cylinder, but in modern high-flying aircraft, the whole cabin is pressurised so that pilot, crew and passengers can still breathe air of similar density, pressure, temperature and composition as that to which they are accustomed at ground level, or at some reasonable height. As for the engines, it has always been preferable to provide extra air rather than oxygen, because although the oxygen is needed for combustion, the nitrogen provides, as it were, the larger part of the working substance which actually drives the engine. In piston engines the extra air is provided by a process known as **supercharging**, which means blowing in extra air by means of a fan or fans; in jet aircraft the principle is fundamentally the same, though simpler because the engine is in itself a supercharger and this, combined with the 'ram effect' at the higher true speeds achieved at altitude, keeps up the supply of air as necessary.

The international standard atmosphere

The reader will have realised that there is liable to be considerable variation in those properties of the atmosphere with which we are concerned – namely, temperature, pressure and density. Since the performance of engine, aeroplane and propeller is dependent on these three factors, it will be obvious that the actual performance of an aeroplane does not give a true basis of comparison with other aeroplanes, and for this reason the **International Standard Atmosphere** has been adopted. The properties assumed for this standard atmosphere in temperate regions are those given in Fig. 2.2, earlier. If, now, the actual performance of an aeroplane is measured under certain conditions of temperature, pressure and density, it is possible to deduce what would have been the performance under the conditions of the Standard Atmosphere, and thus it can be compared with the performance of some other aeroplane which has been similarly reduced to standard conditions.

The altimeter

The instrument normally used for measuring height is the altimeter, which was traditionally merely an aneroid barometer graduated in feet or metres instead of millimetres of mercury or millibars. As a barometer it will record the **pressure** of the air, and since the pressure is dependent on the **temperature** as well as the height, it is only possible to graduate the altimeter to read the height if we assume certain definite conditions of temperature. If these conditions are not fulfilled in practice, then the altimeter cannot read the correct height. Altimeters were at one time graduated on the assumption that the temperature remained the same at all heights. We have already seen that such an assumption is very far from the truth, and the resulting error may be as much as 900 m at 9000 m, the altimeter reading too high owing to the drop in temperature. Altimeters are now calibrated on the assumption that the temperature drops in accordance with the International Standard Atmosphere; this method reduces the error considerably, although the reading will still be incorrect where standard conditions do not obtain.

As a barometer the altimeter will be affected by changes in the pressure of the atmosphere, and therefore an adjustment is provided, so that the scale can be set (either to zero or to the height of the airfield above sea-level) before the commencement of a flight, but in spite of this precaution, atmospheric conditions may change during the flight, and it is quite possible that on landing on the same airfield the altimeter will read too high if the pressure has dropped in the meantime, or too low if the pressure has risen.

Although it is convenient to use SI units for calculations and numerical examples, it should be remembered that knots and feet are still the internationally approved units for speed and altitude respectively when dealing with aircraft operations. If you ever sit at the controls of an aircraft you will almost certainly find that the altimeter is calibrated in feet. We shall therefore use feet (ft) when referring to this instrument.

Modern altimeters are much more sensitive than the old types; some, instead of having one pointer, may have as many as three, and these are geared together like the hands of a clock so that the longest pointer makes one revolution in 1000 ft, the next one in 10 000 ft, and the smallest one in 100 000 ft. Unfortunately this has sometimes proved 'too much of a good thing,' and accidents have been caused by pilots mistaking one hand for another. The more modern tendency is to show the height level in actual figures in addition to one or more pointers. But, like a sensitive watch, the altimeter is of little use unless it can be made free from error, and can be read correctly. In the modern types, if the pilot sets the altimeter to read zero height, which he can do simply by turning a knob, a small opening on the face of the instrument discloses the pressure of the air at that height – in other words, the reading of the barometer. Conversely – and this is the important point – if, while in the air, he finds out by radio the barometric pressure at the airfield at which he wishes to land, he can adjust the instrument so that this pressure shows in the small opening, and he can then be sure that his altimeter is reading the correct height above that aerodrome, and that when he 'touches down' it will read zero. The altimeter may be used in this way for instrument flying and night flying, that is to say when the height above the ground in the vicinity of the aerodrome is of vital importance, but for ordinary cross-country flying during the day it may be preferable to set the **sea-level atmospheric pressure** in the opening. Then the pilot will always know his height above sea-level and can compare this with the height, as shown on the map, of the ground over which he is flying. If this method is used, instead of the altimeter reading zero on landing, it will give the height of the aerodrome above sea-level. There is, however, a snag in this method in that the sea-level atmospheric pressure varies from place to place and so different pilots may set their altimeters differently, thereby increasing the risk of collision; for this reason modern practice for flying above 3000 ft is to set the altimeter to **standard sea-level pressure of**

1013.2 mb, which means, in effect, that all the altimeters may be reading the incorrect height, but that only aircraft flying at the **same** height can have the **same** altimeter readings. Above 3000 ft heights are referred to in terms of **flight levels** (or hundreds of feet), e.g. FL 35 is 3500 ft, FL 40 is 4000 ft, then FL 45, 50, 55, and so on. Increases in flight levels are in fives because of the **quadrantal system** (in operation in the UK) which determines the height at which the pilot must fly for specific compass headings.

The question of altimeter setting has long been a matter for controversy among pilots – and even among nations.

In recent years there have been radical changes in aircraft instruments and displays. Instead of individual instruments there may now be a computer-screen type of display, but the altimeter display still looks quite similar to the traditional instrument. The information on which the display is based may also still be the external pressure, but there are now alternative, more accurate, height-reading devices such as radio or radar altimeters.

The reader who is particularly interested in altimeters and other instruments is referred to *Aircraft Instruments* by E. H. J. Pallett, a companion volume in the Introduction to Aeronautical Engineering Series.

Bernoulli's equation

An aircraft flying through the air causes local changes in both velocity relative to the aircraft and pressure. These changes are linked by Bernoulli's equation. This equation can be written in many forms, but originally it was given by

$$\frac{p}{\rho} + \frac{V^2}{2} + gz = \text{a constant}$$

where z is the height, p is the pressure, ρ is the density, V is the flow speed, and g is the gravity constant.

The equation is sometimes called Bernoulli's integral, because it is obtained by integrating the Euler momentum equation for the case of a fluid with constant density. Since the equation involves a constant density, it should really only be applied to incompressible fluids. Completely incompressible fluids do not actually exist, although liquids are very difficult to compress. Air is definitely compressible, but nevertheless, airflow calculations using Bernoulli's equation give good answers unless the speed of the flow starts going above about half the speed of sound. Bernoulli's equation will also not apply in regions where viscosity is important.

The terms in this equation all have the units of energy per unit mass, and the equation looks temptingly similar to the steady flow energy equation that you will meet if you ever study thermodynamics. The second and third terms do in fact represent kinetic energy per unit mass and potential energy per unit mass respectively. The true energy equation is, however, significantly different, and contains an important extra term, internal energy, which cannot be neglected in compressible airflows. However, whatever Bernoulli's equation is or is not, it remains a useful and simple means for getting approximately correct answers for low-speed flows. Aerodynamicists usually prefer it in the form below:

$$p + \tfrac{1}{2}\rho V^2 + \rho g z = \text{a constant}$$

This is obtained by multiplying the original equation by the density, which makes all of the terms come out in the units of pressure (N/m^2). The last term is usually ignored because changes in height are small in most of the calculations that we perform for airflows around an aircraft, so we write

$$p + \tfrac{1}{2}\rho V^2 = \text{a constant}$$

The first term represents the local pressure of the air, and is called the **static pressure**. The second term $\frac{1}{2}\rho V^2$ is associated with the flow speed and is called **dynamic pressure**. For convenience it is sometimes represented by the letter q, but in this book we will use the full expression.

If we ignore the third term, as above, Bernoulli's equation says that adding the first two terms, the static and the dynamic pressure, produces a constant result. Therefore, if the flow is slowed down so that the dynamic pressure decreases, then to keep the equation in balance the static pressure must increase. If we bring the flow to rest at some point, then the pressure must reach its highest possible value, because the dynamic pressure becomes zero. This maximum value is called the **stagnation pressure** because it occurs at a point where the air has stopped or become stagnant. Using the version of Bernoulli's equation above, we can write

$$p + \tfrac{1}{2}\rho V^2 = \text{a constant} = p\ (\text{stagnation}) + 0$$

This gives the important result that

Static pressure + Dynamic pressure = Stagnation pressure

Air speed measurement

Bernoulli's equation gives rise to a simple method of measuring air speed. You can see that the dynamic pressure is related to the density and the speed, (dynamic pressure = $\frac{1}{2}\rho V^2$), so if we could measure the dynamic pressure and the density we could determine the speed. Fortunately there is a very simple way of measuring the dynamic pressure at least.

If we point a tube directly into the flow of air, and connect the other end to a pressure-measuring device, then that device will read the stagnation pressure. The reason for this is that the tube is full of air and its exit is blocked, so no air can flow down the tube; the oncoming air therefore is brought to rest relative to the tube as it meets the open end of the tube. This type of tube is called a **pitot tube** and provides a means of measuring stagnation pressure. A different result is obtained if we make a hole in the side of a wind tunnel or in the fuselage of an aircraft, and connect this via a tube to a pressure-measuring device. The hole will not impede the flow of air, so the pressure measured will be the local static pressure. A hole used for this purpose is called a **static vent** or tapping. Since static pressure plus dynamic pressure equals stagnation pressure as shown above, it follows that

Stagnation pressure − Static pressure = Dynamic pressure

If instead of connecting the static pressure vent and the pitot tube to two separate pressure-measuring devices, we connect them across one device which measures the **difference** in pressure then, from the above expression, we can see that we will obtain a measurement of the dynamic pressure. Thus we have a simple means of measuring the dynamic pressure, and if only we could find a way of measuring or assessing the density, we could determine the flow speed, but more of this later; let us first concentrate on the measurement of dynamic pressure, which we will show is actually just as important to the pilot as the speed.

The pressure difference measuring device used on aircraft consists of either a diaphragm or a capsule (similar to the type used in an aneroid barometer). The stagnation pressure is applied to one side, and the static pressure is applied to the other. The resulting deflection of the diaphragm can then either be amplified through a series of levers to cause a dial pointer to move, as on older mechanical devices, or can be used to produce a proportional electrical output to be fed into an appropriate electronic circuit. This instrument thus gives a reading that is proportional to the dynamic pressure, but as we shall see, it forms the basis of the **air speed indicator**.

The pitot tube and static pressure hole are located at a suitable convenient position on the aircraft. The location of the static tapping is very important because it is essential to choose a position where the local static pressure is about the same as that in the free stream away from the aircraft. We need, therefore, to find a place where the flow speed is about the same as that in the free stream, and also is not too sensitive to change in the direction that the aircraft is pointing. The pitot and static holes are normally heated to avoid icing at low temperatures.

As an alternative to using separate pitot and static tubes, it is possible to use a combined device called the pitot-static tube which is illustrated in Fig. 2.5. The pitot-static tube consists of two concentric tubes. The inner one is simply a pitot tube, but the outer one is sealed at the front and has small holes in the

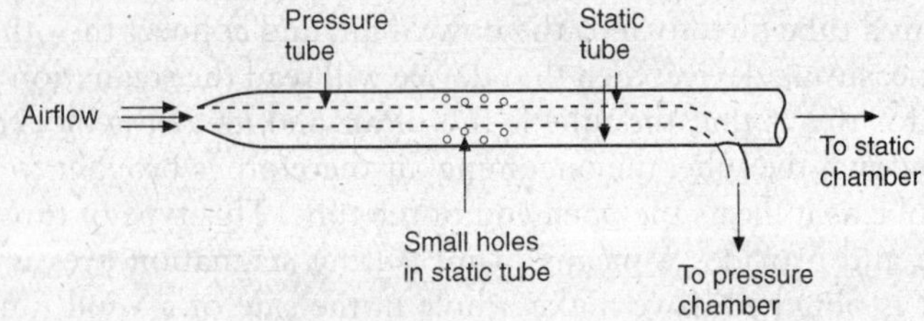

Fig 2.5 Concentric pitot-static tube

side to sense the static pressure. The pitot-static tube is a very convenient device, and by mounting it on the wingtips or the nose it can be arranged so that it is well clear of interference from the flow around the aircraft. Pitot-static tubes are always used for accurate speed measurement on prototypes, but for civil and private aircraft separate pitot and static tappings are normally used. Pitot-static tubes are frequently used in wind tunnels.

The air speed indicator

As stated above, the pitot and static tube combination provides a means of measuring the dynamic pressure. It does not tell us the speed directly, but we can work out the speed if we know the density. Until recently, there was no simple method of measuring density, so all that could be done was to use the dynamic pressure-measuring device described above, and mark on the dial the speed that this dynamic pressure would correspond to at standard sea-level air density. This instrument is called the **Air Speed Indicator (ASI)**. You will see that since the instrument is calibrated assuming the standard sea-level value of air density, it does not give the true speed, unless the aircraft is flying at a height where the density just happens to be equal to the standard sea-level value.

Nowadays, there are devices which can measure the true air speed, but the air speed indicator described above is still an important item on any instrument panel. This is because the lift and other forces on the aircraft are dependent on dynamic pressure, and the air speed indicator gives a reading which is directly related to dynamic pressure. For example, if the dynamic pressure is too small, the wings will not be able to generate enough lift to keep the aircraft in steady level flight. The value of the dynamic pressure and hence the indicated speed at which this occurs will always be the same whatever the height. The pilot just has to remember to keep above this minimum indicated speed. If the pilot had only a true speed indicator, he would have to know what the minimum speed was at every height.

Air speed indicator corrections

The speed indicated by the air speed indicator is called the **Air Speed Indictor Reading (ASIR)**. There are several sources of error in this reading. Firstly, the instrument itself may not have been calibrated correctly, or may be suffering from some wear. This error is called instrument error. By recalibrating the instrument it is possible to determine what the correction should be at every indicated speed. The speed corrected for instrument error is called the **Indicated Air Speed (IAS)**. There will also be errors due to the positioning of the pitot and static tubes on the aircraft. It is virtually impossible to find a position where the static pressure is always exactly the same as the pressure in the free airstream away from the aircraft. To determine the correction for such position errors, the aircraft can be flown in formation with another aircraft with specially calibrated instruments. Once the position error correction has been applied, the speed is known as the **Calibrated Air Speed (CAS)**. Finally, for any aircraft that can fly faster than about 200 mph, it is necessary to apply a correction for compressibility, since Bernoulli's equation only applies to low-speed effectively incompressible flows.

After all the corrections have been applied, the resulting speed is called the **Equivalent Air Speed (EAS)**. Once the equivalent air speed has been obtained it is quite easy to estimate the **True Air Speed (TAS)** which is required for navigation purposes. For a light piston-engined aircraft, the corrections will be relatively small, and for simple navigational estimates, the pilot can assume that the speed read from his instrument the **ASIR** is roughly the same as the equivalent air speed **EAS**. The procedure for calculating the true air speed is as follows.

Suppose that at 6000 m the air speed indicator reads 204 knots, i.e. 105 m/s. Ignoring any instrument or position errors, this means that the pressure on the pitot tube is the same as would be produced by a speed of 105 m/s at standard sea-level density of 1.225 kg/m^3; but this pressure is $\frac{1}{2}\rho V^2$, i.e. $1/2 \times 1.225 \times 105 \times 105$.

Now according to the International Standard Atmosphere the air density at 6000 m is 0.66 kg/m^3, and if the true air speed is V m/s, then the pressure on the pitot tube will be $1/2 \times 0.66 \times V^2$, which must be the same as

$$1/2 \times 1.226 \times 105 \times 105$$

so

$$0.66V^2 = 1.226 \times 105^2$$

or

$$V = \sqrt{1.226/0.66} \times 105 = 1.36 \times 105$$

$$\mathbf{142.8\ m/s.}$$

Thus the indicated air speed is 105 m/s, and the true air speed approximately 143 m/s.

Note that, expressed in symbols, the true air speed (*TAS*) is the equivalent air speed (*EAS*) divided by the square root of the relative density (σ), or

$$TAS = EAS/\sqrt{\sigma}$$

A similar calculation for 12 200 m will reveal the interesting result that at that height the true air speed is slightly more than double the indicated air speed!

We have said that the instrument **indicated** air speed **may sometimes** be more useful to the pilot that the true air speed. For purposes of navigation, however, he must estimate his speed over the ground, and with traditional navigation methods, he must first determine the true air speed, then make corrections to allow for the speed of the atmospheric wind relative to the ground. The true air speed can be determined using the procedure above, but to do this we need to know the air density or the relative density. This can be obtained by using the altimeter reading, and tables for the variation of relative density in the ISA. In practice, as an alternative to calculations, the pilot can use tables showing the relationship between true air speed and indicated air speed at different heights in ISA conditions.

For high-speed aircraft, the indicated air speed has to be corrected for compressibility, and to do this we need a further instrument, one that indicates the speed relative to the local speed of sound. This instrument is called a **machmeter**. However, we are again trying to run before we can walk, and we will leave the treatment of compressible flow to later chapters.

Nowadays navigation has been revolutionised by the introduction of ground-based radio, and satellite navigation systems which can give very accurate indications of position and speed relative to the ground. Despite these advances, however, aspiring pilots still have to learn the traditional methods of navigation in order to qualify for their licence. Old instruments like the air speed indicator and the altimeter are simple and reliable, and will not break down in the event of an electrical failure or a violent thunderstorm. Even on the most advanced modern airliners, an old mechanical air speed indicator and pressure altimeter are fitted, and will continue to work even if all the electrical systems have failed.

The venturi tube

One of the most interesting examples of Bernoulli's Theorem is provided by the venturi tube (Fig. 2.6). This simple but effective instrument is nothing but a tube which gradually narrows to a throat, and then expands even more gradually to the exit. Its effectiveness as a means of causing a decrease of pressure below that of the atmosphere depends very much on the exact shape.

If a photograph is taken, or a diagram made, of the flow of air or water through a venturi tube, it will be observed that the streamlines are closest together at the throat, and this gives an unfortunate impression that the fluid has been compressed at this point. Such an impression is the last thing we want to convey, and what we would like to be able to do is to show a video, or, better still, an actual experiment, which would make it quite clear that while it is true that the streamlines are closer together at the throat, the velocity of flow is also higher. This is the important point: **the dynamic pressure has gone up** and therefore, in accordance with Bernoulli's principle, **the static pressure has gone down**. If a tube is taken from the throat and connected to a U-tube containing water, the suction will be clearly shown.

An interesting experiment with a venturi tube is to place an ordinary pitot tube (without a static) facing the airflow at various positions in the tube. Connect the pitot tube to a U-tube, and leave the other side of the U-tube open to the atmospheric pressure outside the air stream. The pitot tube will record $p + \frac{1}{2}\rho V^2$, i.e. the static pressure in the stream plus the dynamic pressure, and the U-tube will therefore show the difference between this and the atmospheric pressure outside the stream. It will be found that $p + \frac{1}{2}\rho V^2$ is very nearly constant, whether the pitot tube is placed in the free air stream in front of the venturi, or in the mouth, or the throat, or near the exit. This is a convincing proof of Bernoulli's Theorem. The air speed increases from mouth to throat and then decreases again to the exit. The air speed increases very nearly in the same proportion as the area of cross-section of the venturi decreases, and this suggests that there is little or no change in the density of the air. Even more convincing evidence that the density does not change is provided by the flow of water through a venturi tube; the pattern of flow and the results obtained are very similar to those in air, and we know that water is for all practical purposes incompressible.

There are many practical examples of the venturi tube in everyday life, but there is no need to quote them, because we have sufficient examples in flying to illustrate this important principle. The choke tube in a carburettor is one; a wind tunnel is another, the experiments usually being done in the high-speed

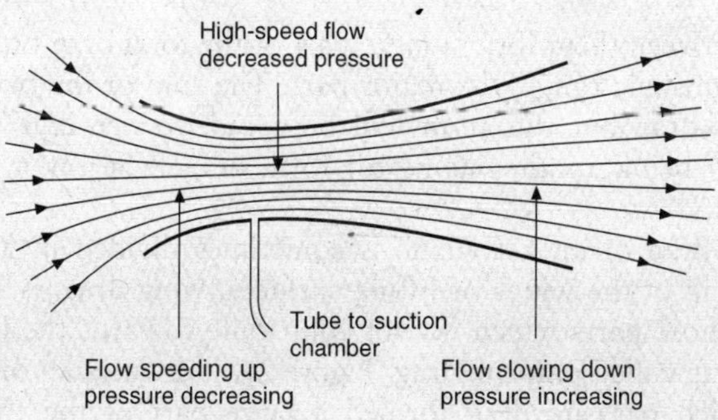

Fig 2.6 Flow through venturi tube

flow at the throat, and the air speed at this point is often measured by a single static hole in the side of the tunnel. A small venturi may be fitted inside a larger one, and the suction at the throat of the small venturi is then sufficient to drive gyroscopic instruments.

Air resistance or drag

Whenever a body is moved through air, or other viscous fluid, a definite resistance to its motion is produced. In aeronautical work this resistance is usually referred to as **drag**.

Drag is the enemy of flight, and efforts must be made to reduce the resistance of every part of an aeroplane to a minimum, provided strength and other essential factors can be maintained. For this reason many thousands of experiments have been carried out to investigate the problems of air resistance; in fact, in this, as in almost every branch of the subject, our knowledge is founded mainly on the mean result of accumulated experimental data. Nowadays, however, there are increasingly more accurate theoretical methods of estimating drag.

Streamlines and form drag

Lines which show the direction of the flow of the fluid at any particular moment are called **streamlines**. A body so shaped as to produce the least possible resistance is said to be of **streamline shape**.

We may divide the resistance of a body passing through a fluid into two parts –

1 **Form drag** or **Pressure drag**

2 **Skin friction** or **Surface friction**

These two between them form a large part of the total drag of an aeroplane – in the high subsonic range, the major part. The sum of the two is sometimes called **profile drag** but this term will be avoided since it is apt to give an impression of being another name for form drag, whereas it really includes skin friction.

The total drag of an aeroplane is sometimes divided in another way in which the drag of the wings or lifting surfaces, **wing drag**, is separated from the drag of those parts which do not contribute towards the lift, the drag of the latter being called **parasite drag**. Figure 2A illustrates an old type of aeroplane in which parasite drag formed a large part of the total. Figure 2B (overleaf) tells another story.

1. **Form Drag.** This is the portion of the resistance which is due to the fact that when a viscous fluid flows past a body, the pressure on the forward-facing part is on average higher than that on the rearward-facing portion. The extreme example of this type of resistance is a flat plate placed at right angles to the wind. The resistance is very large and almost entirely due to the pressure difference between the front and rear faces, the skin friction being negligible in comparison (Fig. 2.7).

Experiments show that not only is the pressure in front of the plate greater than the atmospheric pressure, but that the pressure behind is less than that of the atmosphere, causing a kind of 'sucking' effect on the plate.

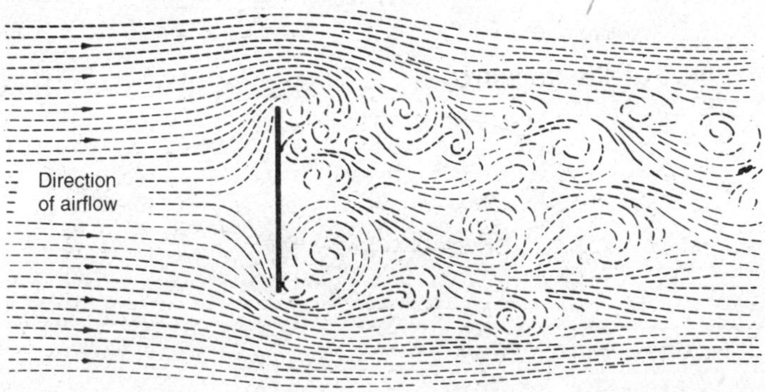

Fig 2.7 Form drag

Fig 2A Parasite drag A flying replica of the Vickers Vimy bomber which was the first aeroplane to make a non-stop flight across the Atlantic (in 1919). The extensive use of bracing wires and struts creates a great deal of parasite drag.

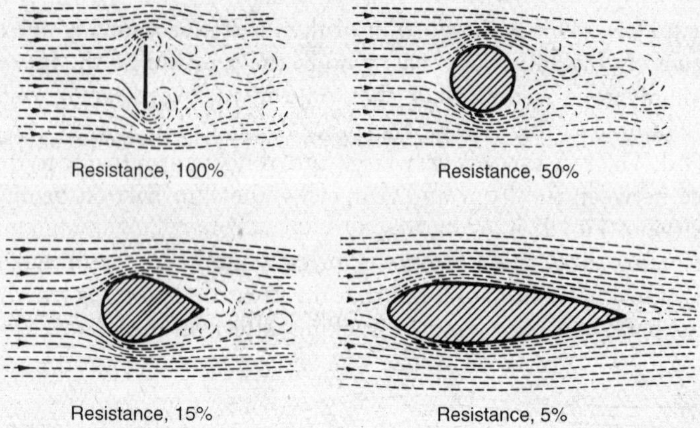

Fig 2.8 Effect of streamlining

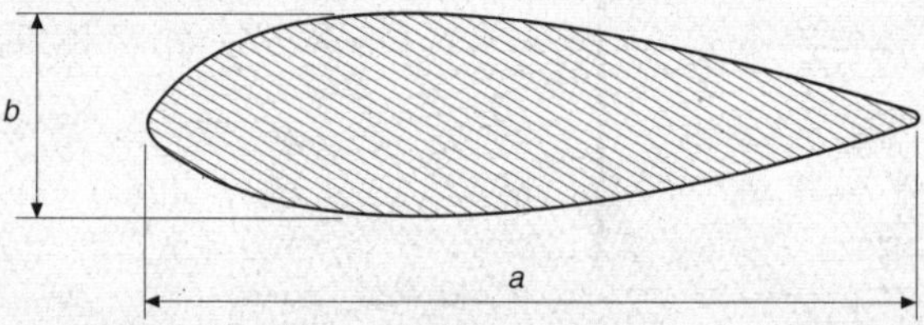

Fig 2.9 Fineness ratio

It is essential that form drag should be reduced to a minimum in all those parts of the aeroplane which are exposed to the air. This can be done by so shaping them that the flow of air past them is as smooth as possible, and much experimental work has been carried out with this in view. The results show the enormous advantage to be gained by the streamlining of all exposed parts; in fact, the figures obtained are so remarkable that they are difficult to believe without a practical demonstration. At a conservative estimate it can be said that a round tube has not much more than half the resistance of a flat plate, while if the tube is converted into the best possible streamline shape the resistance will be only one-tenth that of the round tube or one-twentieth that of the flat plate (Fig. 2.8).

The streamline shapes which have given the least resistance at subsonic speeds have had a **fineness ratio** – i.e. *a*/*b* – of between 3 and 4 (see Fig. 2.9), and the maximum value of b should be about one-third of the way back from the nose. These dimensions, however, may vary considerably without increasing the resistance to any great extent.

Fig 2B Streamlining
(By courtesy of the former British Aircraft Corporation)
Concorde 002 in contrast to the Vickers Vimy of Fig. 2A

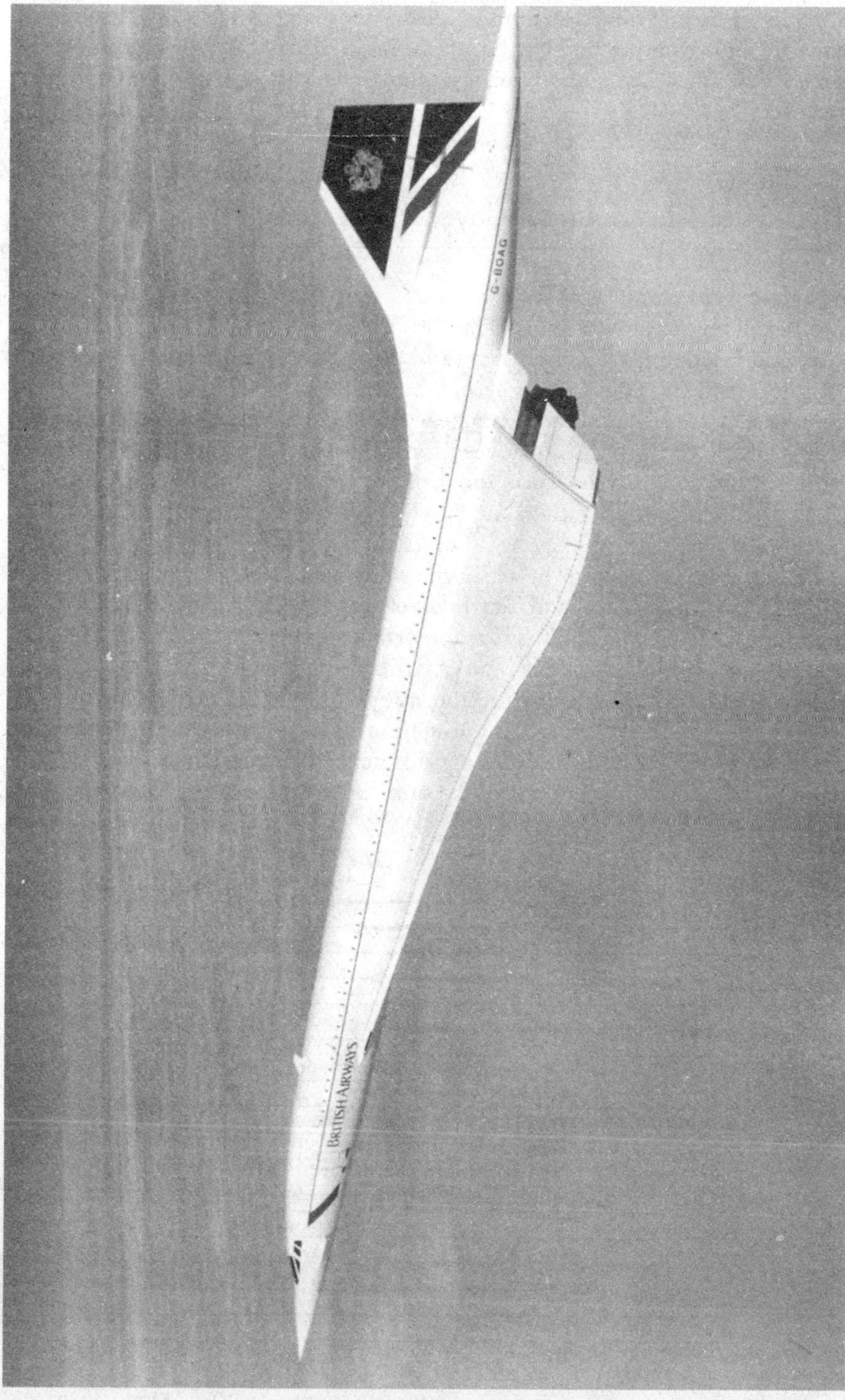
G-BOAG
BRITISH AIRWAYS

It should be mentioned that although we now have a fair idea of the ideal shape for any **separate** body, it by no means follows that two bodies of this shape – e.g. a fuselage and a wing – will give the least resistance when joined together.

Skin friction and boundary layer

Another consideration is that as we decrease the form drag the skin friction becomes of comparatively greater importance.

2. **Skin Friction**. Air is slowed up, and brought to a standstill, very close to a surface. If there is dust on an aeroplane wing before flight, it is usually still there after flight. **The layers of air near the surface retard the layers farther away** – owing to the friction between them, i.e. the viscosity – and so **there is a gradual increase in velocity as the distance from the surface increases** (Fig. 2.10). The distance above the plate in which the velocity regains a value close to that of the free stream may be no more than a few millimetres over a wing.

The layer or layers of air in which the shearing action takes place, that is to say between the surface and the full velocity of the airflow, is called the **boundary layer.** Owing to the great importance of skin friction, and necessity of keeping it within reasonable limits, particularly at high speed, much patient research work has been devoted to the study of the boundary layer.

Now the boundary layer, like the main airflow, may be either **laminar** or **turbulent** (Figs 2.11 and 2C), and the difference that these two types of flow make to the total skin friction is of the same order as the effect of streamlining

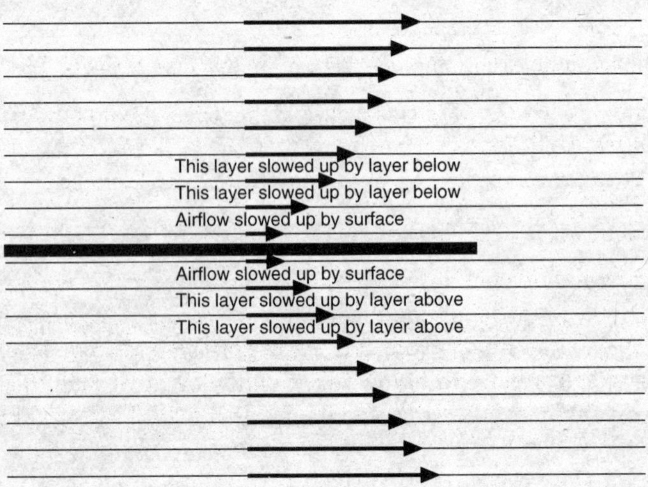

Fig 2.10 Skin friction

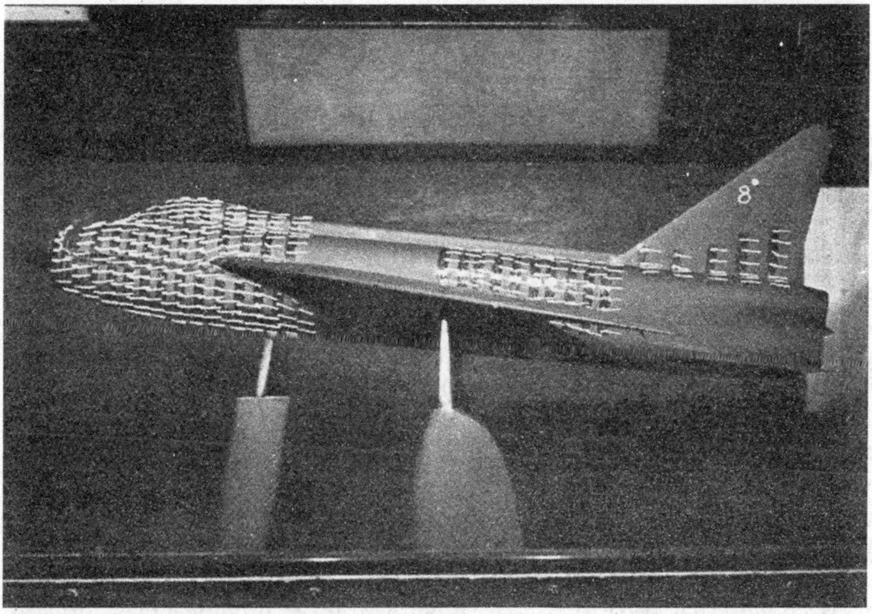

Fig 2C Investigating the boundary layer
(By courtesy of the former British Aircraft Corporation, Preston)
Wool tufts on model of a fuselage

the main flow. It has been stated that if we could ensure a laminar boundary layer over the whole surface of a wing the skin friction would be reduced to about one-tenth of its value.

The turbulent layer is characterised by high frequency eddies superimposed on the average velocity at each distance from the surface, while in the laminar case the 'layers' of air flow smoothly over each other. The turbulent layer, other factors being equal, has a much higher degree of shear at the surface, and it is this which causes the skin friction to be much higher than it is for the

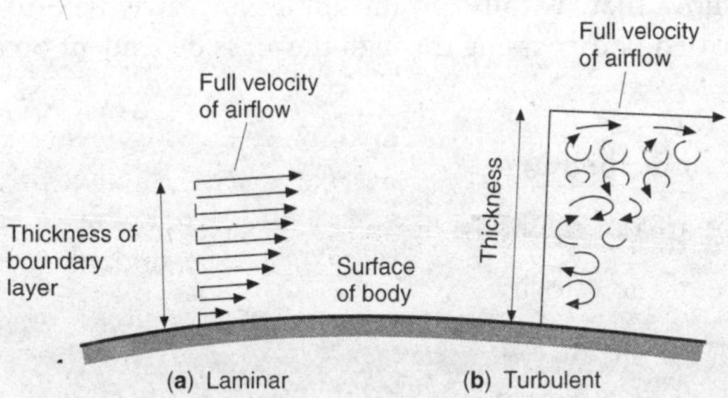

Fig. 2.11 Laminar and turbulent boundary layer
Thickness of layer greatly exaggerated

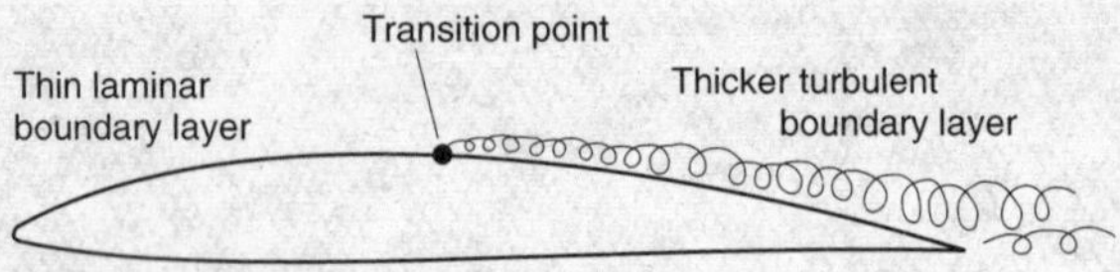

Fig 2.12 Transition point

laminar boundary layer. A smooth surface encourages a laminar layer, although other factors such as viscosity and the flow speed are also important. A smooth surface is also important when the boundary layer is turbulent as in this case the skin friction is reduced by a high degree of surface finish, although it remains considerably above the laminar value.

The usual tendency is for the boundary layer to start by being laminar near the leading edge of a body, but there comes a point, called the transition point, when the layer tends to become turbulent and thicker (Fig. 2.12). As the speed increases the transition point tends to move further forward, so more of the boundary layer is turbulent and the skin friction greater.

If this much is understood it will be obvious that the purpose of much research work has been to discover how the transition point moves forward, and how its movement can be controlled so as to maintain laminar flow over as much of the surface as possible.

But a further complication is that the behaviour of the boundary layer is very dependent on the size of an aerofoil (scale affect); this affects its relative thickness, whether it is laminar or turbulent, and how soon it separates from the surface. This is very important in wind tunnel testing which is discussed later. First let us look at how we can represent the drag produced by skin friction and separation.

Drag coefficient

Experiments show that, within certain limitations, it is true to say that the total resistance of a body passing through the air is dependent on the following factors –

(a) The shape of the body.

(b) The frontal area of the body.

(c) The square of the velocity.

(d) The density of the air.

Of these the velocity squared law is not strictly true at any speed and is definitely untrue at very low and very high speeds: when the speeds are low, matters are complicated by the way in which the boundary layer develops and, at high speeds, by the fact that the air may be compressed.

It is sometimes thought that the air is compressed in front of a body that is moving quite slowly through the air. We know that air is compressible but this does not come into play at speeds well below the speed of sound; at such speeds air behaves very like an almost incompressible liquid such as water. The passage of sound is, of course, caused by compression in the air, and it is only when speeds are reached in the neighbourhood of the speed of sound, about 340 m/s (661 knots), that appreciable compression of the air begins to take place. High-speed aircraft fly in this region, and beyond it. It is also interesting to note that the speed at which sound travels depends on the temperature of the air and becomes appreciably less at high altitudes, and thus the problem of reaching this critical velocity has become an important consideration in high-altitude flying. In such conditions there may be considerable departure from the velocity squared law, but for the low subsonic speeds – say, from 15 to 150 m/s – this law can be taken as accurate enough for practical purposes, so that double the speed means four times the resistance. As regards frontal area, when we are considering bodies of very different dimensions we must remember the scale effect to which we have already referred; we should also notice that, if we have a one-fifth scale model of a body, the frontal area of the full-sized body will be twenty-five times that of the model.

The term 'frontal area' means the maximum projected area when viewed in the direction of normal motion, so the frontal area of an aircraft is the maximum cross-sectional area when viewed from the front. In some instances the surface area would be a more sensible area to take – no general rule can be laid down, and the student should remember that the chief object of experiments on resistance is to compare the resistance of bodies of a similar kind. We are not very much concerned with how the resistance of a wing compares with the resistance of a wheel, but we do wish to compare the resistance of wings of different sizes and shapes, and also the resistances of different types of wheels. Therefore, if we choose one method to measure the area of wings and another to measure the area of wheels (as indeed we do), it does not matter very much. We must, however, agree on which reference area we are using in a particular case if we are to compare results sensibly. It is all a question of convenience. We shall return to this when we look at the lift, drag and pitching moments of an aerofoil in the next chapter.

The law of the variation of the resistance with the density of the air is found to be very nearly correct at ordinary densities, and on first thoughts points to the advantages of flight at high altitudes.

Assuming (a), (b), (c) and (d) to be true, we can express the result by the following formula for bodies of the same shape –

$$R \,\alpha\, \rho V^2 S_F$$

ρV^2 looks suspiciously like the dynamic pressure, $\frac{1}{2}\rho V^2$, and it is, perhaps, not surprising that the force generated by the air stream depends on this quantity.

The general formula for air resistance, or drag, can thus be written as –

$$R = C_D \frac{1}{2} \rho V^2 S_F$$

where C_D is a coefficient (known as the drag coefficient), which depends on the shape of the body and is found by experiment; ρ represents the density of the air, S_F the frontal area of the body, and V the velocity.

The units in this formula will correspond to those adopted in Chapter 1, i.e. the resistance (R) will be in newtons, the density (ρ) in kilograms per cubic metre, the area (S_F) in square metres, velocity (V) in metres per second and the drag coefficient (C_D) merely a number.

From this formula we can estimate the resistance of bodies moving through the air, provided we know the value of C_D for the particular shape concerned. This is usually found by experiment and, in the absence of more accurate information, the following values may be used –

for a flat plate $C_D = 1.2$ (normal to the flow direction)
for a circular tube $C_D = 0.6$ (axis normal to the flow direction)
for a streamline strut $C_D = 0.06$

EXAMPLE 10.1
Find the resistance of a flat plate, 15 cm by 10 cm, placed at right angles to an airflow of velocity 90 km/h. (Assume sea-level air density of 1.225 kg/m^3.)

SOLUTION
Data: $C_D = 1.2$
$\rho = 1.225\ \text{kg/m}^3$
$V = 90\ \text{km/h} = 25\text{m/s}$
$S_F = 15 \times 10 = 150\text{cm}^2 = 0.015\text{m}^2$

$$\text{Resistance} = C_D \frac{1}{2} \rho V^2 S_F = 1.2 \times 0.5 \times 1.225 \times 25 \times 25 \times 0.015 = 6.89\ \text{N}$$

Although we have so far applied this formula to resistance only, it is really of far wider application; it can, in fact, be used to represent any force and, with a small modification, moment, produced by the flow of air and the reader will be well advised to be sure that he understands just what it means. Therefore, let us sum up the position by saying that the aerodynamic drag experienced by any body depends on the shape of the body (represented by the coefficient C_D in the formula), the dynamic pressure of the air when it is of the given density and flowing at the given velocity (represented by $\frac{1}{2}\rho V^2$), and the size of the body, in this case given by the frontal area (represented by S_F).

We can base our coefficient on areas other than S_F, if this is more convenient; this we shall do in the next chapter, when we look at the forces and moments produced by an aerofoil. If we change the reference area, though, C_D will have to be correspondingly changed to keep the total resistance at the correct value; so we must agree on which reference area we are using in each case.

Wind tunnels

Because it is difficult to predict the forces on an aircraft (especially drag) with sufficient precision, scale models must be tested using wind tunnels to provide the necessary information.

In experimental work it is usual to allow the fluid to flow past the body rather than to move the body through the fluid. The former method has the great advantage that the body is at rest, and consequently the measurement of any forces upon it is comparatively simple. Furthermore, since we are only concerned with the **relative** motion of the body and the fluid, the true facts of the case are fully reproduced provided we can obtain a flow of the fluid which would be as steady as the corresponding motion of the body through the fluid.

Many experiments are carried out on models in wind tunnels. There are several types of tunnel, but probably the most commonly used is the closed working-section, closed-return type shown in Figs 2.13 and 2E (overleaf). The model is placed in the narrow working section and air enters through a contraction. The contraction makes the airflow speed in the working section more uniform, and also higher than in the rest of the circuit. Having a high flow speed in the whole circuit would increase the energy losses due to friction. The term closed-return refers to the fact that the air flows round in a complete circuit.

As an alternative, open circuit tunnels are sometimes used. In these, only the working section contraction and fan sections are required as illustrated in Fig. 2.14 (overleaf). The air is simply sucked in from the atmosphere through the contraction to the working section, and then exhausted back into the atmosphere, rather like a large vacuum cleaner. The advantage of the open-return type of tunnel is that it takes up much less space, and costs less than a closed-return type. The principal disadvantages are that dust is drawn in, and the

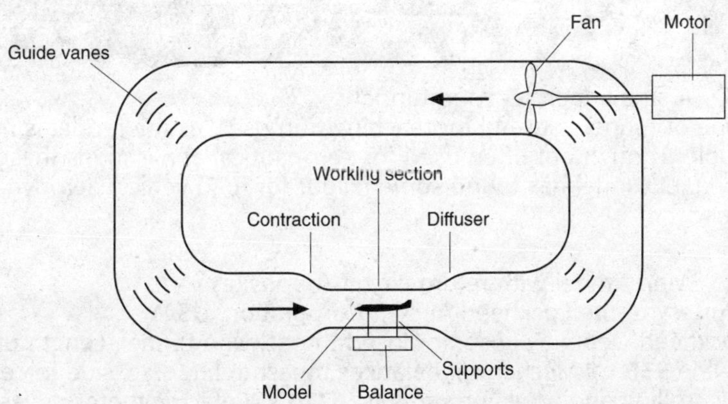

Fig 2.13 A closed working-section closed return wind tunnel

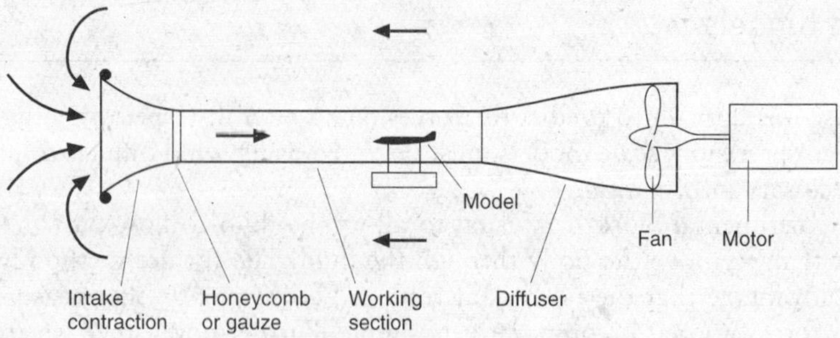

Fig 2.14 A simple open return tunnel

flow may be sensitive to external disturbances. Also, the pressure in the working section must be lower than atmospheric, since the air is drawn in from the atmosphere and speeded up. This means that any small leaks around the working section will pull in a jet of air. This type of tunnel is frequently used in college and university laboratories.

Fig 2D A small open jet wind tunnel
This type of tunnel is useful for teaching purposes, as the model is readily accessible. Though not often used for aeronautical applications these days, the open jet tunnel has found some favour for road vehicle aerodynamic testing.

Fig 2E Wind tunnel with return circuit (opposite)
(By courtesy of the Lockheed Aircraft Corporation, USA)
The Lockheed-Georgia subsonic closed-throat wind tunnel. Length of centre line 238 m. Mechanical balances measure lift, drag, side force and pitching, rolling and yawing moments. 6710 kW electric motor drives fan of 12 m diameter.

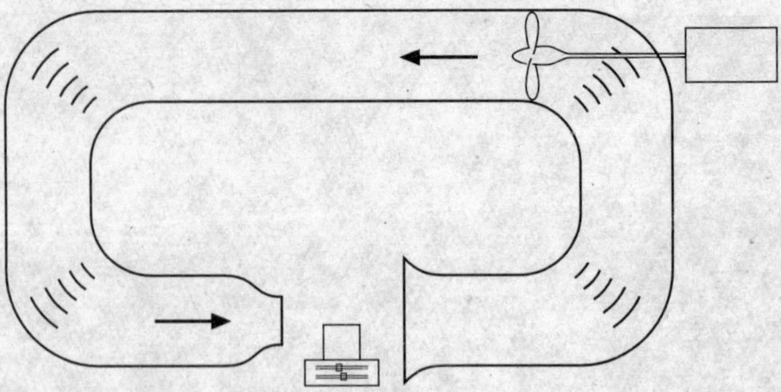

Fig 2.15 An open jet tunnel

Another type of tunnel that is sometimes employed is the open jet type illustrated in Figs 2D (earlier) and 2.15. In this type of tunnel the working section is not enclosed, which gives it its main advantage, accessibility. For teaching purposes, the open working section is particularly useful, and this type of tunnel is also popular for automotive aerodynamic studies where the effects of wall constraint are less predictable than for aircraft. There are other types of tunnel such as the slotted wall, but let us not confuse ourselves with such subtleties at this stage.

Other common types of tunnel are of course the supersonic and transonic, but these are described later in the book, after compressible flow has been explained.

Wind-tunnel balances

To measure the forces exerted by the airflow, the model is normally mounted on to a balance which may be a mechanical or an electrical type. The older mechanical force balance is rather like a simple weighing machine of the type used to weigh babies or patients in a hospital. As illustrated in Fig. 2.16, the force to be measured (such as the lift on the model) is applied via a series of levers and

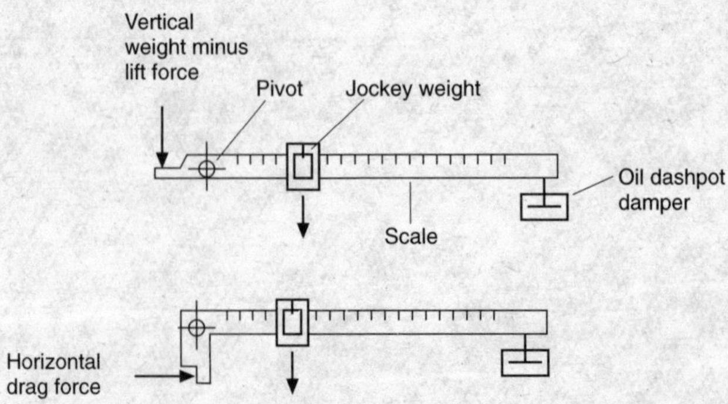

Fig 2.16 A simple balance

pivots or flexures to an arm upon which is placed a jockey weight. The moment of the applied force is balanced by moving the jockey weight along the arm. The position of the jockey weight therefore indicates the magnitude of the force. The further along the arm the jockey weight has to be moved, the greater is the force being balanced. The same principle is used to measure horizontal drag or thrust forces, as shown in Fig. 2.16. One arm has to be used for each of the three forces and three moments that make up the six 'components' illustrated in Fig. 2.17. In small college tunnels it is normal to have only three arms which measure the three most important components, lift, drag and pitching moment. The design of the levers and pivots is very complicated because it is important that a change in the vertical lift force does not affect the arm that is supposed to measure only the horizontal drag force. The balance unit which is large is situated outside the tunnel, and the model is attached to it by means of a number of rods or wires (see Fig. 2D, earlier). Originally the jockey weights were moved by hand, but nowadays they can either be moved automatically by a servo electric motor, or are left fixed, with the force of the arm being measured by an electrical force transducer. A force transducer is an instrument that produces an electrical output that is proportional to the applied force. This brings us to the second form of balance, the electronic type.

The electronic force balance consists of a carefully machined block of metal that is attached to the model at one end and to a supporting structure at the other; this often takes the form of a single rod or 'sting' protruding from the rear of the model. Electrical resistance strain gauges attached to the block produce output voltages proportional to the applied forces. Up to six components can be measured. This type of balance is very small and compact and is normally contained within the model. One potential disadvantage is that there is usually a certain amount of unwanted interference between the different components; changes in lift affect the drag reading, etc. However, the computer-based data acquisition systems to which such balances are invariably attached are able to make corrections and allowances automatically.

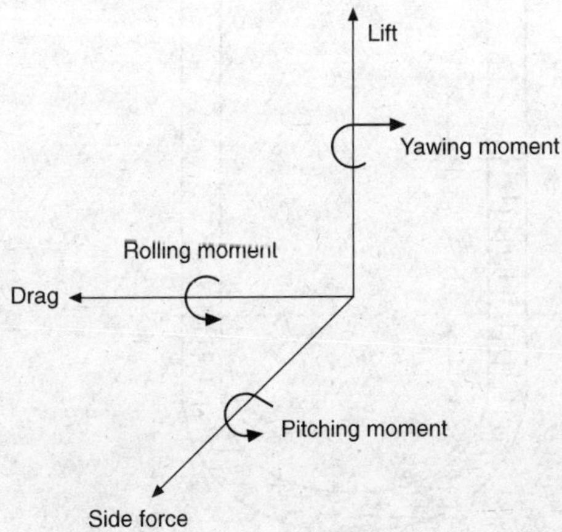

Fig 2.17 The six 'components'

Measuring pressure

We can measure the pressure on the model surface by connecting flexible tubes between small holes (pressure tappings) drilled in the surface and a suitable pressure-measuring device. A simple way to measure the pressure is to use a liquid-filled tube U-tube (Fig 2.18). If the pressure in the tube, connected to the hole in the model, is lower than atmospheric, the liquid level will rise; rather like sucking a drink up a straw. If the pressure is higher than atmospheric (as in Fig 2.18), the liquid will be pushed further down the tube, like blowing down the straw.

The idea is quite simple. The column of liquid between A and B is supported by the difference between the test pressure and atmospheric pressure, which produces an upward force on the column equal to the weight of the liquid between A and B. As we saw for the liquid column in Chapter 1, the upward force is equal to the pressure difference multiplied by the cross-sectional area of the tube. This balances the weight of the liquid, which is equal to its density, ρ, multiplied by the gravitational constant, g, the cross-sectional area of the tube and the height of the column, h.

So what pressure difference is the equivalent to a manometer deflection (pressure 'head' – see Chapter 1) of 1m of water?

Pressure difference $= \rho \times g \times h$

So, putting in the values in SI units, we get

Pressure difference $= 1000 \times 9.81 \times 1 = 9810\ \mathrm{N/m^2}$

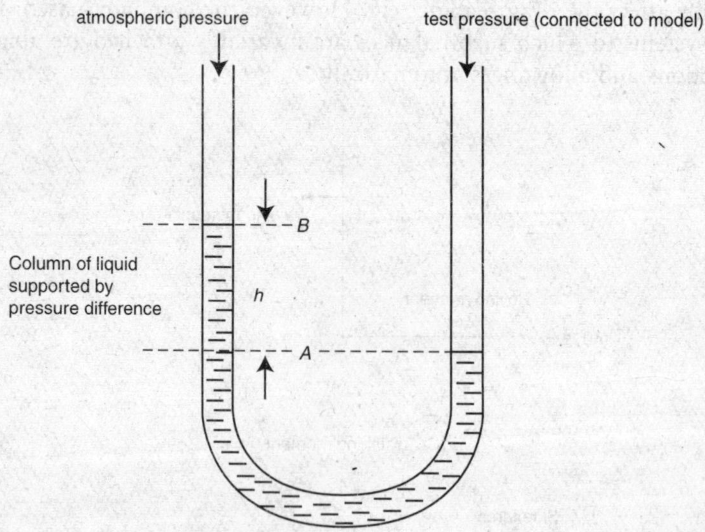

Fig 2.18 Measuring pressure

Absolute and gauge pressure

In the simple U-tube manometer above, the liquid surface at B is left open to the atmosphere, so –

$$\text{Pressure difference} = \text{test pressure} - \text{atmospheric pressure} = \rho \times g \times h$$

Thus –

$$\text{test pressure} = \text{atmospheric pressure} + \rho \times g \times h.$$

This is known as the **absolute pressure.** The pressure difference ($= \rho \times g \times h$) simply gives the pressure change relative to atmospheric pressure. In practice this relative pressure is all we need to know. As this pressure is the pressure obtained directly from the gauge, it has come to be referred to as **gauge pressure.** Sometimes it is convenient to use a different pressure source to act as our reference, instead of atmospheric pressure. We can see that –

$$\text{absolute pressure} = \text{gauge pressure} + \text{atmospheric (or reference) pressure}$$

Nowadays it is generally more convenient to use **pressure transducers** (Fig. 2F) to measure the pressure at the tappings.

Fig. 2F Pressure transducer

The transducer produces an electrical output that is proportional to the pressure being measured and which can be recorded, through a suitable interface, directly on a computer system. Pressure transducers work in a variety of ways. For example, the pressure can be used to deform a diaphragm, which has electrical strain gauges attached to it. They can also be purchased in a wide variety of sizes and pressure ranges. Some pressure transducers measure the pressure relative to a reference source (gauge pressure transducers), others measure relative to a sealed vacuum and thus directly measure absolute pressure.

Other wind tunnel measurements

As well as measuring the forces and pressures on a wind tunnel model, we may want to investigate more detailed features of the flow, so that we can improve the design of our aircraft. We can do this by measuring the velocity at different points in the flow and by using flow visualisation to help us to see what the flow is doing.

One way we can measure speed is to traverse the flow using a miniaturised version of the pitot-static tube (Fig 2.5 earlier), although we must be careful to line it up with the local flow direction. If we want to measure fluctuations in turbulent flow, however, this will not respond fast enough. For this type of speed measurement, we can use a hot wire anemometer or a laser Doppler anemometer. In the hot wire anemometer a very fine wire is heated electrically. The airflow cools the wire and this changes its electrical resistance, which can be measured electronically. Because the wire is so fine (about 5 micro-metres in diameter) it can measure very rapid fluctuations in airspeed. The laser Doppler anemometer (LDA) works by illuminating a very small volume of air at the crossing point of two laser beams. The beams interfere with each other and produce a series of 'fringes' (dark and light stripes). Particles, which are light enough to follow the airstream, are introduced into the flow (seeding) and each reflects a pulse of light as it travels through each light fringe. The frequency of the reflected light is then used to measure the velocity of the particle but some very clever signal processing is needed to sort things out when particles overlap in the illuminated area or when very large particles come through, which are too big to follow the flow. Seeding can also be used to measure the velocity in a 'slice' of the flow illuminated by a light sheet. Pictures are taken, in quick succession, of particles illuminated by the sheet and their velocities are deduced from their changes in position. This is called 'Particle Image Velocimetry' (PIV for short).

Particles in the flow can also be used to make the flow visible. Smoke is often used for this. We can also see what the flow is doing on the surface of a model by using short wool tufts, stuck to the surface at their upstream ends, or by the use of surface oil flow. A common oil flow technique is to mix a white powder such as titanium dioxide with kerosene and a few drops of oleic acid. After running the wind tunnel the titanium dioxide forms streaks showing the pattern of flow on the surface. A note of caution – oil flow needs considerable experience in interpretation as the flow may differ considerably a small distance from the surface, outside the boundary layer.

Sources of error

Wind tunnel experiments on models, even at subsonic speeds, are liable to three main sources of error when used to forecast full-scale results. These are –

1. **Scale Effect.** As mentioned earlier, laws of resistance can be framed which apply well to bodies whose sizes are not very different, but these laws become less accurate when there is a great difference in size between the model and the full scale. A similar effect is noticed when the velocity of the model test differs appreciably from the full-scale velocity.

Corrections can be applied which allow for this 'scale effect' and enable more accurate forecasts to be made. Readers who are interested will find an explanation of scale effect, and of the advantages of the compressed air tunnel, if they refer to Appendix 2. They will also be introduced to the important term Reynolds Number.

2. **Interference from Wind Tunnel Walls** (Fig. 2.19). The second error is due to the fact that in the wind tunnel the air stream is confined to the limits of the tunnel, whereas in free flight the air round the aeroplane is, for all practical purposes, unlimited in extent. In this case too corrections can be applied which considerably reduce the error.

3. **Errors in Model.** The smaller the scale of the model, the more difficult does it become to reproduce every detail of the full-scale body, and since very slight changes of contour may considerably affect the airflow, there will always be errors due to the discrepancies between the model and the full-scale body.

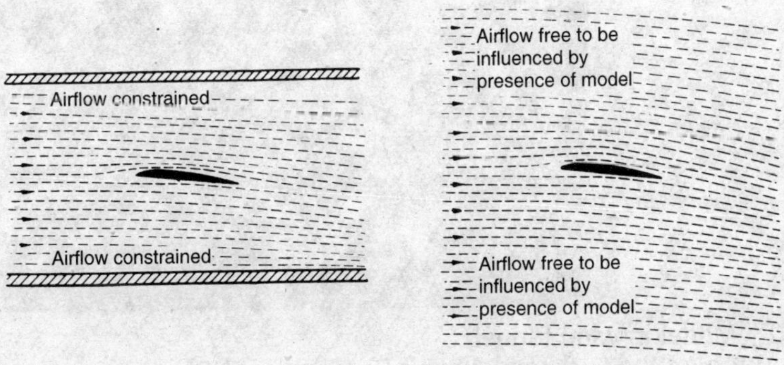

Fig 2.19 Interference from wind tunnel walls

Other experimental methods

Provided these three limitations are fully realised, and due allowances made for them, wind tunnel results can provide us with some very useful experimental data (Fig. 2G).

In addition to wind tunnel tests, experiments may be performed in the following ways –

1. By experiments in water instead of air.
2. By experiments in actual flight.

Computational fluid dynamics (CFD)

Nowadays computers provide a powerful tool for the investigation of the flow around an aircraft or its components. The equations that describe the mechanics of the airflow (the **Navier-Stokes** equations) were known well

Fig 2G Model in wind tunnel
(By courtesy of the Lockheed Aircraft Corporation, USA)
Model being prepared for flutter tests in wind tunnel

before the first flight. However, nobody could solve them, except for a few trivial cases. Even with the advent of computers, capable of performing millions of calculations per second, it is not possible to solve the exact equations but they can be solved approximately. With this powerful tool you might think that there is now no need for wind tunnels. However, there are problems. Data from wind tunnels have to be used in CFD to model turbulence and CFD is not very good at accurately predicting flows where there are large turbulent areas, such as at stall. Predicting flows is therefore still something of a black art. Tunnel results are prone to scale effects, which are difficult to interpret and CFD also has its problems (as does flight testing – it is hard, for example, to sort out effects due to the engine from those due to the airframe). Like most things, there is no substitute for experience and ever more sophisticated ways are being developed to use the power of CFD and wind tunnel testing and flight testing to complement each other.

In the next chapter we look at how we use an aerofoil to produce the more desirable aerodynamic force, lift. Before reading this see if you can answer the following questions. If you can do so you have probably understood most of this chapter and you may proceed with confidence, but you should also try the numerical questions in Appendix 3.

Can you answer these?

1. Why is it difficult to find out the exact height of the atmosphere using a pressure altimeter?
2. What is meant by the density of air?
3. Which falls off more rapidly with height, the density or the pressure?
4. Why does an altimeter have an adjustment so that it can be set before each flight?
5. What are the chief differences between the atmosphere at sea-level and at 30 000 ft, or 10 000 m?
6. What do you understand by the term 'streamline shape'?
7. Distinguish between 'indicated' air speed and 'true' air speed.
8. What is the significance of $\frac{1}{2}\rho V^2$?
9. What is 'position error'?
10. What is the difference between the 'troposphere' and the 'stratosphere'?

11. What is the meaning of (*a*) subsonic, and (*b*) supersonic speeds?
12. What does the symbol 'q' stand for?

For solutions see Appendix 5.
For numerical questions on air and airflow see Appendix 3.

CHAPTER 3

Aerofoils – subsonic speeds

So far we have only considered the resistance, or drag, of bodies passing through the air. In the design of aeroplanes it is our aim to reduce such resistance to a minimum. We now come to the equally important problem of how to generate a force to lift or support the weight of an aircraft.

In the conventional aeroplane this is provided by wings, or aerofoils, which are inclined at a small angle to the direction of motion, the necessary forward motion being provided by the thrust of a rotating airscrew, or by some type of jet or rocket propulsion. These aerofoils are usually slightly curved, but in the original attempts to obtain flight on this system flat surfaces were sometimes used.

Lifting surfaces

If air flows past an aerofoil, a flat plate or indeed almost any shape that is inclined to the direction of flow, we find that the pressure of air on the top surface is reduced while that underneath is increased (Fig. 3.1, overleaf). This difference in pressure results in a net force on the plate trying to push it both upwards and backwards. In the case of a simple flat plate, you might imagine that the net force would act at right angles to the plate. This is not so, because there is also a tangential force caused by the different pressures that act on the small leading and trailing edge face areas. This tangential force, though small, is by no means negligible. Rather surprisingly, the pressure at the leading edge is normally very low, and at small angles of inclination, the tangential force will act in the direction shown in Fig. 3.2 (overleaf). The reasons for the low pressure at the leading edge will be shown later. Note, that although the tangential force may be directed towards the front of the plate, the resultant of the tangential and normal forces must always be tilted back relative to the local flow direction.

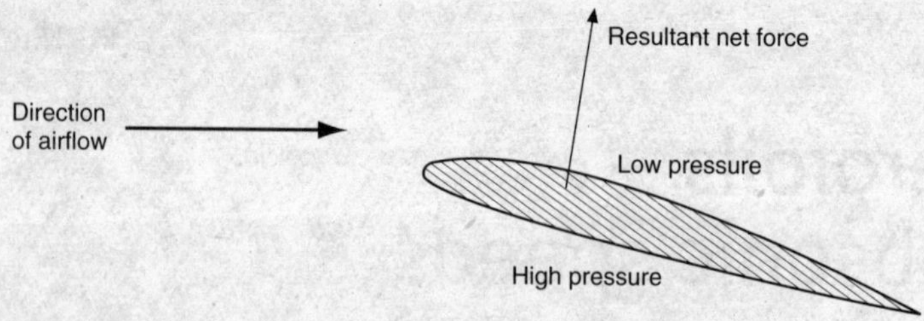

Fig 3.1 Resultant force on an aerofoil due to pressure difference

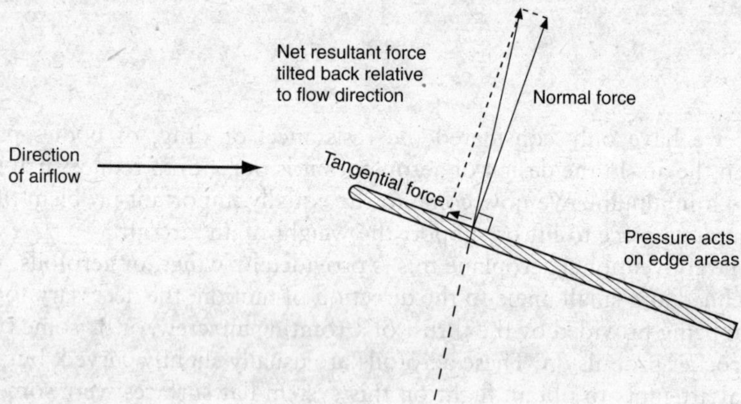

Fig 3.2 Forces due to pressure differences in a flat plate

Lift and drag

The resultant or net force on the lifting surface may be conveniently split into two components relative to the airflow direction as follows –

1. **The component at right angles to the direction of the airflow,** called **LIFT** (Fig. 3.3).
2. **The component parallel to the direction of the airflow,** called **DRAG** (Fig. 3.3)

The use of the term 'lift' is apt to be misleading, for under certain conditions of flight, such as a vertical nose dive, it may act horizontally, and cases may even arise where it acts vertically downwards.

Airflow and pressure over aerofoil

It was soon discovered that a much greater lift, especially when compared with the drag, could be produced by using a curved surface instead of a flat one,

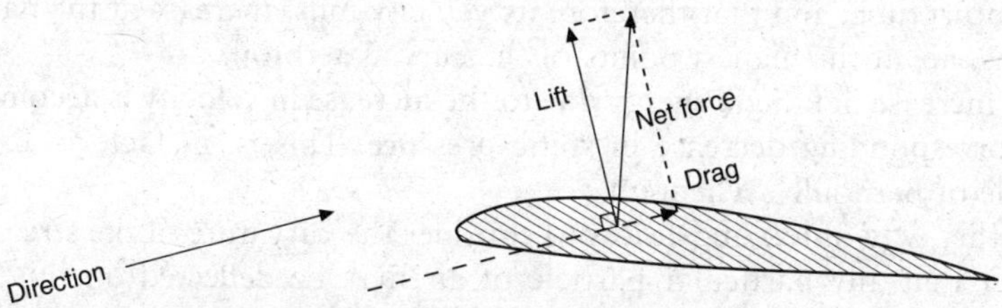

Fig 3.3 Lift and drags shown for the case of a descending aircraft

and thus the modern aerofoil was evolved. The curved surface had the additional advantage that it provided a certain amount of thickness, which was necessary for structural strength.

Experiments have shown that air flows over an aerofoil (Fig. 3.4) much more smoothly than over a flat plate.

In Fig. 3.4, which shows the flow of air over a typical aerofoil, the following results should be noticed –

1. **There is a slight upflow before reaching the aerofoil.**
2. **There is a downflow after passing the aerofoil.** This downflow should not be confused with the downwash produced by the trailing vortices as described later.
3. **The air does not strike the aerofoil cleanly on the nose, but actually divides at a point just behind it on the underside.**
4. **The streamlines are closer together above the aerofoil where the pressure is decreased.**

This last fact is at first puzzling, because, as in the venturi tube, it may lead us to think that the air above the aerofoil is compressed, and that therefore we should expect an increased pressure. The explanation is that the air over the

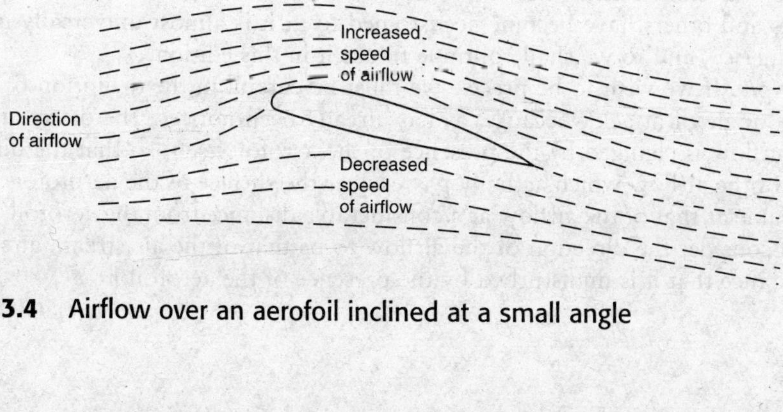

Fig 3.4 Airflow over an aerofoil inclined at a small angle

top surface acts as though it were passing through a kind of bottleneck, similar to a venturi tube, and that therefore its velocity must increase at the narrower portions, i.e. at the highest points of the curved aerofoil.

The increase in kinetic energy due to the increase in velocity is accompanied by a corresponding decrease in static pressure. This is, in fact, an excellent example of Bernoulli's Theorem.

Another way of looking at it is to consider the curvature of the streamlines. In order that any particular particle of air may be deflected on this curved path, a force must act upon it towards the centre of the curve, so it follows that the pressure on the outside of the particle must be greater than that on the inside; in other words, the pressure decreases as we move down towards the top surface of the aerofoil. This point of view is interesting because it emphasises the importance of curving the streamlines.

Chord line and angle of attack

It has already been mentioned that the angle of inclination to the airflow is of great importance. On a curved aerofoil it is not particularly easy to define this angle, since we must first decide on some straight line in the aerofoil section from which we can ensure the angle to the direction of the airflow. Unfortunately, owing to the large variety of shapes used as aerofoil sections it is not easy to define this **chord line** to suit all aerofoils. Nearly all modern aerofoils have a convex under-surface; and the chord must be specially defined, although it is usually taken as the line joining the leading edge to the trailing edge. This is the centre in the particular case of symmetrical aerofoils.

We call the angle between the chord of the aerofoil and the direction of the airflow the angle of attack (Fig. 3.5).

This angle is often known as the angle of incidence; that term was avoided in early editions of this book because it was apt to be confused with the riggers' angle of incidence, i.e. the angle between the chord of the aerofoil and some fixed datum line in the aeroplane. Now that aircraft are no longer 'rigged' (in the old sense) there is no objection to the term angle of incidence; but by the same token there is no objection either to angle of attack – many pilots and others have become accustomed to it; it is almost universally used in America, and so we shall continue to use it in this edition.

[*Note*. If we wish to be precise we must be careful in the definition of the term 'angle of attack', because, as has already been noticed, the direction of the airflow is changed by the presence of the aerofoil itself, so that the direction of the airflow which actually passes over the surface of the aerofoil is not the same as that of the airflow at a considerable distance from the aerofoil. We shall consider the direction of the airflow to be that of the air stream at such a distance that it is undisturbed by the presence of the aerofoil.]

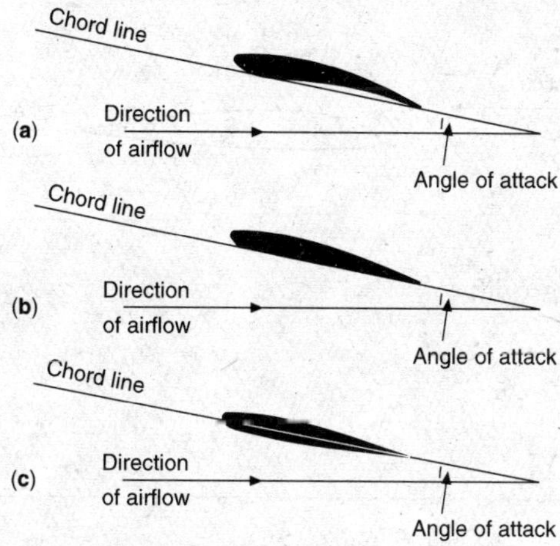

Fig. 3.5 Chord line and angle of attack
(*a*) Aerofoil with concave undersurface.
(*b*) Aerofoil with flat undersurface.
(*c*) Aerofoil with convex undersurface.

Line of zero lift

Now an aerofoil may provide lift even when it is inclined at a slightly negative angle to the airflow. And one may well ask, how can an aerofoil inclined at a negative angle produce lift? The idea seems absurd, but the explanation of the riddle is simply that the aerofoil is not really inclined at a negative angle. Our curious chord may be at a negative angle, but the curved surfaces of the aerofoil are inclined at various angles, positive and negative, the net effect being that of a slightly positive angle, which produces lift.

If we tilt the nose of the aerofoil downwards until it produces no lift, it will be in an exactly similar position to that of a flat plate placed edgewise to the airflow and producing no lift, and if we now draw a straight line through the aerofoil parallel to the airflow (Fig. 3.6, overleaf) it will be the inclination of this line which settles whether the aerofoil provides lift or not.

Such a line is called the line of zero lift or neutral lift line, and would in some senses be a better definition of the chord line, but it can only be found by wind tunnel experiments for each aerofoil, and, even when it has been found, it is awkward from the point of view of practical measurements.

Nor is it of much significance in practical flight, except perhaps in a dive when the angle of attack may approach the no lift condition.

Note that for an aerofoil of symmetrical shape zero lift corresponds to zero angle of attack.

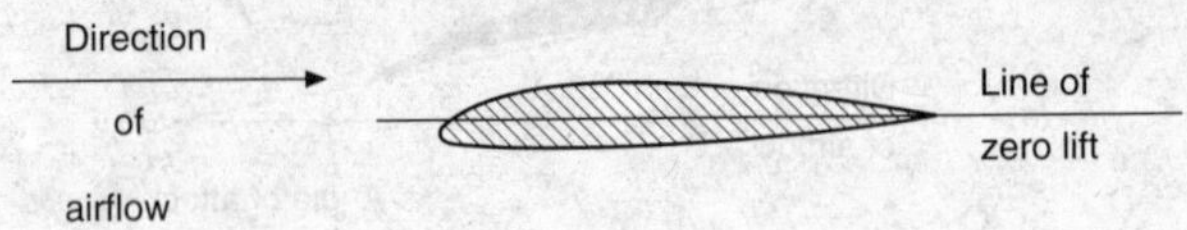

Fig 3.6 Line of zero lift

Pressure plotting

As the angle of attack is altered the lift and drag change very rapidly, and experiments show that this is due to changes in the distribution of pressure over the aerofoil. These experiments are carried out by the method known as 'pressure plotting' (Fig. 3.7), in which a number of small holes in the aerofoil surface (*a*, *b*, *c*, *d*, etc.) are connected to a number of glass manometer tubes (*a*, *b*, *c*, *d*, etc.) containing water or other liquid and connected to a common reservoir. Where there is a suction on the aerofoil the liquid in the corresponding tubes is sucked up; where there is an increased pressure the liquid is depressed. This is really several U-tube manometers connected to a common reservoir (p. 60). Such experiments have been made both on models in wind tunnels and on aeroplanes in flight, and the results are most interesting and instructive.

The reader is advised to work through Example No. 94 in Appendix 3. In this example the results of an actual experiment are given, together with a full explanation of how to interpret the results. In order to follow through to the end of this example it is necessary to have a knowledge of the lift formula given later in this chapter, but the actual 'pressure plotting' can be done without this. Multiple manometers provide a good visual indication of the form of a pressure distribution and are frequently used for teaching. If results are to be recorded, it is more convenient to use pressure transducers coupled to a computer (p. 61).

Pressure distribution

Figure 3.8 shows the pressure distribution, obtained in this manner, over an aerofoil at an angle of attack of 4°. Two points are particularly noticeable, namely –

1. **The decrease in pressure on the upper surface is greater than the increase on the lower surface.**

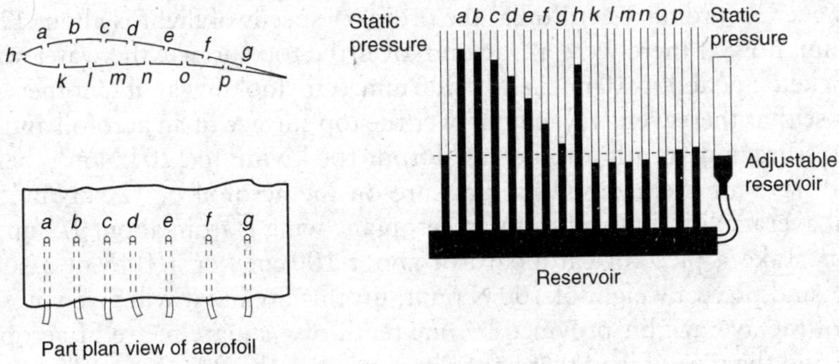

Fig 3.7 Pressure plotting

2. **The pressure is not evenly distributed,** both the decreased pressure on the upper surface and the increased pressure on the lower surface being most marked over the front portion of the aerofoil.

Both these discoveries are of extreme importance.

The first shows that, although both surfaces contribute, it is the **upper surface**, by means of its decreased pressure, which provides the greater part of the lift; at some angles as much as four-fifths.

The student is at first startled by this fact, as this seems contrary to common sense; but, as so often happens, having learnt the truth, he is inclined to exaggerate it, and to refer to the area above the aerofoil as a 'partial vacuum' or even a 'vacuum'. Although, by a slight stretch of imagination, we might allow the term 'partial vacuum', the term 'vacuum' is hopelessly misleading. We find that the greatest height to which water in a manometer is sucked up when air

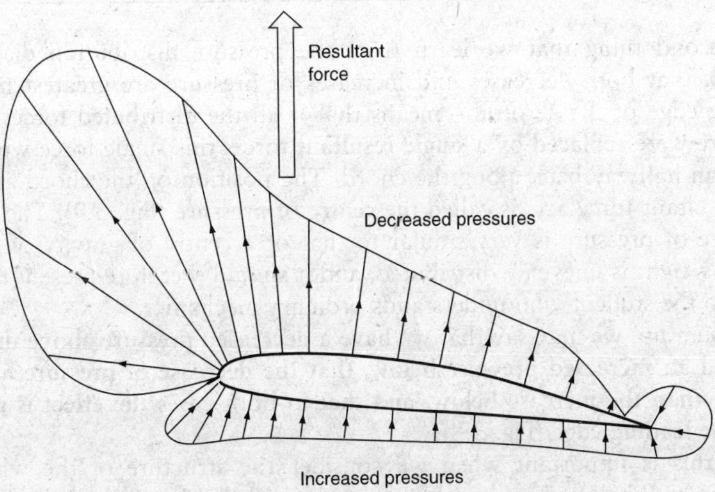

Fig 3.8 Pressure distribution over an aerofoil

flows over an ordinary aerofoil at the ordinary speeds of flight is about 120 to 150 mm; now, if there were a 'vacuum' over the top surface, the water would be sucked up about 10 m, i.e. 10 000 mm. Or, looking at it another way, suppose that there were a 'vacuum' over the top surface of an aerofoil and that the pressure underneath was increased from 100 kN/m^2 to 120 kN/m^2, then we would have an average upward pressure on the aerofoil of 120 kN/m^2. The actual average lift obtained from an aeroplane wing is from about 1/2 up to 5 kN/m^2. Take a piece of cardboard of about 100 cm^2, or 1/10th of a square metre, and place a weight of 100 N on it; lift this up and it will give you some idea of the average lift provided by one-tenth of a square metre of aeroplane wing, and the type of load that has to be carried by the skin. You will not want to repeat the experiment with more than 10 000 N on the cardboard!

The reason why the pressure distribution diagram of Fig. 3.8 has not been completed round the leading edge is because the changes of pressure are very sudden in this region and cannot conveniently be represented on a diagram. The increased pressure on the underside continues until we reach a point head-on into the wind where the air is brought to rest and the increase of pressure is $1/2\,\rho V^2$, or q, as recorded on a pitot tube. The point at which this happens is called the **stagnation point**, and its position round the leading edge varies slightly as the angle of attack of the aerofoil is changed but is always just behind the nose on the underside of positive angles of attack. After the stagnation point there is a very sudden drop to zero, followed by an equally sudden change to the decreased pressure of the upper surface, and rather surprisingly on the nose.

Centre of pressure

The second thing that we learn from the pressure distribution diagram – namely, that both decreases and increases of pressure are greatest near the leading edge of the aerofoil – means that if all the distributed forces due to pressure were replaced by a **single resultant force**, this single force would act less than halfway back along the chord. The position on the chord at which this resultant force acts is called the **centre of pressure** (Fig. 3.9). The idea of a centre of pressure is very similar to that of a centre of gravity of a body whose weight is unevenly distributed, and it should therefore present no difficulty to the student who understands ordinary mechanics.

To sum up, we may say that we have a **decreased pressure above the aerofoil and an increased pressure below, that the decrease of pressure above is greater than the increase below, and that in both cases the effect is greatest near the leading edge** (Fig. 3.8).

All this is important when we consider the structure of the wing; for instance, we shall realise that the top surface or 'skin' must be held down on to the ribs, while the bottom skin will simply be pressed up against them.

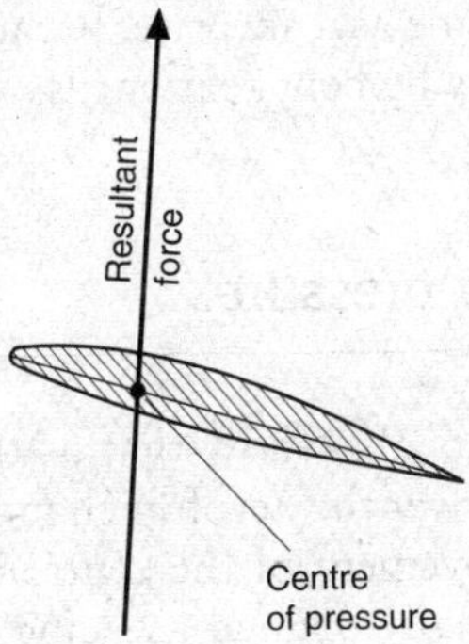

Fig 3.9 Centre of pressure

Total resultant force on an aerofoil

If we add up the distributed forces due to pressure over an aerofoil, and replace it by the total resultant force acting at the centre of pressure, we find that this force is not at right angles to the chord line nor at right angles to the flight direction. Near the tips of swept wings it can sometimes be inclined forward relative to the latter line due to rather complicated three dimensional effects, but over most of the wing, and on average, it must always be inclined backwards, otherwise we would have a forwards component, or negative drag, and hence perpetual motion.

Although the force must on average be inclined backwards relative to the flight direction as in Fig. 3.10 it can often be inclined forwards relative to the chord line normal. Figure 3.10 illustrates the situation. You will see from this figure that there can be a component of the force that is trying to bend the

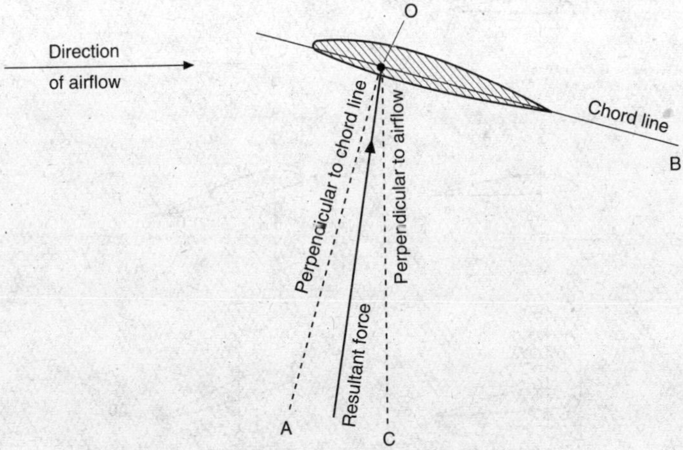

Fig 3.10 Inclination of resultant force

wings forward. This may come as a surprise, because you might have expected that the wings would always be bent rearwards.

Movement of centre of pressure

Pressure plotting experiments also show that as the angle of attack is altered the distribution of pressure over the aerofoil changes considerably, and in consequence there will be a movement of the centre of pressure. The position of the centre of pressure is usually defined as being a certain proportion of the chord from the leading edge. Figure 3.11 illustrates typical pressure distribution over an aerofoil at varying angles of attack. In these diagrams only the lift component of the total pressure has been plotted – the drag component has hardly any effect on the position of the centre of pressure. It will be noticed that at a negative angle, and even at 0°, the pressure on the upper surface near the leading edge is increased above normal, and that on the lower surface is decreased; this causes the loop in the pressure diagram, which means that this portion of the aerofoil is being pushed downwards, while the rear portion is being pushed upwards, so that the whole aerofoil tends to turn over nose first.

So, even at the angle of zero lift, when the upward and downward forces are equal, there is a nose-down pitching movement on the aerofoil; as will be seen later this is a matter of considerable significance. Putting it another way,

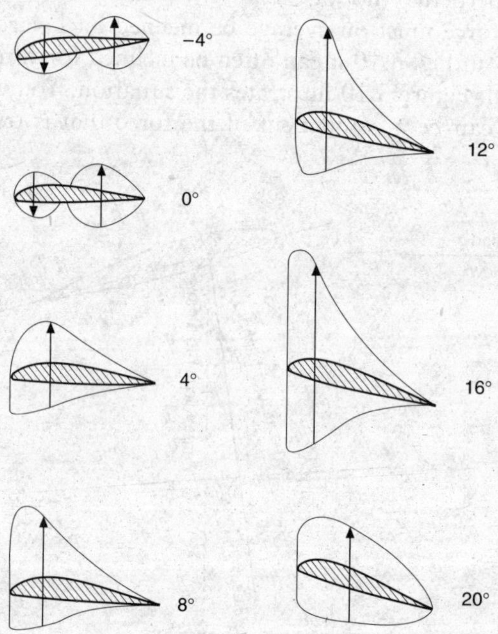

Fig 3.11 How the lift distribution changes with angle of attack

at these negative angles the centre of pressure is a long way back – the only place where we could put one force which would have the same moment or turning effect as the distributed pressure would be a long way behind the trailing edge, in fact at zero lift it could not provide a pitching moment at all unless it were an infinite distance back – which is absurd. Perhaps a more sensible way of putting it is to say that there is a couple acting on the aerofoil, and a couple has no resultant and has the same moment about any point (see Chapter 1).

As the angle of attack is increased up to 16°, the centre of pressure gradually moves forward until it is less than one-third of the chord from the leading edge; above this angle it begins to move backwards again.

Now during flight, for reasons we shall see later, the angle of attack is usually between 2° and 8° and is very rarely below 0° or above 16°. So, for the ordinary angles of flight, as the angle of attack of the aerofoil is increased, the centre of pressure tends to move forward.

Lift a pencil at its centre of gravity and it will lie horizontal; move the position at which you lift it forwards towards the point and the rear end of the pencil will drop: this is because the centre of lift has moved forwards as compared with the centre of gravity. Therefore if the aerofoil is in balance or 'trimmed' at one angle of attack, so that the resultant force passes through the centre of gravity, then the forward movement of the centre of pressure on the aerofoil as the angle of attack is increased will tend to drop still more the trailing edge of the aerofoil; in other words, the angle of attack will increase even more, and this will in turn cause the centre of pressure to move farther forward, and so on. This is called **instability**, and it is one of the problems of flight.

If we were to take the wing off a model aeroplane and try to make it glide without any fuselage or tail, we would find that it would either turn over nose first or its nose would go up in the air and it would turn over on to its back. This is because the wing is unstable, and although we might be able to weight it so that it would start on its glide correctly, it would very soon meet some disturbance in the air which would cause it to turn over one way or the other.

Curiously enough, in the case of a **flat plate, an increase of the angle of attack over the same angles causes the centre of pressure to move backwards;** this tends to dip the nose of the plate back again to its original position, and so makes the flat plate **stable**. For this reason it is possible to take a **flat** piece of stiff paper or cardboard, and, after properly weighting it, to make it glide across the room. If it meets a disturbance the centre of pressure moves in such a way as to correct it. Note that the flat piece of paper will only glide if it is weighted so that the centre of gravity is roughly one-third of the chord back from the leading edge. If it is not weighted the centre of pressure will always be in front of the centre of gravity, and this will cause the piece of paper to revolve rapidly.

The unstable movement of the centre of pressure is a disadvantage of the ordinary curved aerofoil, and in a later chapter we shall consider the steps which are taken to counteract it. Attempts have been made to devise aerofoil

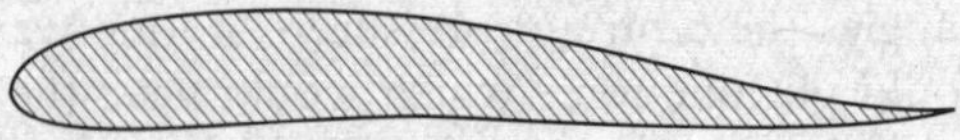

Fig 3.12 Reflex curve near trailing edge

shapes which have not got this unpleasant characteristic, and it has been found possible to design an aerofoil in which the centre of pressure remains practically stationary over the angles of attack used in ordinary flight. The chief feature in such aerofoils is that the under-surface is convex, and that there is sometimes a reflex curvature towards the trailing edge (see Fig. 3.12); nearly all modern aerofoil sections have in fact got convex camber on the lower surface. Unfortunately, attempts to improve the stability of the aerofoil may often tend to spoil other important characteristics.

Lift, drag and pitching moment of an aerofoil

Now the ultimate object of the aerofoil is to obtain the lift necessary to keep the aeroplane in the air; in order to obtain this lift it must be propelled through the air at a definite **velocity** and it must be set at a definite **angle of attack** to the flow of air past it. We have already discovered that we cannot obtain a purely vertical force on the aerofoil; in other words, we can only obtain lift at the expense of a certain amount of drag. The latter is a necessary evil, and it must be reduced to the minimum so as to reduce the power required to pull the aerofoil through the air, or alternatively to increase the velocity which we can obtain from a given engine power. Our next task, therefore, is to investigate how much lift and how much drag we shall obtain from different shaped aerofoils at various angles of attack and at various velocities. The task is one of appalling magnitude; there is no limit to the number of aerofoil shapes which we might test, and in spite of thousands of experiments carried out in wind tunnels and by full-scale tests in the air, it is still impossible to say that we have discovered the best-shaped aerofoil for any particular purpose. However, modern theoretical methods make it possible to predict the behaviour of aerofoil sections. Such methods can even be used to design aerofoils to give specified characteristics.

In wind-tunnel work it is the usual practice to measure lift and drag separately, rather than to measure the total resultant force and then split it up into two components. The aerofoil is set at various angles of attack to the airflow, and the lift, drag and pitching moment are measured on a balance.

The results of the experiments show that within certain limitations the lift, drag and pitching moment of an aerofoil depend on –

(*a*) **The shape of the aerofoil.**

(*b*) **The plan area of the aerofoil.**

(*c*) **The square of the velocity.**

(*d*) **The density of the air.**

Notice the similarity of these conclusions to those obtained when measuring drag, and in all cases there are similar limitations to the conclusions arrived at.

The reader should notice that whereas when measuring drag we considered the **frontal area** of the body concerned, on aerofoils we take the **plan area.** This is more convenient because the main force with which we are concerned, i.e. the lift, is at right angles to the direction of motion and very nearly at right angles to the aerofoils themselves, and therefore this force will depend on the plan area rather than the front elevation. The actual plan area will alter as the angle of attack is changed and therefore it is more convenient to refer results to the **maximum plan area** (the area projected on the plane of the chord), so that the area will remain constant whatever the angle of attack may be. Now we use the symbol *S*, for the plan area of a wing, to replace the frontal areas, which we used when considering drag alone in the previous chapter.

In so far as the above conclusions are true, we can express them as formulae in the forms –

$$\textbf{Lift} = C_L \cdot \frac{1}{2}\rho V^2 \cdot S \text{ or } C_L \cdot q \cdot S$$

$$\textbf{Drag} = C_D \cdot \frac{1}{2}\rho V^2 \cdot S \text{ or } C_D \cdot q \cdot S$$

$$\textbf{Pitching moment} = C_M \cdot \frac{1}{2}\rho V^2 \cdot Sc \text{ or } C_M \cdot q \cdot Sc$$

Since the pitching moment is a moment, i.e. **a force × distance**, and since $\frac{1}{2}\rho V^2 . S$ represents a force, it is necessary to introduce a length into the equation – this is in the form of the chord, c, measured in metres.

The pitching moment is positive when it tends to push the nose upwards, negative when the nose tends to go downwards.

The symbols C_L, C_D and C_M are called the **lift coefficient, drag coefficient** and **pitching moment** coefficient of the aerofoil respectively; they depend on the shape of the aerofoil, and they alter with changes in the angle of attack. The air density is represented by ρ in kilograms per cubic metre, *S* is the plan area of the wing in square metres, *V* is the air speed, in metres per second, *c* the chord of the aerofoil in metres; the method of writing the formulae in terms of $\frac{1}{2}\rho V^2$, or *q*, has already been explained in Chapter 2.

Aerofoil characteristics

The easiest way of setting out the results of experiments on aerofoil sections is to draw curves showing how –

(*a*) **the lift coefficient,**

(*b*) **the drag coefficient,**

(*c*) **the ratio of lift to drag, and**

(*d*) **the position of the centre of pressure, or the pitching moment coefficient,**

alter as the angle of attack is increased over the ordinary angles of flight.

Typical graphs are shown in Figs 3.13, 3.15, 3.16 and 3.17. These do not refer to any particular aerofoil; they are intended merely to show the type of curves obtained for an ordinary general purpose aerofoil.

In Appendix 1 at the end of the book, tables are given showing the values of C_L, C_D, L/D, position of the centre of pressure, and C_M, for a few well-known aerofoil sections. The reader is advised to plot the graphs for these sections, and to compare them with one another (see example Nos. 98 to 101 in Appendix 3). In this way the reader will understand much more clearly the arguments followed in the remaining portion of this chapter.

It is much more satisfactory to plot the **coefficients** of lift, drag and pitching moment rather than the **total** lift, drag and pitching moment, because the coefficients are practically independent of the air density, the scale of the aerofoil and the velocity used in the experiment, whereas the total lift, drag and moment depend on the actual conditions at the time of the experiment. In other words, suppose we take a particular aerofoil section and test it on different scales at different velocities in various wind tunnels throughout the world, and also full-scale in actual flight, we should in each case obtain the same curves showing how the coefficients change with angle of attack.

It must be admitted that, in practice, the curves obtained from these various experiments do not exactly coincide; this is because the theories which have led us to adopt the formula lift = $C_L \cdot \frac{1}{2}\rho V^2 \cdot S$ are not **exactly** true for very much the same reasons as those we mentioned when dealing with drag – for instance, scale effect and the interference of wind-tunnel walls. As a result of the large number of experiments which have been performed, it is possible to make allowances for these errors and so obtain good accuracy whatever the conditions of the experiment.

Now let us look at the curves to see what they mean, for a graph which is properly understood can convey a great deal of information in a compact and practical form.

Lift curve

Let us first see how the lift coefficient changes with the angle of attack (Fig. 3.13).

We notice that when the angle of attack has reached 0° there is already a definite lift coefficient and therefore a definite lift; this is a property of most cambered aerofoils. A flat plate, or a symmetrical aerofoil, will of course give no lift when there is no angle of attack.

Then between 0° and about 12° the graph is practically a straight line, meaning that as the angle of attack increases there is a steady increase in the lift; whereas above 12°, although the lift still increases for a few degrees, the increase is now comparatively small and the graph is curving to form a top, or maximum point.

At about 15° the lift coefficient reaches a maximum, and above this angle it begins to decrease, the graph now curving downwards.

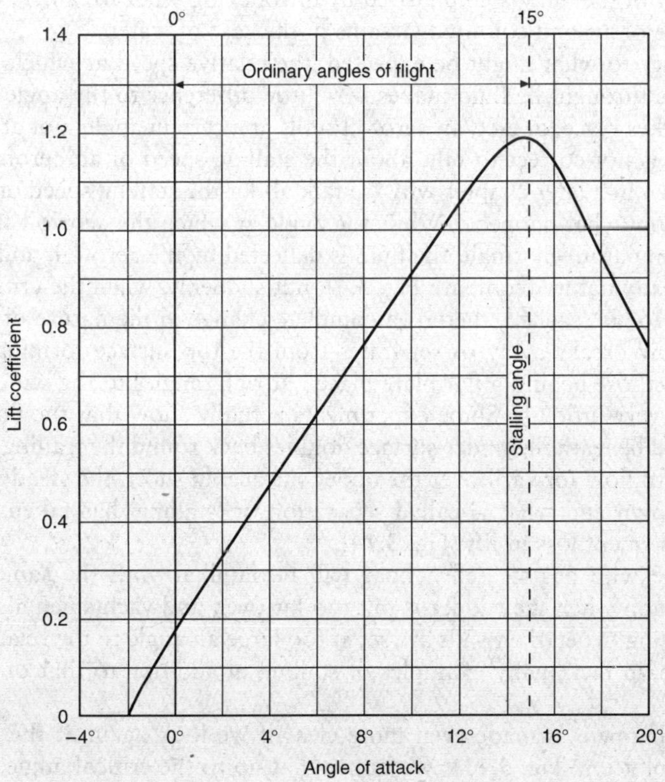

Fig 3.13 Lift curve

Stalling of aerofoil

This last discovery is perhaps the most important factor in the understanding of the why and wherefore of flight. It means that whereas at small angles any increase in the angle at which the aerofoil strikes the air will result in an increase in lift, when a certain angle is reached any further increase of angle will result in a loss of lift.

This angle is called the stalling angle of the aerofoil, and, rather curiously, perhaps, we find that the **shape** of the aerofoil makes little difference to the **angle** at which this stalling takes place, although it may affect considerably the amount of lift obtained from the aerofoil at that angle.

Now, what is the cause of this comparatively sudden breakdown of lift? The student will be well advised to take the first available opportunity of watching, or trying for himself, some simple experiment to see what happens. Although, naturally, the best demonstration can be given in wind tunnels with proper apparatus for the purpose, perfectly satisfactory experiments can be made by using paper or wooden model aerofoils and inserting them in any fairly steady flow of air or water, or moving them through air or water. The movement of the fluid is emphasised by introducing wool streamers or smoke in the case of air and coloured streams in the case of water.

Contrary to what might be expected, the relative speed at which the aerofoil moves through the fluid makes very little difference to the angle at which stalling takes place; in fact, an aerofoil stalls at a certain angle, not at a certain speed. (It is not correct to talk about the stalling speed of an aerofoil, but it will be seen in a later chapter why we talk about the stalling speed of an aeroplane.) Now what happens? While the angle at which the aerofoil strikes the fluid is comparatively small, the fluid is deflected by the aerofoil, and the flow is of a steady nature (compare Fig. 3.4); but suddenly, when the critical angle of about 15° is reached, there is a complete change in the nature of the flow. The airflow breaks away or separates from the top surface forming vortices similar to those behind a flat plate placed at right angles to the wind; there is therefore very little lift. Some experiments actually show that the fluid which has flowed beneath the under-surface doubles back round the trailing edge and proceeds to flow forward over the upper surface. In short, the steady flow has broken down and what is called separation or 'stalling' has taken its place, with consequent loss in lift (Fig. 3.14).

Anyone who has steered a boat will be familiar with the same kind of phenomenon when the rudder is put too far over, and yachtsmen also experience 'stalling' when their sails are set at too large an angle to the relative wind. There are, in fact, many examples of stalling in addition to that of the aerofoil.

What happens is made even more clear if we look again at the results of pressure plotting (Fig. 3.11). We notice that up to the critical angle considerable suction has been built up over the top surface, especially near the leading

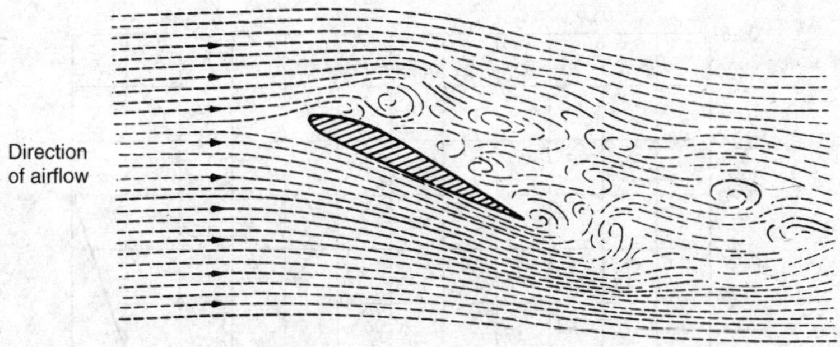

Fig 3.14 Stalling of an aerofoil

edge, whereas when we reach the stalling angle the suction near the leading edge disappears, and this accounts for the loss in lift, because the pressure on other parts of the aerofoil remains much the same as before the critical angle.

Some students are apt to think that **all** the lift disappears after the critical angle; this is not so, as will easily be seen by reference to either the lift curve or to the pressure plotting diagrams. The aerofoil will, in fact, give some lift up to an angle of attack of 90°. Modern interceptor aircraft are sometimes flown at very high angles of attack during violent manoeuvres, so the upper portion of the graph is nowadays quite important.

The stalling angle, then, is that angle of attack at which the lift coefficient of an aerofoil is a maximum, and beyond which it begins to decrease owing to the airflow becoming separated.

The drag curve

Now for the drag coefficient curve (Fig. 3.15, overleaf). Here we find much what we might expect. The drag is least at about 0°, or even a small negative angle, and increases on both sides of this angle; up to about 6°, however, the increase in drag is not very rapid, then it gradually becomes more and more rapid, especially after the stalling angle when the airflow separates.

The lift/drag ratio curve

Next we come to a very interesting curve (Fig. 3.16, overleaf), which shows the relation between the lift and the drag at various angles of attack.

In a former paragraph we came to the conclusion that we want as much lift, but as little drag, as it is possible to obtain from the aerofoil. Now from the lift curve we find that we shall get most lift at about 15°, from the drag curve least drag at about 0°, but both of these are at the extreme range of possible

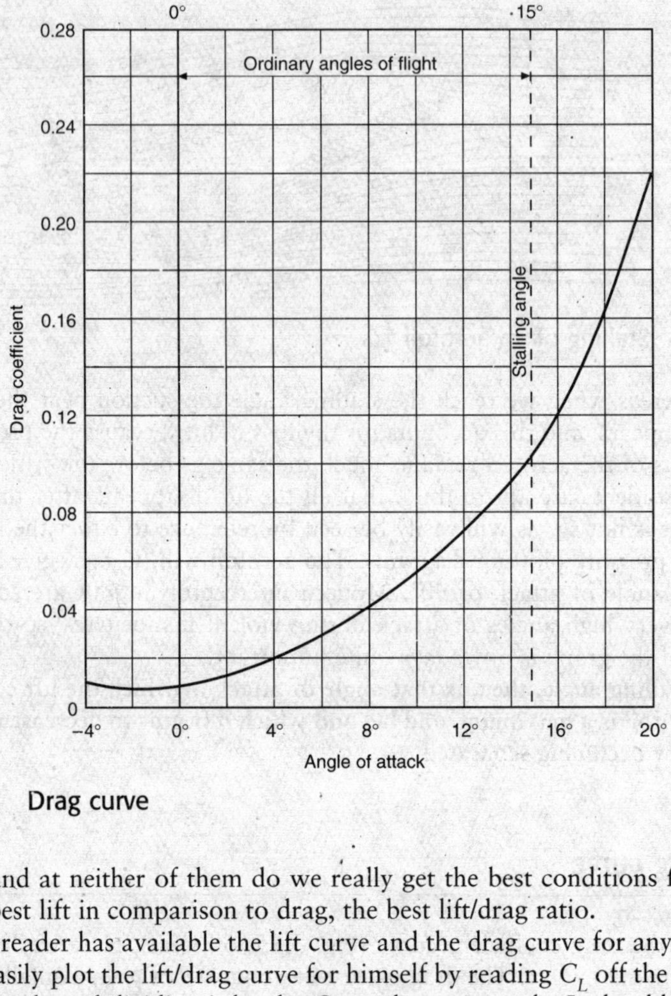

Fig 3.15 Drag curve

angles, and at neither of them do we really get the best conditions for flight, i.e. the best lift in comparison to drag, the best lift/drag ratio.

If the reader has available the lift curve and the drag curve for any aerofoil, he can easily plot the lift/drag curve for himself by reading C_L off the lift curve at each angle and dividing it by the C_D at the same angle. It should be noted that it makes no difference whether we plot L/D or C_L/C_D, as both will give the same numerical value, since $L = C_L \cdot \frac{1}{2}\rho V^2 \cdot S$ and $D = C_D \cdot \frac{1}{2}\rho V^2 \cdot S$.

We find that the lift/drag ratio increases very rapidly up to about 3° or 4°, at which angles the lift is nearly 24 times the drag (some aerofoils give an even greater maximum ratio of lift to drag); the ratio then gradually falls off because, although the lift is still increasing, the drag is increasing even more rapidly, until at the stalling angle the lift may be only 10 or 12 times as great as the drag, and after the stalling angle the ratio falls still further until it reaches 0 at 90°.

The chief point of interest about the lift/drag curve is the fact that this ratio is greatest at an angle of attack of about 3° or 4°; in other words, it is at this angle that the aerofoil gives its best all-round results – i.e. it is most able to do what we chiefly require of it, namely to give as much lift as possible consistent with a small drag.

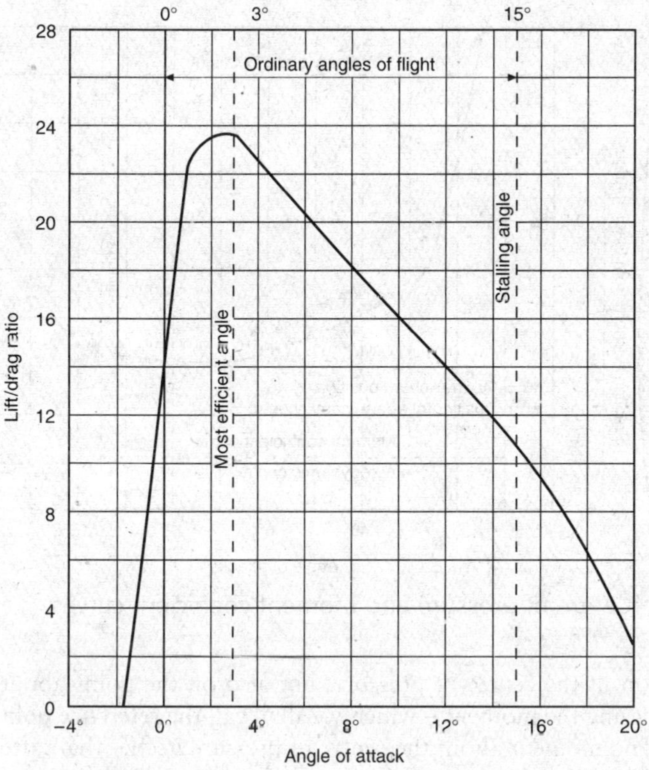

Fig 3.16 Lift/drag curve

The centre of pressure and moment coefficient

Lastly, let us examine the curves (Fig. 3.17, overleaf) which show how the centre of pressure moves, and what happens to the pitching moment coefficient, as the angle of attack is increased.

The centre of pressure curve merely confirms what we have already learnt about the movement of the centre of pressure on an ordinary aerofoil. After having been a long way back at negative angles, at 0° it is about 0.70 of the chord from the leading edge, at 4° it is 0.40 of the chord back, and at 12° 0.30 of the chord; in other words, the centre of pressure gradually moves forward as the angle is increased over the ordinary angles of flight; and this tends towards instability. After 12° it begins to move back again, but this is not of great importance since these angles are not often used in flight.

It is easy to understand the effect of the movement of the centre of pressure, and for that reason it has perhaps been given more emphasis in this book than it would be in more advanced books on the subject.

It is important to remember that the pitching moment, and its coefficient, depend not only on the lift (or more correctly on the resultant force) and on

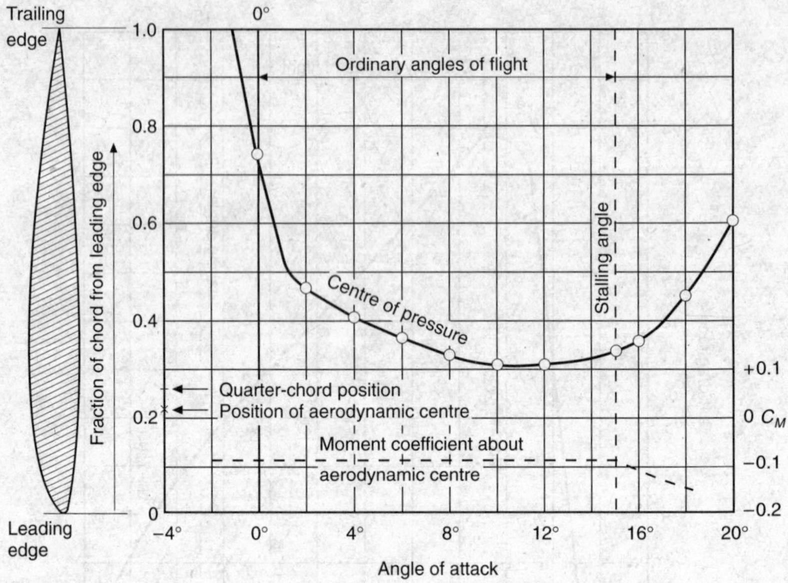

Fig 3.17 Centre of pressure and moment coefficient curves

the position of the centre of pressure, but also on the point about which we are considering the moment – which we shall call the **reference point**. There is, of course, no moment about the centre of pressure itself – that, after all, is the meaning of centre of pressure – but, as we have seen, the centre of pressure is not a fixed point. If we take as our point of reference some fixed point on the chord we shall find that the pitching moment – which was already slightly nose-down (i.e. slightly negative) at the angle of zero lift – increases or decreases as near as matters in proportion to the angle of attack, i.e. the graph is a straight line, like that of the lift coefficient, over the ordinary angles of flight. About the leading edge, for instance, it becomes more and more nose-down as the angle is increased; but about a point near the trailing edge, although starting at the same slightly nose-down moment at zero lift, it becomes less nose-down, and finally nose-up, with increase of angle (Fig. 3.18).

The reader may be surprised at the increasing nose-down moment about the leading edge, because is not the centre of pressure moving forward? Yes, but the movement is small and the increasing lift has more effect on the pitching moment. The intelligent reader may be even more surprised to hear that an increasing nose-down tendency is a requirement for the pitching stability of the aircraft, for have we not said that the movement of the centre of pressure was an unstable one? Yes, this is a surprising subject, but the answer to the apparent paradox emphasises once again the importance of the point of reference; in considering the stability of the whole aircraft our point of reference must be the **centre of gravity**, and the centre of gravity is always, or nearly

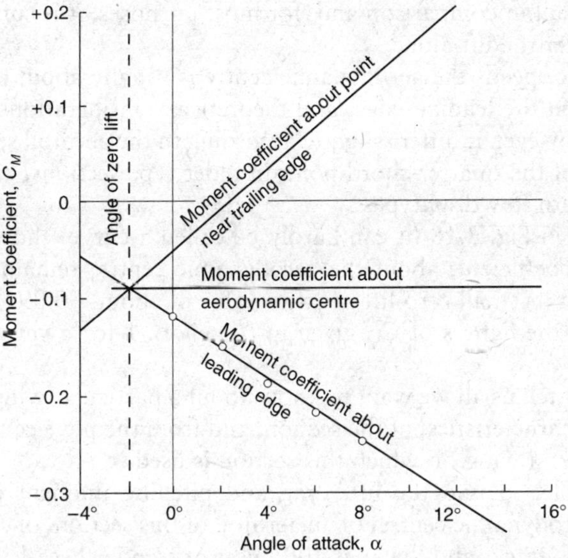

Fig 3.18 Moment coefficient about different reference points

always, behind the leading edge of the wing, so the change of pitching moment with angle of attack is more like that about the trailing edge – which is definitely unstable.

Aerodynamic centre

But something else of considerable importance arises from the differing effects of different reference points. For if about the leading edge there is a steady increase, and about a point near the trailing edge a steady decrease in the nose-down pitching moment, there must be some point on the chord about which there is no change in the pitching moment as the angle of attack is increased, about which the moment remains at the small negative nose-down value that it had at the zero lift angle (Figs 3.17 and 3.18).

This point is called the **aerodynamic centre** of the wing.

So we have two possible ways of thinking about the effects of increase of angle of attack on the pitching moment of an aerofoil, or later of the whole aeroplane; one is to think of the lift changing, and its point of application (centre of pressure) changing; the other is to think of the point of application (aerodynamic centre) being fixed, and only the lift changing (Fig. 3.19, overleaf). Both are sound theoretically; the conception of a moving centre of pressure may sound easier at first, but for the aircraft as a whole it is simpler to consider the lift as always acting at the aerodynamic centre. In both methods we really ought to consider the total force rather than just the lift, but the drag is small in comparison and, for most purposes, it is sufficiently accurate to consider the lift alone.

At subsonic speeds the aerodynamic centre is usually **about one-quarter of the chord from the leading edge**, and theoretical considerations confirm this. In practice, however, it differs slightly according to the aerofoil section, usually being ahead of the quarter-chord point in older type sections, and slightly aft in more modern low drag types.

The graph in Fig. 3.18 (it can hardly be called a curve) shows how nearly the moment coefficient, about the aerodynamic centre, remains constant on our aerofoil at its small zero-lift negative value of about -0.09. This is further confirmed by the figures of C_M given in Appendix 1 for a variety of aerofoil shapes.

The graphs tell us all we want to know about a particular wing section; they give us the '**characteristics**' of the section, and from them we can work out the effectiveness of a wing on which this section is used.

For example, to find the lift, drag and pitching moment per unit span (about the aerodynamic centre) of an aerofoil of this section, of chord 2 metres at 6° angle of attack, and flying at 100 knots at standard sea-level conditions.

From Figs 3.13, 3.15 and 3.17, we find that at 6° –

$$C_L = 0.6$$

$$C_D = 0.028$$

$$C_M = -0.09 \text{ about aerodynamic centre}$$

$$100 \text{ knots} = 51.6 \text{ m/s}$$

Since $\frac{1}{2}\rho V^2$ (or q) is common to the lift, drag and moment formulae, we can first work out its value –

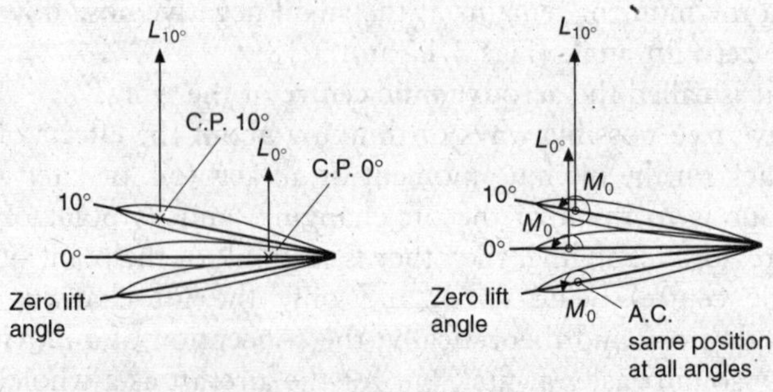

Fig. 3.19 Centre of pressure and aerodynamic centre
Two ways of looking at it.

$$q = \frac{1}{2}\rho V^2 = \frac{1}{2} \times 1.225 \times 51.6 \times 51.6 = 1631 \text{ N/m}^2$$

So lift $= C_L . q . S = 0.6 \times 1631 \times 2 = \mathbf{1957\ N}$

drag $= C_D . q . S = 0.028 \times 1631 \times 2 = \mathbf{91.3\ N}$

pitching moment $= C_M . q . Sc = -0.09 \times 1631 \times 20 \times 2$
$= \mathbf{-5872\ N\text{–}m}$

But where is the aerodynamic centre on this aerofoil?

At zero lift there is only a pure moment, or couple, acting on the aerofoil, and since the moment of a couple is the same about any point, this moment, and its coefficient, must be equal to that about the aerodynamic centre, which we shall call $C_{M\,.\,AC}$ (sometimes written as C_{MO}), and this by definition will remain the same whatever the angle of attack.

For all practical purposes we can assume that the aerodynamic centre is on the chord line, though it may be very slightly above or below. So let us suppose that it is on the chord line, and at distance x from the leading edge, and that the angle of attack is $\alpha°$ (Fig. 3.20).

The moment about the aerodynamic centre, i.e. $C_{M\,.\,AC} . q . Sc$, will be equal to the moment about the leading edge (which we will call $C_{M\,.\,LE} . q . Sc$) plus the moments of L and D about the aerodynamic centre; the leverage being $x \cos \alpha$ and $x \sin \alpha$ respectively.

So

$$C_{M\,.\,AC} . q . Sc = C_{M\,.\,LE} . q . Sc + C_L . q . S . x . \cos \alpha + C_D . q . S . x . \sin \alpha$$

and, dividing all through by $q . S$,

$$C_{M\,.\,AC} . c = C_{M\,.\,LE} . c + C_L . x . \cos \alpha + C_D . x . \sin \alpha$$

$$\therefore x = c . (C_{M\,.\,AC} - C_{M\,.\,LE})/(C_L \cos \alpha + C_D \sin \alpha)$$

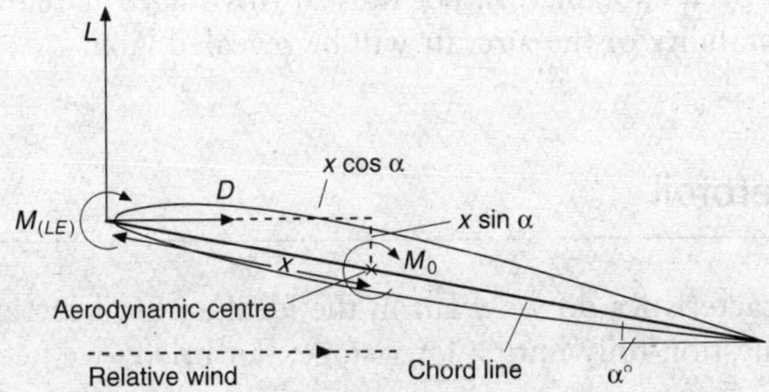

Fig 3.20 To find aerodynamic centre

or, expressed as a fraction of the chord,

$$x/c = (C_{M\,.\,AC} - C_{M\,.\,LE})/(C_L \cos\alpha + C_D \sin\alpha)$$

But the moment coefficient about the leading edge for this aerofoil at 6° is −0.22 (see Fig. 3.18), and $C_{M\,.\,AC}$ is −0.09 (Fig. 3.17),

$$C_L = 0.6,\ \cos 6° = 0.994,\ C_D = 0.028,\ \sin 6° = 0.10$$

So

$$x/c = (-0.09 + 0.22)/(0.6 \times 0.994 + 0.028 \times 1.10)$$

$$= 0.13/(0.60 + 0.003)$$

$$= \mathbf{0.216}$$

which means that the aerodynamic centre is 0.216 of the chord, or 0.432 metres, behind the leading edge, and so in this instance is forward of the quarter-chord (0.25) point.

Notice that at small angles, such as 6°, cos α is approx 1, sin α is nearly 0, so we can approximate by forgetting about the drag and saying that

$$x/c = (C_{M\,.\,AC} - C_{M\,.\,LE})/C_L$$

About the centre of pressure there is no moment, so

$$\text{(Distance of C.P. from L.E.)}/c = -C_{M\,.\,LE}/(C_L \cos\alpha + C_D \sin\alpha)$$

$$= -C_{M\,.\,LE}/C_L \text{ approx}$$

$$= +0.22/0.60 = \mathbf{0.37}$$

thus confirming the position of the C.P. as shown in Fig. 3.17.

All this has been explained rather fully at this stage; its real significance in regard to the stability of the aircraft will be revealed later.

The ideal aerofoil

But what characteristics do we want in the ideal aerofoil section? We cannot answer that question fully until a later stage, but briefly we need –

1. **A High Maximum Lift Coefficient.** In other words, the top part of the lift

curve should be as high as possible. In our imaginary aerofoil it is only about 1.18, but we would like a maximum of 1.6 or even more. Why? Because we shall find that the higher the maximum C_L, the lower will be the landing speed of the aeroplane, and nothing will contribute more towards the safety of an aircraft than that it shall land at a low speed.

2. **A Good Lift/Drag Ratio.** If we look again at Fig. 3.16, we can see that at a particular angle of attack, the lift/drag ratio of the aerofoil has a maximum value. This ratio does not occur at the angle of attack for minimum drag (Fig. 3.15) or at that for maximum lift coefficient (Fig. 3.13), but somewhere in between. Why is this ratio important? Because to get the smallest possible resistance to motion for a given weight we must operate at this angle of attack, and the higher the maximum lift/drag ratio, the smaller the air resistance that will be experienced.

   The real importance of both high lift/drag ratio and high $C_L^{3/2}/C_D$, discussed below, will become clearer when we talk about aircraft performance (Chapter 7). Let us just note here that both are important from the point of view of aerofoil design.

3. **A High Maximum Value of $C_L^{3/2}/C_D$.** The power required to propel an aeroplane is proportional to drag × velocity, i.e. to DV. For an aeroplane of given weight, the lift for level flight must be constant (being equal to the weight). If L is constant, D must vary inversely as L/D (or C_L/C_D). From the formula $L = C_L . \frac{1}{2}\rho V^2 S$ it can be seen that if L, ρ and S are constant (a reasonable assumption), then V is inversely proportional to $\sqrt{C_L}$ (or $C_L^{1/2}$). Thus power required is proportional to DV, which is inversely proportional to $(C_L/C_D) \times C_L^{1/2}$, i.e. to $C_L^{3/2}/C_D$. In other words, the greater the value of $C_L^{3/2}/C_D$, the less the power required, and this is especially important from the point of view of climbing and staying in the air as long as possible on a given quantity of fuel and as we have seen, getting the best economy from a piston-engined aircraft. If the reader likes to work out the value of this fraction for different aerofoils at different angles, and then compares the best value of each aerofoil, it will be possible to decide the best aerofoil from this point of view.

4. **A Low Minimum Drag Coefficient.** If high top speed rather than economical cruise is important for an aircraft, then we will need low drag at small lift coefficient, and hence small angles of attack. The drag coefficient at these small angles of attack will be related to the minimum drag coefficient (Fig. 3.15).

5. **A Small and Stable Movement of Centre of Pressure.** The centre of pressure of our aerofoil moves between 0.75 and 0.30 of the chord during ordinary flight; we would like to restrict this movement because if we can rely upon the greatest pressures on the wing remaining in one fixed pos-

ition we can reduce the weight of the structure required to carry these pressures. We would also like the movement to be in the stable rather than in the unstable direction.

Looking at this another way: as we have explained, the moment coefficient at zero lift is slightly negative on most aerofoils, and about the leading edge becomes more nose-down as the angle of attack is increased, and this tends towards stability. Yes, but our real reference point should be about the centre of gravity and, as we have also explained, this is usually not only behind the leading edge but also behind the aerodynamic centre, and may even be behind the trailing edge. So, in fact, this is not what we want for stability about the centre of gravity. On the contrary, we would prefer the exact opposite, i.e. a slight positive (nose-up) moment coefficient at zero lift, and this decreasing to negative as the angle of attack is increased. Most aerofoil sections do not give this; but later we shall find that there are means of achieving it.

6. **Sufficient Depth to enable Good Spars to be Used.** Here we are up against an altogether different problem. Inside the wing must run the spars, or other internal members, which provide the strength of the structure. Now the greater the depth of a spar, the less will be its weight for a given strength. We must therefore try to find aerofoils which are deep and which at the same time have good characteristics from the flight point of view.

Compromises

So much for the ideal aerofoil. Unfortunately, as with most ideals, we find that no practical aerofoil will meet all the requirements. In fact, attempts to improve an aerofoil from one point of view usually make it worse from other points of view, until we are forced either to go all out for one characteristic, such as maximum speed, or to take a happy mean of all the good qualities – in other words, to make a compromise, and all compromises are bad! It is perhaps well that we have introduced the word 'compromise' at this stage, because the more one understands about aeroplanes the more one realises that an aeroplane is from beginning to end a compromise. We want an aeroplane which will do this, we want an aeroplane which will do that; we cannot get an aeroplane which will do both this and that, therefore we make an aeroplane which will half do this and half do that – a 'half and half affair'. And of all the compromises which go to make up that final great compromise, the finished aeroplane, the shape of the aerofoil is the first, and perhaps the greatest, compromise.

Camber

How can we alter the shape of the aerofoil section in an attempt to obtain better results? The main changes that we can make are in the curvature, or camber, of the centre line, i.e. the line equidistant from the upper and lower surfaces, and in the position of the maximum camber along the chord.

In symmetrical sections, some of which have been very successful, there is of course no camber of the centre line; other sections have centre line cambers of up to 4 per cent or more of the chord.

Generally speaking we get good all-round characteristics and a smooth stall when the maximum camber is situated about 40 per cent of the chord back. Aerofoils with the maximum camber well forward, say at 15 per cent to 20 per cent of the chord, may have low drag but are apt to have poor stall characteristics – a rather sudden breakaway of the airflow.

The other main features that can be varied are the maximum thickness, the variation of thickness along the chord and the position of maximum thickness – not necessarily the same as that of maximum camber. There is considerable variation of maximum thickness (Fig. 3.21) even in commonly used aerofoils, from very thin sections with about 6 per cent of the chord to thick sections of 18 per cent or more. Reasonably thick sections are best at low speed, and for pure weight carrying, thin ones for high speed. Remember that it is the thickness compared with the chord that matters, thus the Concorde with its large chord of nearly 30 metres achieved a remarkable thickness/chord ratio of 3 per cent.

The greater the camber of the centre line the more convex will be the upper surface, while the lower surface may be only slightly convex, flat or even slightly concave (though this is rare in modern types). Sometimes there is a

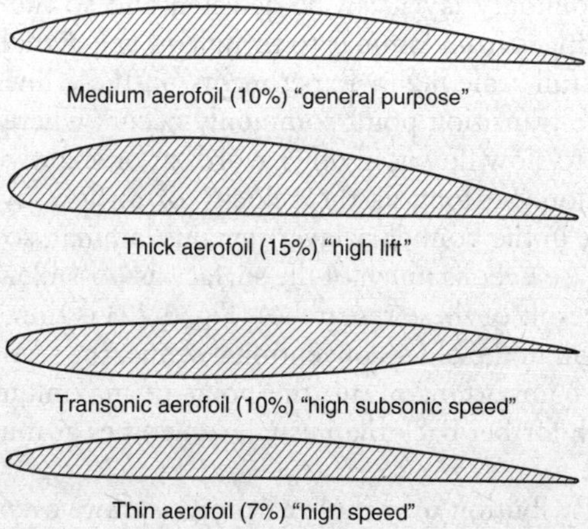

Fig 3.21 Thickness of aerofoil sections

reflex curve of the centre line towards the trailing edge (Fig. 3.12); this tends to reduce the movement of the centre of pressure and makes for stability.

Laminar flow aerofoils

The attainment of really high speeds, speeds approaching and exceeding that at which sound travels in air, has caused new problems in the design and in the flying of aeroplanes. Not the least of these problems is the shape of the aerofoil section.

Speed is a comparative quantity and the term 'high speed' is often used rather vaguely; in fact, the problem changes considerably at the various stages of high speed. In general, we may say that we have so far been considering aerofoil sections that are suitable for speeds up to 400 or 500 km/h (say 220 to 270 knots) – and we must remember that although these speeds have now been far exceeded they can hardly be considered as dawdling. Furthermore all aeroplanes, however fast they may fly, must pass through this important region. At the other end of the scale are speeds near and above the so-called 'sound barrier', shall we say from 800 km/h (430 knots) up to – well, what you will! The problems of such speeds will be dealt with in later chapters. Notice that there is a gap, from about 500 to 800 km/h (say 270 to 430 knots), and this gap has certain problems of its own; among other things, it is in this region that the so-called laminar flow aerofoil sections have proved of most value.

The significance of the boundary layer was explained in Chapter 2. Research on the subject led to the introduction of the laminar flow or low drag aerofoil, so designed as to maintain laminar flow over as much of the surface as possible. By painting the wings with special chemicals the effect of turbulent flow in the boundary layer can be detected and so the transition point, where the flow changes from laminar to turbulent, can actually be found both on models and in full-scale flight. Experiments on these lines have led to the conclusion that the transition point commonly occurs where the airflow over the surface begins to slow down, in other words at or slightly behind the point of maximum suction. So long as the velocity of airflow over the surface is increasing the flow in the boundary layer remains laminar, so it is necessary to maintain the increase over as much of the surface as possible. The aerofoil that was evolved as a result of these researches (Fig. 3.22) is thin, the leading edge is more pointed than in the older conventional shape, the section is nearly symmetrical and, most important of all, the point of maximum camber (of the centre line) is much farther back than usual, sometimes as much as 50 per cent of the chord back.

The pressure distribution over these aerofoils is more even, and the airflow is speeded up very gradually from the leading edge to the point of maximum camber.

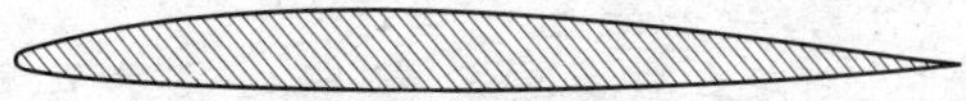

Fig 3.22 Laminar flow aerofoil section

There are, of course, snags – and quite a lot of them. It is one thing to design an aerofoil section that has the desirable characteristics at a small angle of attack, but what happens when the angle of attack is increased? As one would expect, the transition point moves rapidly forward! It has been found possible, however, to design some sections in which the low drag is maintained over a reasonable range of angles. Other difficulties are that the behaviour of these aerofoils near the stall is inferior to the conventional aerofoil and the value of $C_L$max is low, so stalling speeds are high. Also, the thin wing is contrary to one of the characteristics we sought in the ideal aerofoil.

But by far the most serious problem has been that wings of this shape are very sensitive to slight changes of contour such as are within the tolerances usually allowed in manufacture. The slightest waviness of the surface, or even dust, or flies, or raindrops that may alight on the surface, especially near the leading edge and, worst of all, the formation of ice – any one of these may be sufficient to cause the transition point to move right up to the position where the irregularity first occurs, thus causing all the boundary layer to become turbulent and the drag due to skin friction to be even greater than on the conventional aerofoil. This is a very serious matter, and led to the tightening up of manufacturing and maintenance tolerances.

With swept wings, the flow along the leading edge towards the tip usually causes transition to occur very near the leading edge and nullifying the effect of any laminar flow section. Because of scale effect, this may not happen on a wind tunnel model, making testing all the more difficult.

Another and more drastic method of controlling the boundary layer is to provide a source of suction, with the object of 'sucking the boundary layer away' before it goes turbulent.

This has the advantage that a much thicker wing section can be used (Fig. 3.23). The practical difficulty is in the power and weight involved in providing a suitable source of suction. Laminar boundary layers separate from the surface more easily than turbulent layers and suction may also be applied just before the point of separation to prevent this happening. Both suction and blowing (Fig. 3.24, overleaf) may also be used to prevent the separation of the turbulent boundary layer on an ordinary aerofoil.

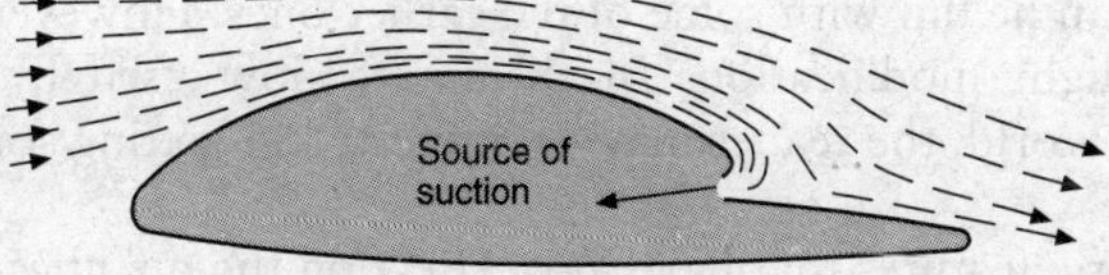

Fig 3.23 Control of boundary layer by suction (schematic drawing)

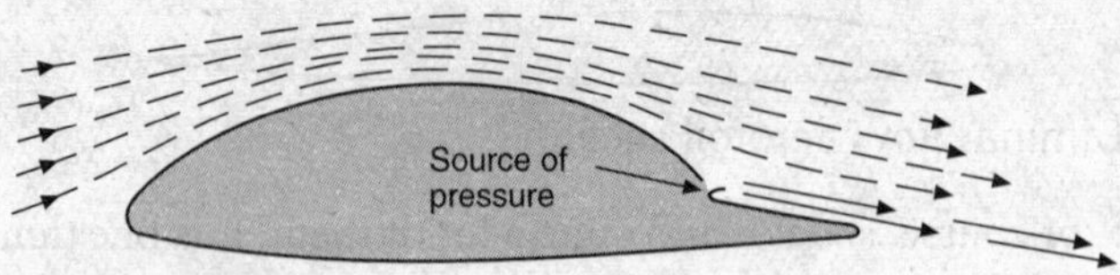

Fig 3.24 Control of boundary layer by pressure (schematic drawing)

Design and nomenclature of aerofoil sections

In the early days, in fact until the late 1930s, very few aerofoil shapes were suggested by theory; the usual method was to sketch out a shape by eye, give it a thorough test and then try to improve on it by slight modifications. As a result of this method we had a mass of experimental data obtained under varying conditions in the various wind tunnels of the world. The results were interpreted in different ways, and several systems of units and symbols were used, so that it was difficult for the student or aeronautical engineer to make use of the data available.

It is true that this hit-and-miss method of aerofoil design produced a few excellent sections but it was gradually replaced by more systematic methods. The first step in this direction was to design and test a 'family' of aerofoils by taking a standard symmetrical section and altering the curvature, or camber, of its centre line. An early example of this resulted in the RAF series of aerofoils in the UK (RAF referred to the Royal Aircraft Factory). In Germany similar investigations were made with series named after the Gottingen Laboratory, and in America with the Clark Y series.

Later sections have been based on theoretical calculations but, whatever the basis of the original design, we still rely on wind tunnel tests to decide the qualities of the aerofoil.

The naming and numbering of sections has also been rather haphazard. At first the actual number, such as RAF 15, meant nothing except perhaps that it was the 15th section to be tried. But the National Advisory Committee for Aeronautics in America soon attempted to devise a system whereby the letters and numbers denoting the aerofoil section served as a guide to its main features; this meant that we could get a good idea of what the section was like simply from its number. Unfortunately the system has been changed from time to time, and this has caused confusion; while the modern tendency to have more figures and letters in a number has resulted in such complication that the student finds it more difficult to get information about the section from the number than he did with some of the earlier ones. However since NACA sections, or slight modifications of them, are now used by nearly every country in the world, the reader may be interested in getting some idea of the systems.

The geometric features that have most effect on the qualities of an aerofoil section are –

(*a*) **the camber of the centre line;**

(*b*) **the position of maximum camber;**

(*c*) **the maximum thickness, and variation of thickness along the chord;**

and, perhaps rather surprisingly –

(*d*) **the radius of curvature of the leading edge;**

(*e*) **whether the centre line is straight, or reflexed near the trailing edge; and the angle between the upper and lower surfaces at the trailing edge.**

The NACA sections designed for comparatively low speed aircraft are based on either the four- or five-digit system; laminar flow sections for high subsonic speeds on the 6, 7 or 8 systems (the 6, 7 or 8 being the first figure, not the number of digits).

In each system there are complicated formulae for the thickness distribution, the radius of the leading edge and the shape of the centre line, but we need not worry about these; what is easier to understand is the meaning of the digits or integers, for instance, in the four-digit system –

(*a*) the **first digit** gives the **maximum camber** as a percentage of the chord;

(*b*) the **second digit** gives the **position of the maximum camber**, i.e. distance from the leading edge, in **tenths** of the chord;

(*c*) the **third** and **fourth** digits indicate the **maximum thickness** as a percentage of the chord.

Thus NACA 4412 has a maximum camber of 4 per cent of the chord, the position of this maximum camber is 40 per cent of the chord back, and the maximum thickness is 12 per cent of the chord. In a symmetrical section there is of course no camber so the first two digits will be zero; thus NACA 0009 is a symmetrical section of 9 per cent thickness.

Notice that these are all geometric features of the section, but in later systems attempts are made to indicate also some of the aerodynamic characteristics, for instance, in the **five-digit** system –

(*a*) the '**design lift coefficient**' (in tenths) is three-halves of the **first digit;**

(*b*) the **second** and **third digits** together indicate twice the distance back of the **maximum camber**, as a percentage of the chord;

(*c*) and the last two once again the **maximum thickness.**

The 'design lift coefficient' is the lift coefficient at the angle of attack for normal level flight, usually at about 2° or 3°.

Most of these sections have a 2 per cent camber, and in fact there is some relationship between the design lift coefficient and the maximum camber which has sometimes led to confusion about the meaning of the first digit; also the point of maximum camber is well forward at 15 per cent, 20 per cent or 25 per cent of the chord (which accounts for the doubling of the second and third digits to 30, 40 or 50). In fact the most successful, and so the most common of these sections, begins with the digits 230, followed by the last two indicating the thickness. Thus NACA 23012, as used on the Britten-Norman Islander, has a design lift coefficient of 0.3 (it also has 2 per cent camber), the maximum camber is at 15 per cent of the chord, while the maximum thickness is 12 per cent.

The forward position of the maximum camber in the five-digit sections results in low drag, but poor stalling characteristics, which explains why, when these sections are used near the root of a wing, they are often changed to a four-digit one (which gives a smooth stall) near the tip.

It should be noted that the position of maximum thickness (not indicated in either of these systems) is not necessarily the same as that of maximum camber, and in one British system eight digits were used so that this too could be indicated; two pairs of digits gave the thickness and its position, two other pairs the maximum camber and its position. Figure 3.25 illustrates 1240/0658 based on this system. For a symmetrical section the last four figures are omitted since they would all have been zero.

The reader may like to sketch for himself such sections as NACA 4412 and 23012, but he will have to judge the position of maximum thickness by eye.

In the NACA 6, 7 and 8 series, as in nearly all the NACA series, the last two digits again indicate the percentage thickness, but the other figures, letters, suffixes, dashes and brackets become so complicated that it is necessary to refer to tables. Most of these sections are particularly good for high subsonic speeds.

Many aircraft now use "tailor made" sections. This is particularly the case with transonic transport aircraft, which are designed to very fine limits to improve economy.

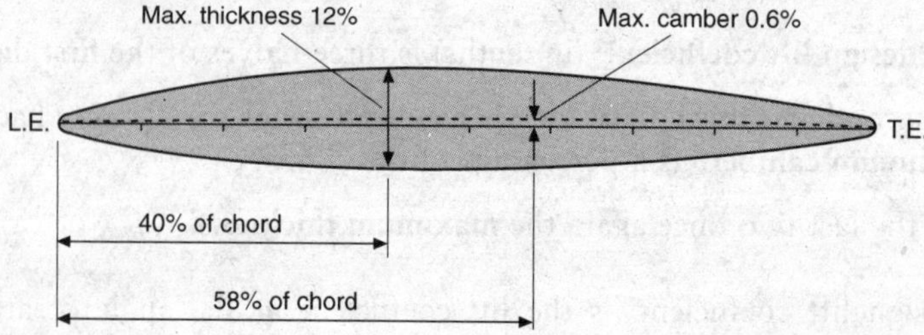

Fig 3.25 Aerofoil section 1240/0658

Aspect ratio

We have so far only considered aerofoils from the point of view of their cross-section, and we must now consider the effect of the plan form. Suppose we have a rectangular wing of 12 m^2 plan area; it could be of 6 m span and 2 m chord, or 8 m span and 1.5 m chord, or even 16 m span and 0.75 m chord. In each case the cross-sectional shape may be the same although, of course, to a different scale, depending on the chord. Now according to the conclusions at which we have already arrived, the lift and drag are both proportional to the area of the wing, and therefore since all of these wings have the same area they should all have the same lift and drag. Experiments, however, show that this is not exactly true and indicate a definite, though small, advantage to the wings with larger spans, both from the point of view of lift and lift/drag ratio.

The ratio span/chord is called aspect ratio (Fig. 3.26), and the aspect ratios of those wings which we have mentioned are therefore 3, 5.33 and 21.33 respectively, and the last one, with its 'high aspect ratio', gives the best results (at any rate at subsonic speeds which is what we are concerned with in this chapter). Why? It is a long story, and some of it is beyond the scope of this book; but the reader has the right to ask for some sort of explanation of one of the most interesting and, in some ways, one of the most important, problems of flight. So here goes!

Induced drag

Experiments with smoke or streamers show quite clearly that the air flowing over the top surface of a wing tends to flow inwards (Fig. 3.27, overleaf). This is because the decreased pressure over the top surface is less than the pressure outside the wing tip. Below the under-surface, on the other hand, the air flows outwards, because the pressure below the wing is greater than that outside the

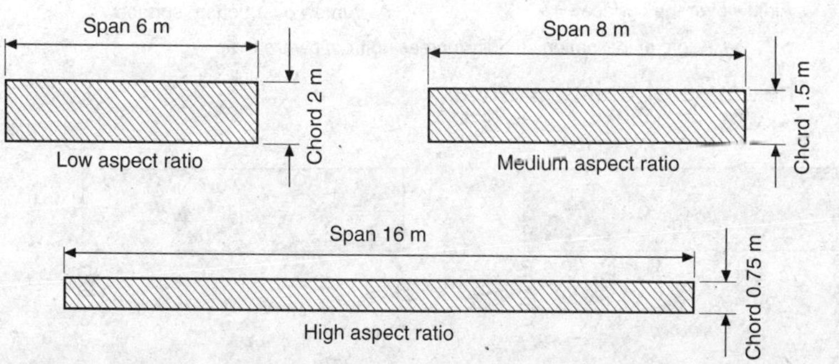

Fig. 3.26 Aspect ratio
The area of each wing is 12 m^2.

wing tip. Thus there is a continual spilling of the air round the wing tip, from the bottom surface to the top. Perhaps the simplest way of explaining why a high aspect ratio is better than a low one is to say that the higher the aspect ratio the less is the proportion of air which is thus spilt and so is ineffective in providing lift – the less there is of what is sometimes called 'tip effect' or 'end effect'.

When the two airflows, from the top and bottom surfaces, meet at the trailing edge they are flowing at an angle to each other and cause vortices rotating clockwise (viewed from the rear) from the left wing, and anti-clockwise from the right wing. All the vortices on one side tend to join up and form one large vortex which is shed from each wing tip (Fig. 3.28). These are called **wing-tip vortices.**

All this is happening every time and all the time an aeroplane is flying, yet some pilots do not even know the existence of such vortices. Perhaps it is just as well, perhaps it is a case of ignorance being bliss. In earlier editions of this book it was suggested that if only pilots could see the vortices, how they would talk about them! Well, by now most pilots have seen the vortices or, to be more correct, the central core of the vortex, which is made visible by the condensation of moisture caused by the decrease of pressure in the vortex (Figs 3A and 3B, overleaf). These visible (and sometimes audible!) trails from the wing tips should not be confused with the vapour trails caused by condensation taking place in the exhaust gases of engines at high altitudes (Fig. 3C, overleaf).

Now if you consider which way these vortices are rotating you will realise that there is an upward flow of air outside the span of the wing and a down-

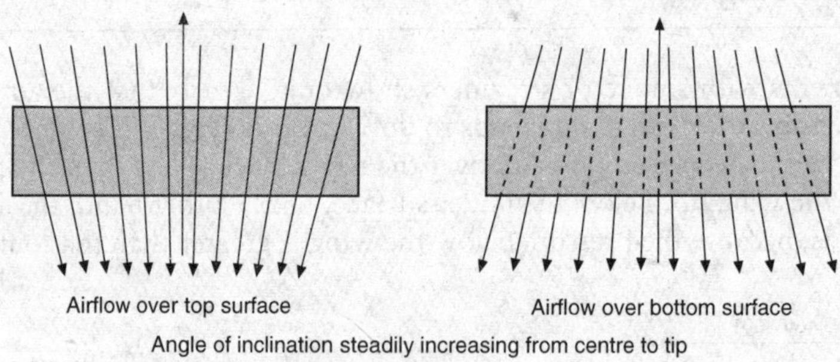

Fig 3.27 The cause of trailing vortices

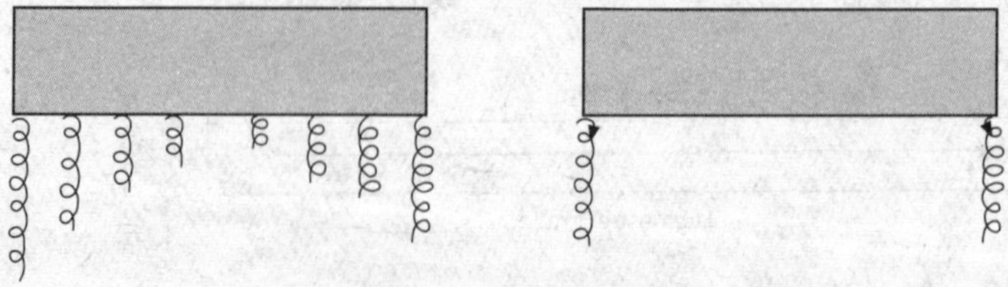

Fig 3.28 Trailing vortices which become wing-tip vortices

ward flow of air behind the trailing edge of the wing itself. This means that the net direction of flow past a wing is pulled downwards. Therefore the lift – which is at right angles to the airflow – is slightly backwards, and thus contributes to the drag (Fig. 3.29). This part of the drag is called **induced drag.**

In a sense, induced drag is caused by the lift; so long as we have lift we must have induced drag, and we can never eliminate it altogether however cleverly the wings are designed. But the greater the aspect ratio, the less violent are the wing-tip vortices, and the less the induced drag. If we could imagine a wing of infinite aspect ratio, the air would flow over it without any inward or outward deflection, there would be no wing tip vortices, no induced drag. Clearly such a thing is impossible in practical flight, but it is interesting to note that an aerofoil in a wind tunnel may approximate to this state of affairs if it extends to the wind-tunnel walls at each side, or outside the jet stream in an open jet type of tunnel. The best we can do in practical design is to make the aspect ratio as large as is practicable. Unfortunately a limit is soon reached – from the structural point of view. The greater the span, the greater must be the wing strength, the heavier must be the structure, and so eventually the greater weight of structure more than counterbalances the advantages gained. Again it is a question of compromise. In practice, aspect ratios for flight at subsonic speeds vary from 6 to 1 up to about 10 to 1 for ordinary aeroplanes, but considerably higher values may be found on sailplanes, and even more in man-powered aircraft, where aerodynamic efficiency must take precedence over all other considerations (see Fig. 3D, later) and very low values for flight

Fig 3A Wing-tip vortices
(By courtesy of the former British Aircraft Corporation, Preston)
The low pressure at the core of the vortex causes a local condensation fog on a damp day.

Fig 3B Rolling up of vortices
(By courtesy of the former British Aircraft Corporation, Preston)
A unique demonstration of the rolling up process; the wing tip and flap tip are each shedding vortices that are strong enough to cause condensation, and the pair roll around one another.

at transonic and supersonic speeds (see Fig. 3E, later). Fig. 3.30 shows how aspect ratio affects the lift curve, not only in the maximum value of C_L but in the slope of the curve, the stalling angle actually being higher with low values of aspect ratio. Notice that the angle of no lift is unaffected by aspect ratio.

The theory of induced drag can be worked out mathematically and experiment confirms the theoretical results. The full calculation involved would be

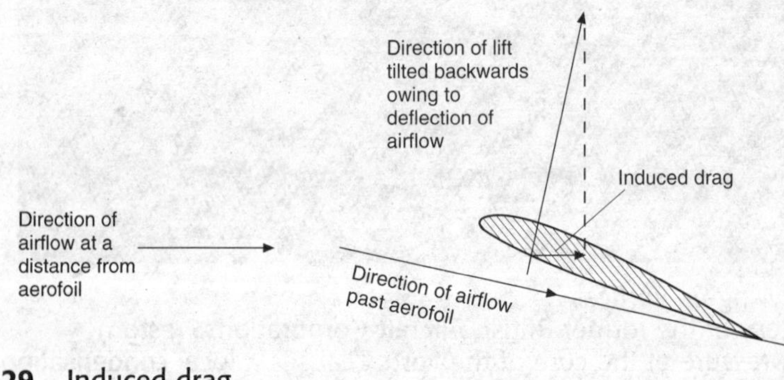

Fig 3.29 Induced drag

Fig 3C Exhaust trails
(By courtesy of General Dynamics Corporation, USA)
A modified B36 with 6 piston engines and 4 jet engines, and carrying an atomic reactor as an experiment on the shielding of crews and aircraft components.

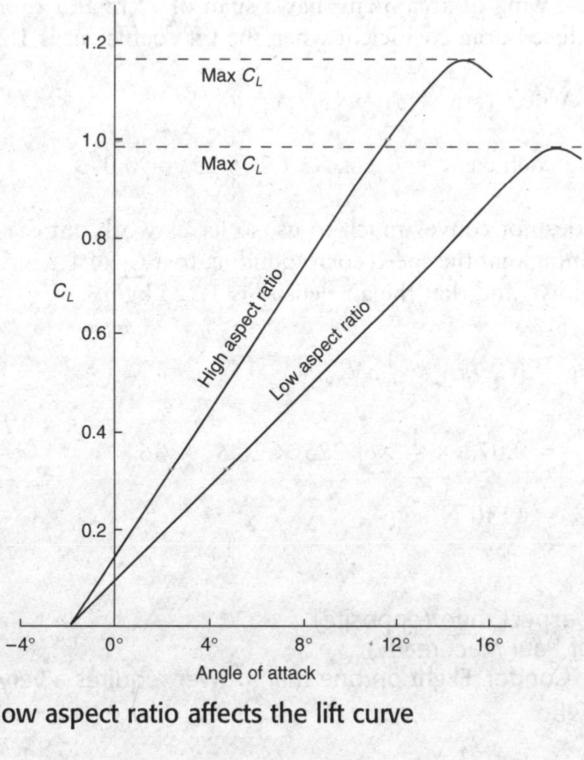

Fig 3.30 How aspect ratio affects the lift curve

out of place in a book of this kind, but the answer is quite simple and the reader may like to know it, especially since it helps to give a clearer impression of the significance of this part of the drag.

The coefficient of induced drag is found to be $C_L{}^2/\pi A$ for wings with an elliptical planform, where A is the aspect ratio and C_L the lift coefficient (the value for tapered wings may be 10 per cent to 20 per cent higher, depending on the degree of taper). This means that the actual drag caused by the vortices is $(C_L{}^2/\pi A)\,.\,\frac{1}{2}\rho V^2\,.\,S$, but since the $\frac{1}{2}\rho V^2\,.\,S$ applies to all aerodynamic forces, it is sufficient to consider the significance of the coefficient, $C_L{}^2/\pi A$. In the first place, the fact that A is underneath in the fraction confirms our previous statement that the greater the aspect ratio, the less the induced drag; but it tells us even more than this, for it shows that it is a matter of simple proportion: if the aspect ratio is doubled, the induced drag is halved. The significance of the $C_L{}^2$ is perhaps not quite so easy to understand. C_L is large when the angle of attack is large, that is to say when the speed of the aircraft is low; so induced drag is relatively unimportant at high speed (probably less than 10 per cent of the total drag), more important when climbing (when it becomes 20 per cent or more of the total) and of great importance for taking off (when it may be as high as 70 per cent of the total). In fact, the **induced drag is inversely proportional to the square of the speed**, whereas all the remainder of the drag is **directly proportional** to the square of the speed.

It is easy to work out simple examples on induced drag, e.g. –

A monoplane wing of area 36 m^2 has a span of 15 m and chord of 2.4 m. What is the induced drag coefficient when the lift coefficient is 1.2?

$$\text{Aspect ratio} = A = 15/2.4 = 6.25$$

$$\text{Induced drag coefficient} = C_L{}^2/\pi A \; 5 \; 1.2^2/6.25\pi = \mathbf{0.073}$$

Perhaps this does not convey much to us, so let us work out the actual drag involved, assuming that the speed corresponding to a C_L of 1.2 is 52 knots, i.e. 96 km/h (26.5 m/s), and that the air density is 1.225 kg/m^3.

$$\text{Induced drag} = (C_L{}^2/\pi A)\,.\,\frac{1}{2}\rho V^2\,.\,S$$

$$= 0.073 \times \frac{1}{2} \times 1.225 \times 26.5^2 \times 36$$

$$= \mathbf{1130\ N}$$

Fig 3D High aspect ratio (opposite)
(By courtesy of Paul MacCready)
The Gossamer Condor. Flight on one man-power requires a very high value of lift/drag.

Let us take it even one step further and find the power required to overcome this induced drag –

Power = DV = 1130 3 26.5 = **30 kW (about 40 horse-power)**.

This example will help the reader to realise that induced drag is something to be reckoned with; it is advisable to work out similar examples, which will be found at the end of the book.

Circulation

An interesting way of thinking about the airflow over wings and wing-tip vortices is the theory of circulation. The fact that the air is speeded up over the upper surface, and slowed down on the under surface of a wing, can be considered as a circulation round the wing superimposed upon the general speed of the airflow (this does not mean that particles of air actually travel round the wing). This circulation is, in effect, the cause of lift. If we now consider this circulation as slipping off each wing tip, and continuing downstream, we have the wing-tip vortices; and they rotate, as already established, downwards behind the wing and upwards outside the wing tips.

But this is not all. When the wing starts to move, or when the lift is increased, the wing sheds and leaves behind a vortex rotating in the opposite direction to the circulation round the wing – sometimes called the starting vortex – so there is a complete system of vortices, round the wing, then the wing-tip vortices, and finally the starting vortex. The wing-tip vortices and the starting vortex are gradually damped out with time – owing to viscosity – but the exertion of engine power (which ultimately is what creates the vortices, and so the lift and induced drag) keeps renewing the circulation round the wing, and the wing-tip vortices which result from it.

This is not just a theory; the flow over the wing can be clearly seen in experiments, as can the wing-tip vortices, while the starting vortex is easily demonstrated by starting to move a model wing, or even one's hand, through water. But perhaps the most extraordinary example of the reality of the effect of aspect ratio on circulation and wing-tip vortices is that by clever formation flying of say three or five aircraft, with the centre one leading, and the outer ones with their wing tips just behind the opposite wing tips of the

Fig 3E Low aspect ratio (opposite)
(By courtesy of the Grumman Corporation, USA)
For high-speed flight, the wings of the F14 are swung back producing an aspect ratio of less than 1. For low-speed flight they can be swung forward giving a higher aspect ratio.

leading aircraft, it is possible to achieve something of the same result (which is illustrated in flying for maximum range) as with an aircraft of three or five times the span! This is hardly a practical proposition for flying across the Atlantic, but it has been illustrated by careful experiment, and geese and other birds used the technique long before we discovered it!

Taper and shape of wing tips

In addition to changes of aspect ratio, the plan form of the wing may be tapered from centre to wing tip; this is often accompanied by a taper in the depth of the aerofoil section (Fig. 3.31) and also by a 'wash-out', or decrease of angle of incidence, towards the wing tip – sometimes too a different aerofoil section is used near the tips. The tapered wing has advantages both from the structural and aerodynamic points of view. This is a feature in which we were slow to accept the teachings of nature, for the wings of most birds have a decided taper. Where the chord is not constant along the span, the numerical value of the aspect ratio is usually taken as the fraction (span/mean chord), or $span^2$/area.

Taper in plan form means a sweepback of the leading edge, or a sweepforward of the trailing edge, or both. Considerable sweepback of the whole wing

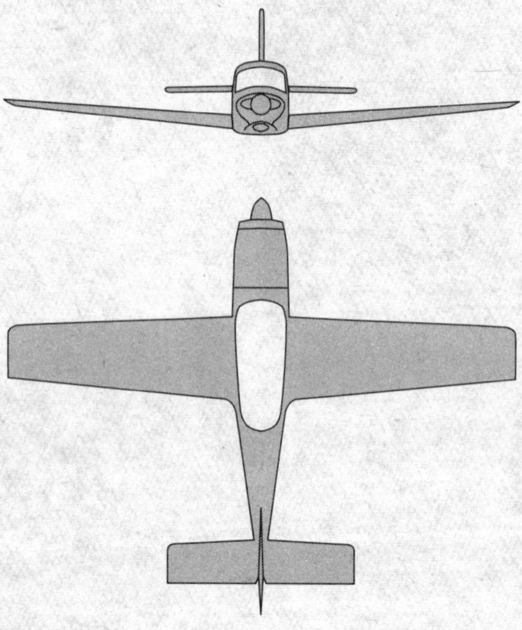

Fig 3.31 Tapered wing

is sometimes used, but this is usually more for consideration of stability or for very high-speed flight, and discussion of the problem from these points of view is deferred to later chapters.

Variable camber

Many attempts have been made to provide aerofoils with some kind of variable camber so that the pilot might be able to alter his aerofoil from a high-lift type to a high speed type at will. Owing to the tremendous advantage to be gained by such a device, it is not surprising to find that much ingenuity has been expended, many patents have been taken out, and it is not easy to compare the rival merits of the various slots, flaps, slotted flaps, and so on. Figure 3.32 (overleaf) shows some of the devices with the increase in maximum lift claimed for each, but we must not take these figures as the only guide to the usefulness or otherwise of each device, because there are other points to be considered besides maximum lift. For instance, we may want a good ratio of maximum C_L to minimum C_D (which indicates a good speed range), or an increase in drag as well as lift, the flaps acting as an air brake, which may be useful in increasing the gliding angle (explained later). Another important consideration is the simplicity of the device; anything which needs complicated operating mechanism will probably mean more weight, more controls for the pilot to work, something more to go wrong.

Flaps and slots

Although there is a large variety of high-lift devices nearly all of them can be classed as either **slots or flaps** – or a combination of the two (Fig. 3.32).

Slots may be subdivided into –

(*a*) **Fixed slots.**

(*b*) **Controlled slots.**

(*c*) **Automatic slots.**

(*d*) **Blown slots.**

Flaps may be subdivided into –

(*a*) **Camber flaps.**

(*b*) **Split flaps.**

(*c*) **Slotted flaps.**

(*d*) **Lift flaps.**

(*e*) **Blown flaps.**

(*f*) **Jet flaps.**

(*g*) **Nose flaps.**

(*h*) **Spoilers.**

(*i*) **Lift dumpers.**

(*j*) **Air brakes.**

We can also classify the effects of both slots and flaps on the characteristics of an aerofoil by saying that their use may cause one or more of the following –

(*a*) **Increase of Lift.**

(*b*) **Increase of Drag.**

(*c*) **Change of Stalling Angle.**

(*d*) **Decrease of Lift.**

(*e*) **Change of Trim.**

Slots

If a small auxiliary aerofoil, called a slat, is placed in front of the main aerofoil, with a suitable gap or slot in between the two (Fig. 3F, overleaf), the maximum lift coefficient of the aerofoil may be increased by as much as 60 per cent (Fig. 3.33, overleaf). Moreover the stalling angle may be increased from 15° to 22° or more, not always an advantage as we shall discover when we consider the problems of landing. An alternative to the separate slat, simpler but not so effective, is to cut one or more slots in the basic aerofoil itself, forming as it were a slotted wing.

The reason behind these results is clearly shown in Fig. 3.34 (later). Stalling is caused by the breakdown of the steady streamline airflow. On a slotted wing the air flows through the gap in such a way as to keep the airflow smooth, following the contour of the surface of the aerofoil, and continuing to provide lift until a much greater angle is reached. Numerous experiments confirm this conclusion. It is, in effect, a form of boundary layer control as described earlier.

The extra lift enables us to obtain a lower landing or stalling speed, and this was the original idea. If the slots are permanently open, i.e. fixed slots, the extra drag at high speed is a disadvantage, so most slots in commercial use are

High-lift devices	*Increase of maximum lift*	*Angle of basic aerofoil at max. lift*	*Remarks*
 Basic aerofoil	–	15°	Effects of all high-lift devices depend on shape of basic aerofoil.
 Plain or camber flap	50%	12°	Increase camber. Much drag when fully lowered. Nose-down pitching moment.
 Split flap	60%	14°	Increase camber. Even more drag than plain flap. Nose-down pitching moment.
 Zap flap	90%	13°	Increase camber and wing area. Much drag. Nose-down pitching moment.
 Slotted flap	65%	16°	Control of boundary layer. Increase camber. Stalling delayed. Not so much drag.
 Double-slotted flap	70%	18°	Same as single-slotted flap only more so. Treble slots sometimes used.
 Fowler flap	90%	15°	Increase camber and wing area. Best flaps for lift. Complicated mechanism. Nose-down pitching moment.

Fig 3.32 High lift devices

Note. Since the effects of these devices depend upon the shape of the basic aerofoil, and the exact design of the devices themselves, the values given can only be considered as approximations. To simplify the diagram the aerofoils and the flaps have been set at small angles, and not at the angles giving maximum lift.

High-lift devices	*Increase of maximum lift*	*Angle of basic aerofoil at max. lift*	*Remarks*
 Double-slotted Fowler flap	100%	20°	Same as Fowler flap only more so. Treble slots sometimes used.
 Krueger flap	50%	25°	Nose-flap hinging about leading edge. Reduces lift at small deflections. Nose-up pitching moment.
 Slotted wing	40%	20°	Controls boundary layer. Slight extra drag at high speeds.
 Fixed slat	50%	20°	Controls boundary layer. Extra drag at high speeds. Nose-up pitching moment.
 Movable slat	60%	22°	Controls boundary layer. Increases camber and area. Greater angles of attack. Nose-up pitching moment.
 Slat and slotted flap	75%	25°	More control of boundary layer. Increased camber and area. Pitching moment can be neutralised.
 Slat and double-slotted Fowler flap	120%	28°	Complicated mechanisms. The best combination for lift; treble slots may be used. Pitching moment can be neutralised.
 Blown flap	80%	16°	Effect depends very much on details of arrangement.
 Jet flap	60%	?	Depends even more on angle and velocity of jet.

Fig 3.32 continued

Fig 3F Leading edge slat and slot
(By courtesy of Fiat Aviazione, Torino, Italy)

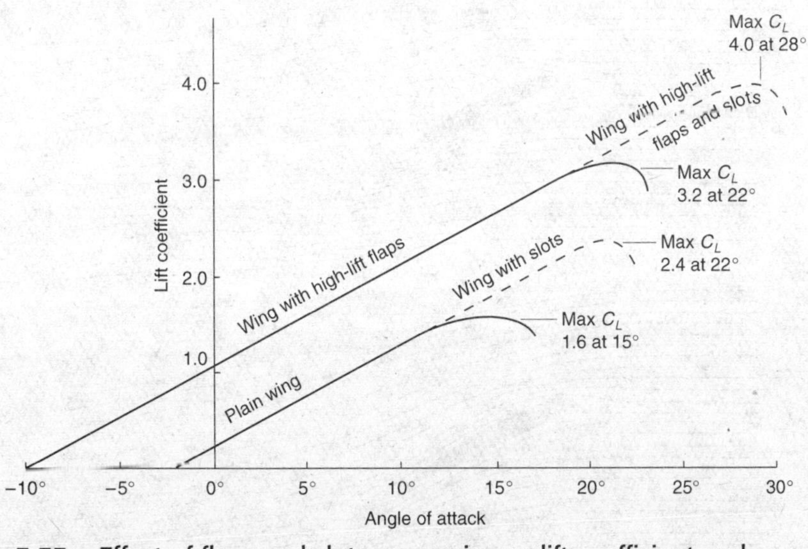

Fig 3.33 Effect of flaps and slots on maximum lift coefficient and stalling angle

controlled slots, that is to say, the slat is moved backwards and forwards by a control mechanism; and so can be closed for high-speed flight and opened for low speeds. In the early days experiments were made which revealed that, if left to itself, the slat would move forward of its own accord. So automatic slots came into their own; in these the slat is moved by the action of air pressure, i.e. by making use of that forward and upward suction near the leading edge. Figure 3.35 shows how the force on the slat inclines forward as the stalling angle is reached. The opening of the slot may be delayed or hastened by 'vents' at the trailing or leading edge of the slat respectively (Fig. 3.36), and there may be some kind of spring or tensioning device to prevent juddering, which may be otherwise likely to occur. It is also important to ensure that the slots open on both wings at the same time!

Before leaving the subject of slots – for the time being, at any rate – there are a couple of interesting points which may be worth mentioning. Firstly, the value of the slot in maintaining a smooth airflow over the top surface of the wing can be materially enhanced by blowing air through the gap between slat and wing; this may be called a blown slot. Secondly, what might be called the 'slot idea' may be extended to other parts of the aircraft. Specially shaped cowlings can be used to smooth the airflow over an engine, and fillets may be used at exposed joints, and other awkward places, to prevent the airflow separating.

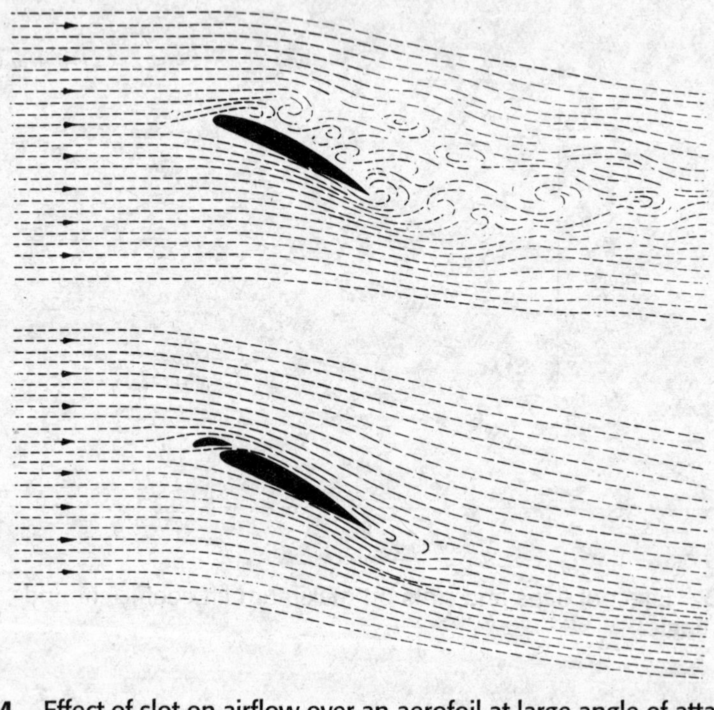

Fig 3.34 Effect of slot on airflow over an aerofoil at large angle of attack

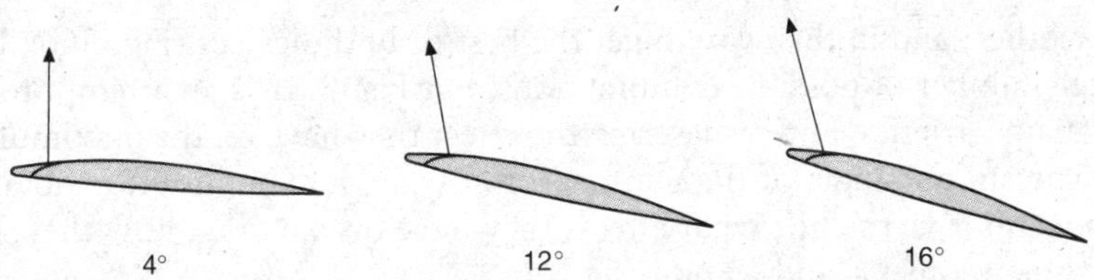

Fig 3.35 Direction of force on slat at varying angles of attack

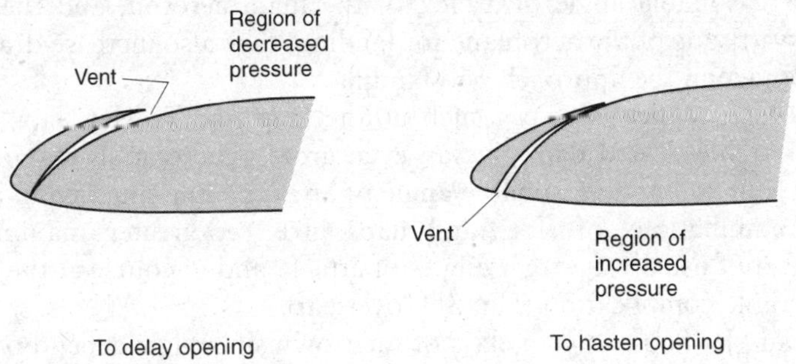

Fig 3.36 Effect of vents on opening of automatic slots

Flaps

The history of flaps is longer, and just as varied, as that of slots. The **plain** or **camber flap** works on the same principle as an aileron or other control surface; it is truly a 'variable camber'. Such flaps were used as early as the 1914–1918 war, and the original idea was the same as with slots, to decrease landing speed with flaps down, and retain maximum speed with flaps up. Their early use was almost exclusively for deck-landing purposes. It seemed at first as though the invention of slots, which followed a few years after that war, might sound the death-knell of flaps. Far from it – if anything it has been the other way round, for flaps have become a necessity on modern aircraft. Flaps, like slots, can increase lift – honours are about even in this respect so far as the plain (or camber) flap, or split flap is concerned. But these flaps can also increase drag – not, like slots, at high speed when it is not wanted, but at low speed when it is wanted. But the main difference between the effects of flaps and slots is shown in Fig. 3.33; from this it will be seen that whereas slots merely prolong the lift curve to higher values of the maximum lift coefficient, when the angle of attack of the main portion of the aerofoil is beyond the normal stalling angle, the high-lift type of flap increases the lift coefficient available throughout the whole range of angles of attack.

However it is no longer appropriate to compare the relative merits of slots and flaps because in modern aircraft it is usual to combine the two in some form or other; and in this way to get the best of both devices (Fig. 3G). There is a large number of possible combinations, but Fig. 3.32 is an attempt to sum up the main varieties, and to describe the effect they have on the maximum lift coefficient, on the angle of the main aerofoil when maximum lift is obtained, why they improve the lift, what effects they have on the drag, how they affect the pitching moment, and so on.

From this figure it will be seen that the simpler flaps such as the camber flap, split flap and single slotted flap give a good increase in maximum lift coefficient at a reasonable angle of attack of the main aerofoil, and therefore a reasonable attitude of the aeroplane for landing; they also increase drag which is an advantage in the approach and landing.

The more complicated types such as the **Zap** and **Fowler flap**, and the **double- or treble-slotted flap**, give an even greater increase in maximum lift coefficient, but still at a reasonable angle of attack; while the even more complicated combinations of slots and flaps give yet greater maximum lift coefficients, but usually at larger angles of attack, and of course at the expense of considerable complication (Fig. 3H, overleaf).

Blown and **jet flaps** are in a class of their own since they depend on power to produce the blowing, and this may be a serious disadvantage in the event of power failure. The true jet flap isn't a flap at all, but simply an efflux of air, or a jet stream in the form of a sheet of air ejected under pressure at or near the trailing edge of the aerofoil. This helps to control the boundary layer, and if the sheet of air can be deflected the reaction of the jet will also contribute directly to the lift.

The **Krueger** and other types of nose flap are used mainly for increasing lift for landing and take-off on otherwise high-speed aerofoils.

Spoilers, air brakes, dive brakes, lift dumpers and suchlike are a special category in that their main purpose is to increase drag, or to destroy lift, or both; moreover, they need not necessarily be associated with the aerofoils (Fig. 3I, overleaf). They are used for various purposes on different types of aircraft; to spoil the L/D ratio and so steepen the gliding angle on high-performance sailplanes and other 'clean' aircraft; to check the speed before turning or manoeuvring; to assist both lateral and longitudinal control; to 'kill' the lift and provide a quick pull-up after landing; and on really high-speed aircraft to prevent the speed from reaching some critical value as in a dive. They will be considered later as appropriate to their various functions.

Fig 3G Flaps and slats (opposite)
Double-slotted flaps and leading edge slats are used on the Tornado. Because the flaps extend across the entire span, there is no room for ailerons, instead, the slab tailplane surfaces can move differentially as well as collectively, and this 'taileron' serves both for roll and pitch control.

AJ
ZE199

Fig 3H Multi-element slotted flaps
Three-element slotted Fowler-type flaps extend rearwards and down as this Boeing 737 prepares to land.

Icing

All these mechanical devices are designed to vary the characteristics of an aerofoil according to our needs, but there is one important form of variable camber which is the work of nature and over which we have little control, namely the formation of ice. Brief mention has already been made of this problem in connection with laminar flow aerofoils, but the effects of icing may be far wider than this, affecting as they do, not only the wings, but many parts of the aircraft, the engine intakes and even the propeller. Icing conditions can arise in various conditions of atmospheric humidity and temperature, but they become worse at regions of low pressure such as on the upper surface of wings near the leading edge, and at engine intakes – just the places where any alteration of contour can be most serious. Apart from the effect on shape, the actual weight of accumulation of ice can be considerable and this alone has been the cause of accidents, as has also the breaking off of lumps of ice which may enter the engine or strike other parts of the aircraft.

Many methods both of prevention and cure have been used to combat the ice problem, and they may be divided into three main categories – mechanical methods (such as rubber overshoes alternately inflated and deflated) designed to break up the ice; heating methods (using the heat of the engines or separate heaters) designed to melt the ice on the leading edges of the wings, fins, engine intakes, etc.; and the use of special anti-icing fluid (about the only method suitable for propellers where it is flung out from the hub). All these, necessary though they may be, mean extra weight and complication, and some of them absorb part of the engine power.

Can you answer these?

If you understand aerofoils you have broken the back of the problems of flight – so test yourself with the following questions.

1. How does the pressure distribution over an aerofoil change as we increase the angle of attack from negative values to beyond the stalling angle?
2. What is meant by the centre of pressure of an aerofoil?
3. Why is it more convenient to speak of the lift coefficient and drag coefficient rather than the lift and drag of an aerofoil?
4. What is meant by the aerodynamic centre of an aerofoil section?
5. What do you understand by the stalling angle of an aerofoil? Why should one not talk about the stalling speed of an aerofoil?
6. What is aspect ratio and what is its significance?

For solutions see Appendix 5.
For numerical examples see Appendix 3.

Fig 31 Speed brakes
Speed brakes on the wings of the last Vulcan bomber (now sadly retired). The cables of a braking parachute can also just be seen trailing from the rear.

CHAPTER 4

Thrust

Introduction

In Chapter 2 we made a study of drag – the force that tries to hold the aeroplane back. In this chapter we shall deal with thrust – the force that opposes drag and keeps the aeroplane going forward. In steady level flight the thrust must be equal to the drag, in order to accelerate the aeroplane it must be greater than the drag, and in climbing it must also be greater than the drag because it will have to support some proportion of the weight. The actual conditions of balance of the forces will be dealt with in the next chapter; it is sufficient at this stage to realise that we must provide the aeroplane with considerable thrust, and that the performance that we can achieve from the aeroplane will be largely dependent upon the amount of thrust that we can provide.

Once the aeroplane is clear of the ground, the only reasonable way of obtaining thrust is to push air or something else backwards and to rely on the reaction to push the aeroplane forwards. This is, in fact, what is done, and to save complication the same system is usually used while still on the ground. The precise physical process by which this reaction is produced and transmitted to the aircraft depends on the type of propulsion system used.

The thrust-provider, of whatever kind it may be, must be supplied with energy. This will usually be in the form of a fuel, which is fed into some kind of 'engine' where, in burning, its chemical energy is changed into thermal energy, which in turn is converted into the mechanical work done in propelling the aeroplane against the drag. Methods of providing thrust differ only in the way in which these various conversions are effected, and in the efficiency of the conversion, that is to say in the proportion of useful work got out to the energy supplied.

Propulsion systems

This is a book on the mechanics of flight, and it would be out of place to go into the thermodynamics of propulsion systems, but we will give a brief description of the basic principles of the more important methods of propulsion. The oldest system, the piston-engine and propeller, is still in common use for light aircraft, and involves a large number of mechanical parts. The propeller itself is, of course, an almost entirely aerodynamic device. In the ramjet and rocket, described later, there are few significant mechanical parts, apart from fuel pumps, and the entire system relies entirely on aerodynamic and thermodynamic principles. Between the two extremes there are the jet and turboprop engines which involve a combination of mechanical and aero-thermodynamic principles.

Turbojet propulsion

Nowadays, the most common form of aircraft engine is the turbojet, which is normally just referred to as the **jet engine**. The turbojet is in principle, a very simple form of propulsion unit based on the **gas turbine engine**. A basic version is shown schematically in Fig 4.1. Air enters at the front and its pressure and temperature are raised by the action of the compressor. Heat is added in the combustion chamber by burning fuel (usually kerosene), and the heated air leaves at high speed. Part of the energy that it has gained is used to drive a turbine, which in turn drives the compressor. The air leaves as a high-speed jet. As the speed of the air has increased, the momentum has increased, and the reaction to this momentum change is a thrust force pushing the engine forward. The intake faces forward, in flight, so the air is effectively "rammed" in. This so-called **ram effect** helps to compress the air, and as the forward speed increases, less and less work has to be done by the compressor, leaving more of the energy increase to be used to generate thrust, so the efficiency of the propulsion system improves.

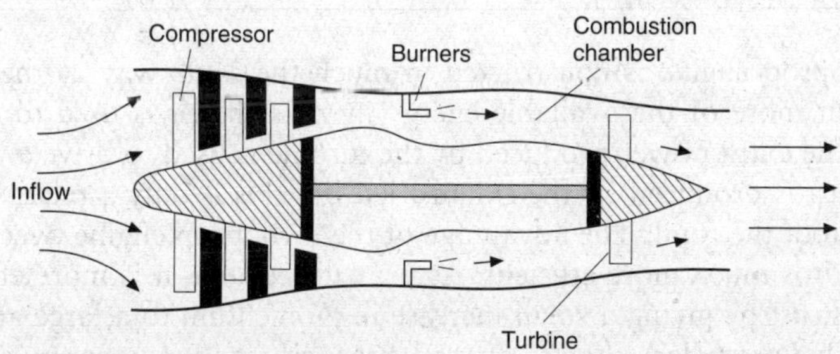

Fig 4.1 Principle of the turbojet

Fig 4A Turbojet propulsion
The hot end of a MiG-29. Two Tumansky R-33D low-bypass turbojet engines, each producing 81.4 kN thrust (with reheat). Note the complex variable area outlet nozzles.

The turbojet is much simpler than the "piston" engines that it has almost entirely displaced, and it has no reciprocating parts to wear and cause vibrations. Most importantly, it produces very much more thrust for a given weight at high speed. An added advantage is that it will work efficiently close to and beyond the speed of sound, where propellers cannot be used, as described later.

The simple type of jet engine shown in Fig. 4.1 has now largely been replaced by the more efficient high by-pass and fan-jet engines described later in this chapter.

Turboprop and turboshaft engines

The turboprop engine is constructed in much the same way as the turbojet engine, but more of the available energy in the exhaust is used to drive the turbine. The extra power produced by the turbine is used to drive a propeller. Some thrust is produced by the exhaust jet, but this is only a relatively small proportion of the total. The advantage of the turboprop engine over the pure jet is that it is much more efficient. As we will see later, it is more efficient to produce thrust by giving a small increase in momentum to a large amount of air (as with a propeller), than to give a large increase in momentum to a small amount (as with a pure turbojet). However, as mentioned previously, propellers produce serious problems at high speed. They are also noisy, and require high maintenance, and require the addition of a heavy gearbox to reduce the turbine speed, which can be up to 100 000 rpm or more, down to the few thousand rpm required for a propeller. They are normally used for relatively low-speed aircraft such as small airliners or heavy-lift transport types (Fig. 4B).

Instead of driving a propeller, the gas-turbine may be used to drive the rotor-blades of a helicopter, and in this application it is normally known as a turbo-shaft engine. Most military helicopters use turboshaft engines. Turboshaft engines are also used for purposes such as the production of auxiliary electrical power.

The ramjet and scramjet

The ramjet engine will be mentioned here, because, although hardly ever used for aircraft, it is important for missiles and potentially important for the early stages of spacecraft flight. In principle, the ramjet is extremely simple, and, as

Fig 4B Turboprop propulsion
The Lockheed Orion with four turboprops. Note the huge wide-chord 'paddle-blade' propellers.

shown in Fig. 4.2, consists of nothing more than a duct or tube of a special shape which faces the airflow in a moving aircraft. It relies on the forward speed to collect and compress the air, the so-called **ram effect** described in the preceding section. The air thus compressed flows past, a source of heat, such as a jet of burning fuel, as in a jet engine. From this, the air gains energy, flowing out faster than it entered. As with the jet engine, the reaction to this increase in air flow speed and momentum is a thrust force, which pushes the engine forward. This simple engine has no moving parts and there is almost nothing to go wrong. The exhaust air jet does not need to drive a turbine, and can be used entirely for direct thrust production. It should be noted that the shape of the duct is important. You cannot make a ramjet by burning fuel in a simple constant-bore tube; all you would get from that would be drag and some hot air.

So if ramjets are so simple, why have they rarely been used on aircraft? The answer is that there is a major snag, and that is that they will only generate thrust efficiently at high speeds. In fact, they only start to become as efficient as a jet engine at several times the speed of sound. At zero forward speed they will not work at all, so any vehicle to be propelled by a ramjet must first be accelerated by some other type of engine or propulsion system. Ramjets have been used for missiles, as shown in Fig 4C. In this case, a simple rocket is used to propel the vehicle to a high speed, whereafter, the more efficient ramjet takes over. The rocket and ramjet combination thus gives a greater range than would be possible with the highly inefficient rocket engine alone. Ramjets have rarely been used on aircraft except for a few experimental types, but the SR71 'Blackbird' spy-plane used a special engine which functioned as a turbojet engine at low speeds, and as a ramjet at high speeds.

The scramjet (supersonic combusting ramjet) works on the same principle as the ramjet, but in this case, the air flow in the engine is everywhere supersonic. In very high speed **hypersonic** flight this is necessary, because the thermodynamic efficiency becomes too low if the air speed is reduced to less than the speed of sound anywhere in the engine. The problem is that in a supersonic flow, the flame produced by burning simple fuels simply blows out. One solution is to use two fuels that react violently as they are mixed. This may be acceptable for military aircraft and missiles, but fare-paying passengers

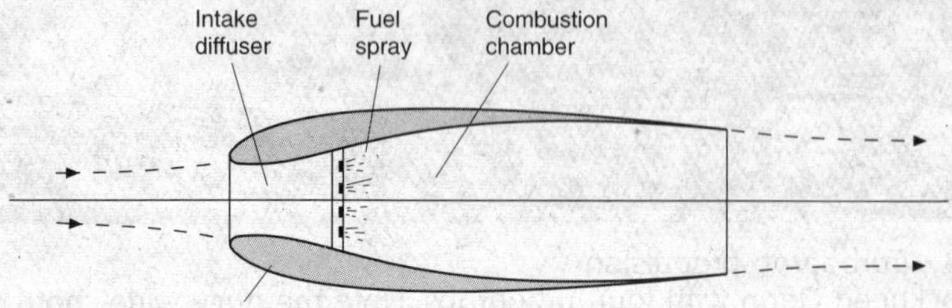

Fig 4.2 Principle of the ramjet

might not relish the idea of sitting in an aircraft containing highly reactive fuels that must never meet except in the combustion chamber!

Rocket propulsion

Rocket propulsion is also very straightforward in principle. As illustrated in Fig 4.3 (overleaf), in the simplest motors, fuel is burned in a combustion chamber, to create heat and a high pressure gas. The hot gases then flow out through the specially-shaped throat and outlet nozzle at high speed. The main difference between the rocket and other forms of propulsion is that air is not used as the oxidant in the burning process, and the gases that are emitted from the outlet are all derived from the fuels.

Apart from its use in some high speed research aircraft, and one highly dangerous German interceptor aircraft of Word War II (the Messerschmitt Me 163), the rocket is mainly used for missiles and spacecraft. The strange vehicle shown in Fig. 4D (overleaf), which used a rocket both for propulsion and lift, never developed into a practical means of transport.

Rocket motors come in two basic types, a solid fuel type, where the fuel and oxidiser are combined in a stable solid form, as in a simple firework, and a liquid propellant type, where two chemicals, one usually the fuel and the other a powerful oxidiser, are mixed together and burned in a combustion chamber.

Fig 4C Ramjet propulsion
The BAC Bloodhound missile used two ramjets: the large tubular shapes with silvery ends which can be seen above and below the body. The smaller tubes are solid-fuel rockets which were used to boost the missile to high speed; they were jettisoned once the ramjets had taken over.

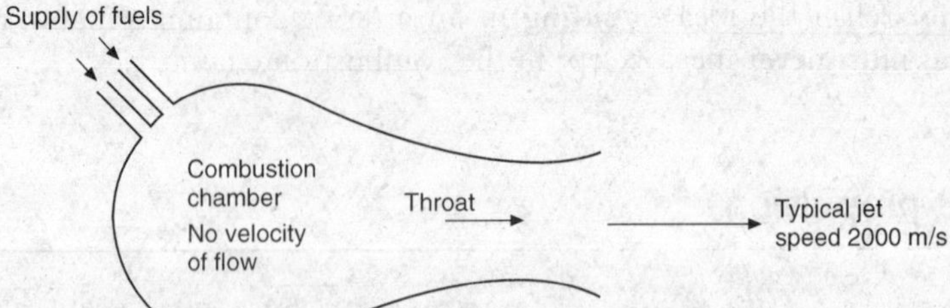

Fig 4.3 Principle of rocket propulsion

Fig 4D Rocket propulsion
(By courtesy of Bell Aerospace Division of Textron Inc, USA)
Flying the two-man Pogo from the rear platform is the rocket belt operator; on the front deck is the rocket technician.

Rocket motors have a very high thrust to weight ratio, and are essentially very simple. However they use fuel and oxidiser at a very high rate, and so have relatively short duration. The need to carry oxidiser as well as fuel means that the total weight of expendable chemicals carried is much higher than in engines that use air as the oxidant. Recently, engines have been developed which use air as the oxidant at low altitudes and pure rocket propulsion at high altitudes or in space.

Apart from the simple chemically fuelled rockets, several more advanced types have been developed. These include ones in which ionised particles or plasma is accelerated to very high speeds by electrostatic or electromagnetic forces. Some of these have been used on spacecraft, normally as low-thrust control jets.

Engine and propeller propulsion

Finally, we come to the old and well-tried system of a propeller driven by an internal combustion engine (Figs 4.4 and 4E, overleaf). Here there is the clear dividing line between the propeller and the engine. We shall consider the propeller in more detail later in this chapter. There are, of course, some problems of airflow even in a reciprocating engine, and we may often use the ram effect of a forward-facing intake as an aid to raising the pressure of the incoming air, just as we may use the backward exhaust as a partial form of jet propulsion. In the cooling system we may even emulate the ramjet by collecting the air in ducts, using the otherwise wasted heat of the engine to give it energy, and ejecting it through a venturi tube – another little bit of jet propulsion. Or, of course, the engine that drives the propeller may itself be a gas turbine, and in this case we can allot almost at will the proportion of the power that we take from the propeller and from the jet respectively as in the **turboprop system.**

Thrust and momentum

All these systems have the common feature that they provide thrust as a result of giving momentum to the air, or other gases. In accordance with the principles of mechanics the **amount of thrust provided will be equal to the rate at which momentum is given to the air.**

In symbols, **if *m* kilograms is the mass of air affected per second,** and if it is given an extra velocity of v **metres per second** by the propulsion device, then the **momentum given to the air per second is *mv*,** so

$$T = mv$$

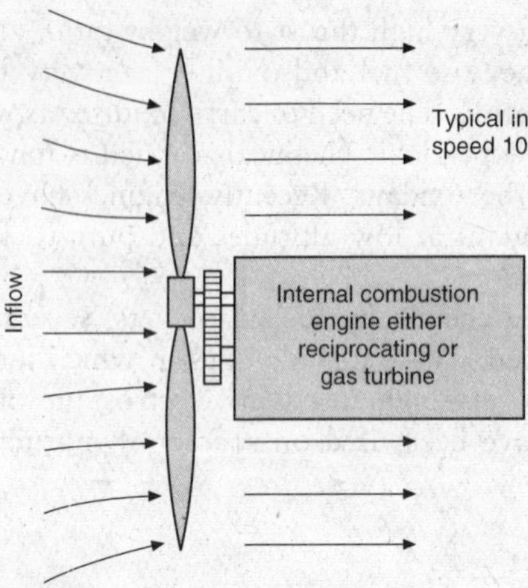

Fig 4.4 Principle of propeller propulsion

Fig 4E Piston engine and propeller
The piston engine and propeller is still the most common arrangement for light general aviation aircraft.

Now clearly the same thrust could be provided by a large *m* and a small *v*, or by a small *m* and a large *v*; in other words, by giving a large mass of air a small extra velocity or a small mass of air a large extra velocity. Which will work best in practice?

Let us now work a step further in symbols and figures. To make life easy, we will choose to consider the case of a stationary aircraft, perhaps just about to start its take-off run. Now the rate at which kinetic energy is given to *m* kg of air per second to produce a slipstream or jet speed of v m/s is $\frac{1}{2}mv^2$ watts. So while 1 kg/s given 10 m/s has the same rate of momentum change and therefore produces the same thrust as 10 kg/s given 1 m/s, the rate of change of energy of the former is

$$\frac{1}{2} \times 1 \times 10^2 = 50 \text{ watts}$$

and of the latter

$$\frac{1}{2} \times 10 \times 1^2 = 5 \text{ watts}$$

It is clear therefore that the latter will require less work and that there will be less waste of energy; in other words, it will be more efficient than the former as a means of producing thrust.

From this point of view the propeller comes first because it throws back a large mass of air at comparatively low velocity, the jet engine comes next, and the rocket a bad third in that it throws back a very small mass at a very high velocity.

You may wonder therefore why propellers ever started to go out of fashion. The problem is that it is difficult to make them work well at high speed. Since the propeller has a rotational as well as a forward speed, it follows that the blade tips will start to move through the air faster than the speed of sound long before the rest of the aircraft. The occurrence of supersonic flow at the blade tips causes all sorts of problems, and although great advances in propeller design have been made, jet propulsion provides the only practical alternative for high-speed flight.

High by-pass and turbofan engines

In order to make a jet engine more efficient, we need to arrange it so that a larger mass of air is somehow given a smaller increase in speed. The method used is to increase the size of the compressor fan and to allow a proportion of the air to pass round the outside of the engine. The momentum given to this 'by-pass' air contributes to the thrust. There are also a number of secondary advantages, the most significant being a reduction in noise. Another function of the by-pass air is to help cool the engine and to make use of some of the otherwise wasted heat to increase the thrust.

By increasing the amount of by-pass air, the so-called fan jet (see Fig. 4.5, overleaf) was evolved. The fan is not really part of the gas turbine compressor, and may sometimes be mounted at the rear of the engine.

Attempts to increase the efficiency still further lead to even larger fans until they become ducted propellers, or eventually unducted advanced turboprops, so that after many stages of development we will have come full circle back to the propeller! Lower by-pass engines will still however be required for very high-speed flight.

After so much talk of efficiency it is as well to remember that **efficiency is not everything!** We sometimes want value at all costs rather than value for money. The thrust given by a jet engine is almost independent of speed, while the thrust of a propeller, especially if it is of fixed pitch, falls off badly both above and below a certain speed. It is thrust that enables us to fly and gives us performance, and sometimes we may be more than willing to pay the price provided we get the thrust.

This seems an appropriate point at which to mention yet another difference between jet propulsion and propeller propulsion, one that is related to the fact just mentioned that the thrust of a jet is almost independent of speed; **so the power developed by a jet engine**, i.e. thrust × speed, **varies with the speed** and there is no satisfactory way of measuring it, either on the ground or in flight; when the aircraft is stationary on the ground, and the engine is running, there is no forward velocity – so the power is nil, but the thrust may be considerable, and can be measured. That is why the performance capability of a jet engine – or of a rocket – is given **in terms of thrust and not of power**. But when an engine drives a propeller, and this applies whether the engine is of the turbine or piston type, the thrust, as we have said, is variable, but the power produced at the propeller shaft may be considerable even when the aircraft is stationary, and what is more it can be measured – the propeller acts as a brake on the engine, and the power is sometimes measured by other kinds of brake, and is sometimes called brake power – so these engines are compared according to the power they produce, and not by the thrust which would be meaningless.

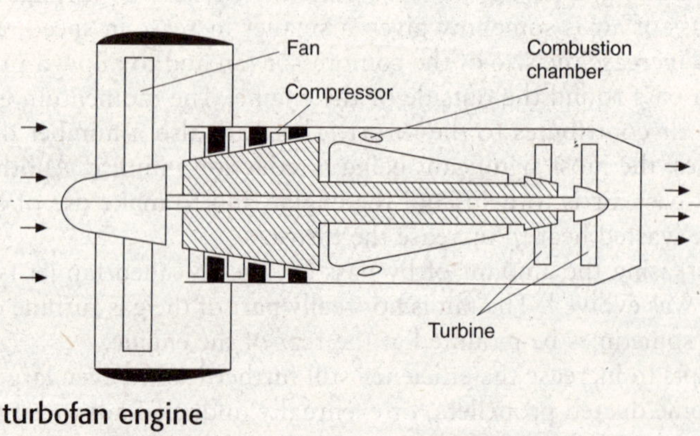

Fig 4.5 A turbofan engine

The propeller or airscrew

Of the various systems of propulsion, the propeller has been most used in the past, and for many types of aircraft it is likely to be a long time in dying. More and more gas turbines, rather than reciprocating engines, are being used for driving propellers but that does not in any way affect the aerodynamic problems involved. It is right, therefore, that we should give brief consideration to those problems. Some of them also are common to those of the helicopter, some too to the blades of compressors and fans and turbines, and these are further reasons why we should consider them.

The object of the propeller is to convert the **torque**, or turning effect, given by the power of the engine, into a straightforward pull, or push, called **thrust**.

If an airscrew is in front of the engine it will cause tension in the shaft and so will **pull** the aeroplane – such an airscrew is called a **tractor**. If, on the other hand, it is behind the engine, it will **push** the aeroplane forward, and it is called a **pusher** (Figs 4F and 4G, later). In Fig. 4H (later) there is an unusual combination of both pusher and tractor propellers.

How it works

Each part of a propeller blade has a cross-section similar to that of an aerofoil; in fact, in some cases exactly the same shape of section has been used for both purposes. The thrust of the propeller is obtained because the chord at each part of the blade is inclined at a small angle (similar to the angle of attack of an aerofoil) to its direction of motion. Since, however, the propeller is both rotating and going forward, the direction of the airflow against the blade will be at some such angle as is shown in Figs 4.6 (overleaf) and 4.7 (later). This will result in lift and drag on the blade section, just as it does on an aerofoil. Actually in a propeller we are not so much concerned with the forces perpendicular and parallel to the airflow, i.e. lift and drag, as the force acting along the axis of the aeroplane (**the thrust force**) and at right angles to the rotation (**the resistance force**). So the total force on the blade must be resolved into thrust and resistance forces, as in Fig. 4.7. The difference between these and lift and drag is clearly seen by comparing Figs 4.6 and 4.7.

The total torque force on the propeller blades will cause a turning moment or torque which opposes the engine torque, and also tends to rotate the complete aeroplane in the opposite direction to that in which the propeller is revolving. When the propeller is revolving at a steady number of revolutions per minute, then the **propeller torque and the engine torque will be exactly equal and opposite.**

Helix angle and blade angle

Why is the theory of the propeller more involved than that of the aerofoil? Chiefly because the local direction of motion of the blade is along a helix rather than a straight line, and, what is more, every section of the propeller blade travels on a different helix (Fig. 4.8, overleaf). The angle (ϕ) between the resultant direction of the airflow and the plane of rotation (Fig. 4.6) is called the **angle of advance** or **helix angle**, and it is a different angle at each section of the blade. The sections near the tip move on a helix of much greater diameter, and they also move at a much greater velocity than those near the boss.

Since all the sections must be set at a small **extra** angle to give the angle of attack, and since for maximum efficiency this extra angle should be approximately the same at all parts of the blade, it is clear that the **blade angle,** or **pitch angle,** must vary like the helix angle from boss to tip. Figure 4.9 (later) shows a typical variation of blade angle.

The **blade angle** is best defined as the angle which the chord of the propeller section at any particular place makes with the horizontal plane when the pro-

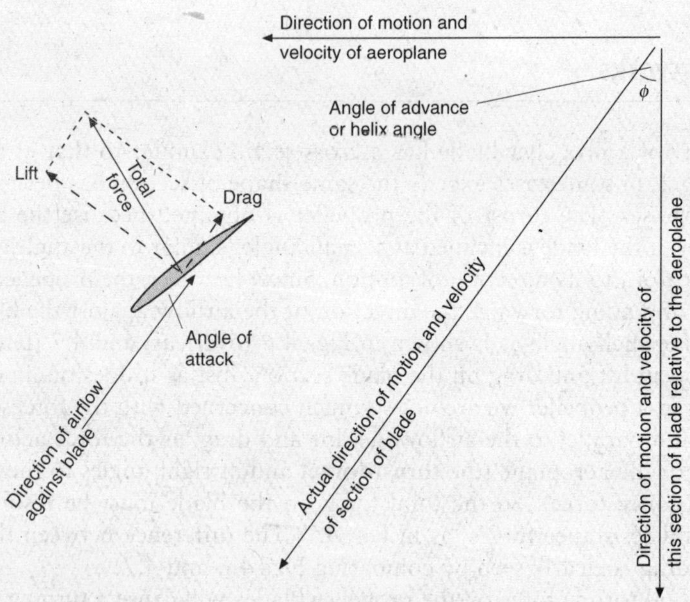

Fig 4.6 Motion of a section of propeller blade
Showing resolution of total force into lift and drag

Fig 4F Pusher – old type (opposite)
The Bristol Boxkite. One blade of the small pusher propeller can be seen protruding behind and below the minimal wooden box fuselage with its barrel-shaped fuel tank.

peller is laid flat on its boss on this horizontal plane, its axis being vertical (Fig. 4.10, overleaf). The figure shows how the blade angle is made up of the helix angle plus the angle of attack.

Advance per revolution

In a propeller, the blade angle at each section is greater than the helix angle and, what is more important, the distance moved forward in one revolution (called the **advance per revolution**) is not by any means a fixed quantity, as it depends entirely on the forward speed of the aeroplane.

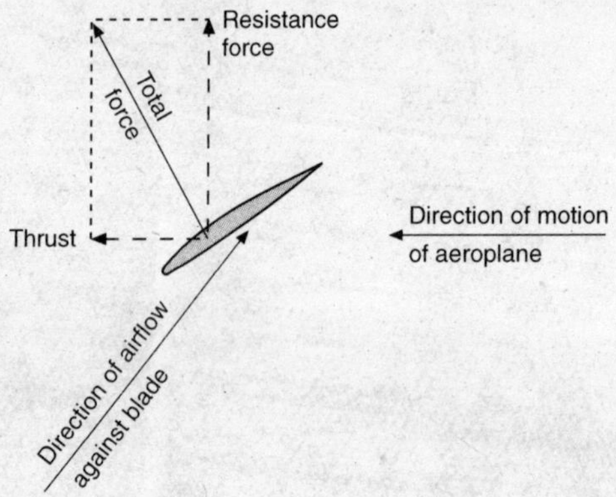

Fig 4.7 Motion of a section of a propeller blade
Showing resolution of total force into thrust and resistance

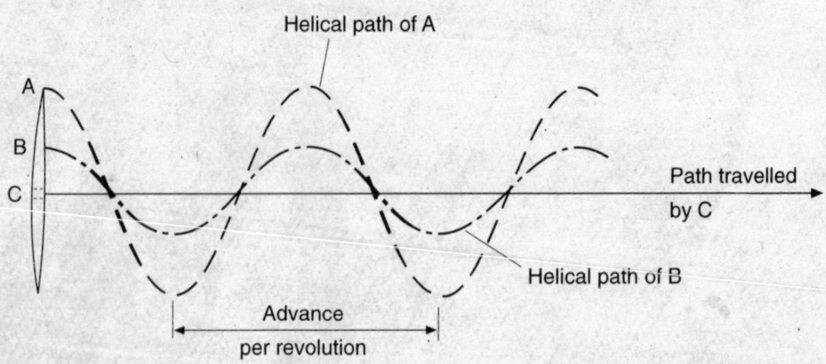

Fig 4.8 Helical paths travelled by various sections of propeller blade

Fig 4G Pusher – new type (opposite)
(By courtesy of the Beech Aircraft Corporation, USA)
The Beech Starship. Twin pusher turboprop with many other advanced features including tail-first 'canard' configuration.

N2000S

Fig 4H Pusher and tractor
(By courtesy of Cessna Aircraft Company, USA)
Unique 4/6 seater with two tandem horizontally opposed air-cooled engines, each driving a 2-blade feathering constant-speed propeller.

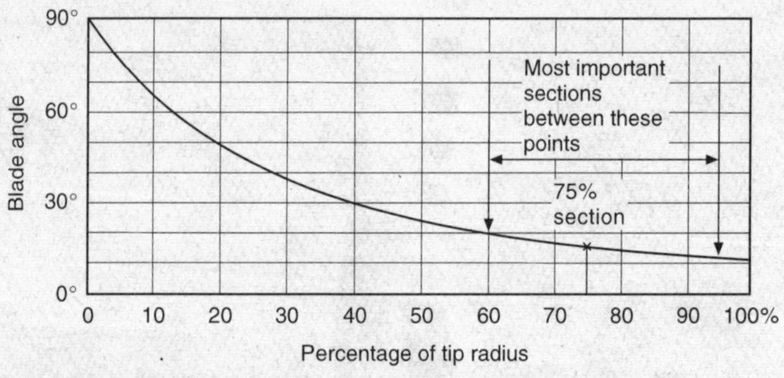

Fig 4.9 Variation of blade angle

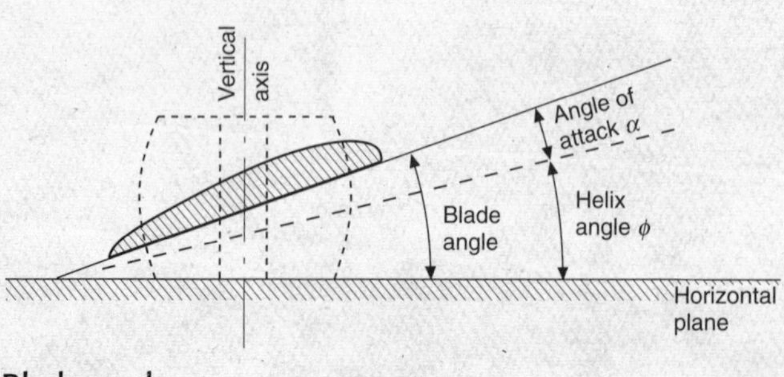

Fig 4.10 Blade angle

For instance, if an aeroplane is flying at 100 m/s and the propeller is making 1200 rpm – i.e. 20 revs per second – then the advance per revolution will be 100/20 = 5 metres. But the same aeroplane may fly at 80 m/s, with the same revolutions of the propeller, and the advance per revolution will be only 4 metres; while when the engine is run up on the ground and there is no forward motion, the advance per revolution will obviously be 0.

Considering first a fixed-pitch propeller, if the angle of a blade section at a radius of r metres is $\theta°$, and if this particular blade section were to move parallel to its chord – i.e. so that its angle of attack was 0° – while at the same time it made one complete revolution, then the distance travelled forward, p metres, would be a definite quantity and would correspond to the pitch of an ordinary screw, the relation $p = 2\pi r \tan\theta$ being true. This is best seen graphically by setting off the blade angle θ from the distance $2\pi r$ drawn horizontally, p being the vertical height. If the same operation is carried out at different distances from the axis of the propeller (Fig. 4.11), it will be found that the value of p is practically the same for all sections of the blade since as the radius r increases there is a corresponding decrease in the blade angle θ, and $2\pi r \tan\theta$ remains constant.

Pitch of a propeller

This quantity, p, is called the **geometric pitch**, since it depends only on the geometric dimensions and not on the performance of the propeller. The value of the geometric pitch of a fixed-pitch propeller may vary from about 1 metre for a slow type of aeroplane to the 5 or 6 metres that was used on Schneider Trophy and other racing aircraft (Fig. 4I).

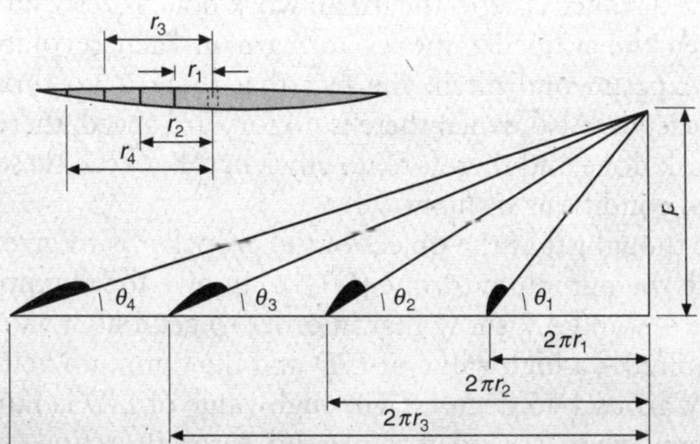

Fig 4.11 Geometric pitch

Fig 4I Fixed pitch propeller
(By courtesy of what was the Fairey Aviation Co Ltd)
Two-blader with very large pitch angle, as used in the Schneider Trophy contest, 1931.

The designer of a propeller may find it convenient to consider the pitch from a different viewpoint. When the advance per revolution reaches a certain value, the thrust becomes zero, the reason being that the angle of attack of each part of the blade has become so small that the aerofoil section of the blade provides no thrust. (Notice how this corresponds to the small negative angle at which an aerofoil ceases to give lift.) The **experimental mean pitch** is defined as the **distance the propeller will move forward in one revolution when it is giving no thrust.**

Efficiency

Now the **efficiency** of a propeller is the ratio of the useful work given out by the propeller to the work put into it by the engine. Mechanical work done is measured by the force multiplied by the distance moved, and so when either the force or the distance is zero, the useful work done is zero, and the efficiency nil. Thus when the propeller moves forward in each revolution a distance equal to the experimental pitch, the fact that there is no thrust means that there is no efficiency. Also, when there is no forward speed, there is no distance moved, no work done and therefore no efficiency. Between these two extremes are the normal conditions of flight.

It might be thought that the object of the propeller is to give the maximum thrust (T) with the minimum torque (Q), i.e. to give **the maximum *T/Q* ratio.** However, Figs 4.6 and 4.7 show that in order to get a high value of T/Q, two things are required – **a high value of *L/D*** and the optimum helix angle, which is theoretically around 45 degrees. The high value of L/D is fairly easy, and is an old problem; what is needed is a good aerofoil section, set at near the correct small angle to the relative air flow, and this means twisting the blade, as already explained.

The provision of the optimum helix angle is more difficult, as this would require matching the rotational speed to the forward speed. In practice, this is impractical, and the propeller is normally run at near constant speed, as described later. In any case, the optimum helix angle can only be obtained at one position along the blade, since the blade is twisted. However, the tip of the propeller is moving faster than the inboard sections, and thus tends to produce a high proportion of the thrust, so it is the angle of the tip that is most important. With fixed pitch propellers, a compromise on the pitch angle has to be made between high efficiency cruising, and high thrust for take-off.

Under conditions of maximum efficiency the advance per revolution is usually considerably less than the experimental pitch. The experimental pitch is sometimes called the **ideal pitch**, while the advance per revolution is the actual **practical pitch**. The difference between the two is called the **slip**, and is usually expressed as a percentage.

The calculation of propeller efficiency is quite straightforward. For example, if the total drag of an aeroplane at 65 m/s is 4.22 kN and the power developed by the engine when the aeroplane is flying at this speed is 336 kW, then –

Work given to propeller per second $= 336\,000$ joules

Work done by propeller per second $= 4220 \times 65$

$= 274\,300$ joules

So efficiency of the propeller = Work got out/Work put in $\times$ 100 per cent

$= (274\,300/336\,000) \times 100$ per cent

= 81.6 per cent

This represents the approximate value of the efficiency obtainable from a good propeller, although in some instances it may rise as high as 85 or even 90 per cent. The best efficiency is obtained when the slip is of the order of 30 per cent.

For those who prefer to examine this question in terms of mathematical symbols the efficiency of a propeller can be deduced as follows –

Let v = forward velocity in m/s

T = thrust of propeller in newtons

n = revolutions per second of engine

Q = torque exerted by engine in N-m

Work done by thrust T at v metres per second $= Tv$ joules per second, or watts

Work given by engine $= 2\pi Q$ joules per rev

$= 2\pi nQ$ joules per second, or watts

Efficiency of propeller $= (Tv/2\pi nQ) \times 100$ **per cent**

Tip speed

The power developed by a piston engine depends upon the pressures attained during combustion in the cylinders and on the revolutions per minute. The greatest power in most engines is developed at a fairly high number of revolutions per minute; and if the propeller rotates at the same speed as the engine crankshaft, the tip speed of the propeller blades is liable to approach or exceed the speed of sound (about 340 m/s in air at ground level, and less at higher altitudes). This causes compressibility effects (see Chapter 11), which, in turn, mean an increase in torque and decrease in thrust; in other words, **a loss of efficiency**. It is clearly of little purpose to design an engine to give high power, if at such power the propeller is to become less efficient and so transfer a lower proportion of the engine power to the aircraft. In the early stages of compressibility some improvement can be effected by changing the blade section near the tip to a thin laminar-flow type and by washing out the blade angle slightly; if this is done the loss is not serious so long as the actual speed of the tip does not exceed the speed of sound. As a further help a reduction gear is often introduced between the engine crankshaft and the propeller; the reduction is not usually very large, perhaps 0.7 or 0.8 to 1, but is just sufficient to reduce the tip speed to a reasonable margin below the speed of sound.

The tip speed, of course, depends not only on the revolutions per minute, but also on the forward speed of the aeroplane and the diameter of the propeller. The high forward speed of modern aeroplanes is such that it is becoming very difficult to keep the tip speed down below the speed of sound, and it would seem that at forward speeds of 350 knots or more some loss in efficiency must be accepted. At 430 knots the loss in efficiency is serious and has spread to a larger proportion of the propeller blades so that it affects not only the tips but what should be the most efficient sections. At this stage there is nothing for it but for the propeller to retire gracefully and hand over supremacy to jet propulsion.

A further objection to high tip speed is that the noise caused by the propeller (incidentally a large proportion of the total noise) is much intensified, especially in the plane in which the propeller is rotating. This can be annoying both outside and inside the aircraft, and in severe cases, structural damage can result.

Variable pitch

For low-speed aeroplanes the thrust of a fixed-pitch propeller is usually found to be greatest when there is no forward speed, i.e. when the aeroplane is stationary on the ground. The thrust developed under these conditions is called the **static thrust**, and its large value is very useful since it serves to give the aeroplane a good acceleration when starting from rest and thus reduces the run required for taking off. But in high-speed aircraft a fixed-pitch propeller designed for maximum speed would have such a large pitch, and, therefore, such steep pitch angles, that some portions of the blades would strike the air at angles of as much as 70° or more when there is no forward speed, the efficiency and static thrust would be very poor, and great difficulty would be experienced in taking off. The only remedy is variable pitch.

This requirement led to the development of the so-called **constant-speed propeller** (Fig. 4J) in which the pitch is automatically adjusted so that the propeller revolves at a given rate decided by the pilot, and remains at that rate irrespective of throttle opening or manoeuvres of the aeroplane. Thus engine and propeller can work at high efficiency irrespective of conditions, such as take-off, climb, maximum speed, altitude, and so on.

Fig 4J Constant-speed propeller
A classic constant-speed propeller design. The hydraulically-actuated speed control unit is housed in the small domed unit on the front of the hub.

It has already been stated that there are problems common to propeller and helicopter blades, and one of these is that of variable pitch, which in helicopters is a virtual necessity as a means of control.

An extension of the idea of variable pitch leads to a propeller with the pitch variable not only over the range of blade angles that will be required for normal conditions of flight but beyond these angles in both directions.

If the blade can be turned beyond the normal fully-coarse position until the chord lies along the direction of flight, thus offering the minimum resistance, the propeller is said to be **feathered**. This condition is very useful on a multi-engined aircraft for reducing the drag of the propeller on an engine that is out of action. It has another advantage too in that it is a convenient method of stopping the propeller and so preventing it from 'windmilling'; this reduces the risk of further damage to an engine that is already damaged.

The turning of the blade beyond the fully-fine position makes the propeller into an effective **air brake**; it has exactly the opposite effect to feathering by causing the maximum drag, which occurs when the blade angle is approximately 2° or 3°.

If the blade angle is still further reduced, i.e. to negative angles, then instead of allowing the blades to windmill, we can run the engine and produce negative thrust or drag. This produces an excellent brake for use in slowing up the aeroplane after landing since it gives a high negative thrust at low forward speeds.

Number and shape of blades

The propeller must be able to absorb the power given to it by the engine; that is to say, it must have a resisting torque to balance the engine torque, otherwise it will race, and both propeller and engine will become inefficient.

The climbing conditions are particularly difficult to satisfy since high power is being used at low forward speeds; and if we do satisfy these conditions – by any of the methods suggested below – it will be difficult to get efficiency in high-speed flight. Thus, the propeller becomes a compromise like so many things in an aeroplane.

The ability of the propeller to absorb power may be increased by –

1. Increasing the blade angle and thus the angle of attack of the blades.
2. Increasing the length of the blades, and thus the diameter of the propeller.
3. Increasing the revolutions per minute of the propeller.
4. Increasing the camber of the aerofoil section of which the blade is made.
5. Increasing the chord (or width) of the blades.
6. Increasing the number of blades.

With so many possibilities one might think that this was an easy problem to solve, but in reality it is one that has caused considerable difficulty. First, the blade angle should be such that the angle of attack is that giving maximum efficiency; there is, therefore, little point in trying to absorb more power if, in so doing, we lose efficiency. The second possibility is to increase the diameter, in other words, to increase the blade aspect ratio, but quite apart from the bogey of tip speed, with large propellers there is the problem of providing enough ground clearance. The third would mean high tip speed and consequent loss of efficiency. The fourth, as with aerofoils, would simply mean a less efficient section; it would seem, too, that we must face even thinner aerofoil sections to avoid loss of efficiency at high speed. So we are left with the last two, and fortunately they provide some hope. Either will result in an increase in what is called the **solidity** of the propeller. This really means the ratio between that part of the propeller disc which, when viewed from the front, is solid and the part which is just air. The greater the solidity, the greater the power that can be absorbed.

Of the two methods of increasing solidity, increase of chord and increase of number of blades, the **former** is the **easier**, the **latter** the **more efficient**. The so-called paddle blades are examples of the former method. But there is a limit to this, first, because the poor aspect ratio makes the blades less efficient.

So, all in all, an increase in the number of blades is the most attractive proposition, and that is why we saw, first, the two-blader (yes, there has been a one-blader! – but only one); then, in turn, three, four, five, and six blades; and we might have gone to eight- and ten-bladers had not jet propulsion come along at the critical time.

After four or, at the most, five blades, it becomes inconvenient to fit all the blades into one hub, and it is, in effect, necessary to have two propellers for each engine. If we are going to have two propellers, we may as well rotate them in opposite directions (Fig. 4K, overleaf) and so gain other advantages which will become more apparent when we have considered the effects of the propeller on the aeroplane.

The slipstream

The propeller produces thrust by forcing the air backwards, and the resultant stream of air which flows over the fuselage, tail units, and other parts of the aeroplane is called the **slipstream**.

The extent of the slipstream may be taken roughly as being that of a cylinder of the same diameter as the propeller. Actually there is a slight contraction of the diameter a short distance behind the propeller.

The velocity of the slipstream is greater than that at which the aeroplane is travelling through the air; the increase in velocity may be as much as 100 per cent, or even more, at the stalling speed of the aeroplane. This means that the

Fig 4K Contra-rotating propellers
Four sets of these six-bladed contra-rotating propellers were employed to propel the old Shackleton patrol aircraft. The noise inside the fuselage, which had no padding or sound absorption material, made the long duration flights decidedly arduous for the crew.

velocity of the air flowing over all those parts in the slipstream is twice that of the airflow over the other parts, and so the drag is four times as great as corresponding parts outside the slipstream. At higher forward speeds the difference is not as great, being only about 50 per cent at normal speeds, and as little as 10 per cent at high speeds. The extra velocity of the slipstream may be beneficial in providing more effective control for rudder and elevators, especially when the aeroplane is travelling slowly through the air, e.g. when taxying, or taking off, or flying near the stalling speed. With jet propulsion, however, it is not advisable for the hot jet to strike the tail plane which, in consequence, is often set very high (Fig. 5F).

In addition to increased velocity, the propeller imparts a **rotary motion** to the slipstream in the same direction as its own rotation; so it will strike one side only of such surfaces as the fin, and so may have considerable effects on the directional and lateral balance of the aeroplane. If these effects are compensated for in normal flight – e.g. by offsetting the fin so that it does not lie directly fore and aft – then the balance will be upset when the engine stops and the slipstream ceases to exert its influence.

Gyroscopic effect

The rotating mass of the propeller or the compressor in the case of a jet engine may cause a slight **gyroscopic effect**. A rotating body tends to resist any change in its plane of rotation, and if such change does take place there is superimposed a tendency for the plane of rotation to change also in a direction at right angles to that in which it is forced. This can easily be illustrated with an ordinary bicycle wheel; if the wheel, while rapidly rotating, is held on a horizontal shaft and the holder attempts to keep the shaft horizontal while he turns, the shaft will either tilt upwards or downwards according to whether he turns with the opposite or the same sense of rotation as that of the wheel. Thus if the propeller rotates clockwise when viewed from the pilot's cockpit (the usual method of denoting the rotation), the nose will tend to drop on a right-hand turn and the tail to drop on a left-hand turn. It is only in exceptional cases that this effect is really appreciable, although it used to be very marked in the days of rotary engines when the rotating mass was considerable.

Swing on take-off

There is often a tendency for an aeroplane to swing to one side during the take-off run. This must be due to some asymmetric feature of the aircraft, and it is an interesting problem to try to track down the real villain that is causing the swing.

The pilot should be the first suspect. He himself is not symmetrical, he may be right-handed (or left-handed), he probably looks out on one side of the aeroplane and may even sit on one side. Certain it is that some aircraft which have swung violently when the pilot has tried to keep them straight have gone as straight as a die when left to themselves!

The second and main suspect is undoubtedly the propeller. But which of its asymmetric effects is the chief cause of swing in any particular aircraft is not so easy to determine. If the propeller rotates clockwise, the **torque reaction** will be anti-clockwise, the left-hand wheel will be pressed on the ground and the extra friction should tend to yaw the aircraft to the left. But let us not forget that the torque reaction may be compensated and, in that case, the behaviour of the aeroplane will depend on how it is compensated.

The slipstream – assuming the same clockwise propeller – will itself rotate clockwise and will probably strike the fin and rudder on the left-hand side, again tending to yaw the aircraft to the left. But the slipstream too may be compensated.

The **gyroscopic effect** will only come in when the tail is being raised. Again the tendency will be to swing to the left if the propeller rotates clockwise. Try it with the bicycle wheel.

Apart from the compensating devices already mentioned the tendency to swing can be largely, if not entirely, eliminated by opposite rotating propellers on multi-engined aircraft (Fig. 4K), by contra-rotating propellers on single-engined aircraft and by jet propulsion or rocket propulsion instead of propellers.

Contra-rotating propellers not only give the greater blade area, or solidity, that is required to absorb large power, but they eliminate or very nearly eliminate all the asymmetrical effects of slipstream, propeller torque, and gyroscopic action. It is curious that the average pilot hardly realised the existence of these asymmetrical effects – until he lost them. Pilots who flew behind contra-rotating propellers for the first time reported that the aircraft was easy to handle and nice to fly. This is hardly surprising; what perhaps is surprising is that the previous ill-effects of one-way rotation had been so little noticed. The second propeller straightens the slipstream created by the first and so causes a straight high-speed flow of air over wings and tail; this improves the control and there is little or no resultant torque tending to roll the aircraft in one direction, and therefore no need to counteract such tendency; the gyroscopic effects are also neutralised. All this means that there should be no tendency to swing to one side during take-off, no roll or yaw if the throttle is suddenly opened or closed, no difference in aileron or rudder trim, whether the engine is on or off – in short, the aircraft should be easy to handle and nice to fly.

Can you answer these?

Some simple questions about thrust and propellers –

1. What is a ramjet?
2. What is meant by the blade angle of a propeller, and why does this angle decrease from boss to tip?
3. Distinguish between the 'advance per revolution', the 'geometric pitch' and the 'experimental pitch' of a propeller.
4. What is slip?
5. What are the advantages of a variable-pitch propeller?
6. Why is the tip speed an important factor in propeller design?
7. Why is solidity important, and how can it be increased?
8. What methods of propulsion can be used outside the earth's atmosphere?

For solutions see Appendix 5.
Turn to Appendix 3 for a few simple numerical examples on thrust.

CHAPTER 5

Level flight

Introduction

The flight of an aeroplane may be considered as consisting of various stages. First, **the take-off**, during which the aircraft is transferred from one medium to another; then **the climb**, during which the pilot gains the height at which the level part of the flight will be made; then a period of this **steady flight at a constant height**, interrupted in certain cases by periods of manoeuvres, or aerobatics; **the approach** back towards the earth; and finally **the landing**.

On long distance flights the main portion may consist of **a long slow steady climb**, which is more economical than maintaining the same height as fuel is consumed, and the weight of the aircraft is reduced, and so it is often only during a small portion of each flight that the aeroplane may be considered as travelling in straight and level flight at uniform velocity (Fig. 5A, overleaf).

The four forces

Now, what are the forces which keep the aeroplane in its state of steady level flight? First the **lift**, which will be vertically upwards since the direction of motion is horizontal. This we have created with the express object of keeping the aeroplane in the air by opposing the force of gravity, namely, the **weight**. But we can only produce lift if the aeroplane is moved forward and for this we need the **thrust** provided by the propeller or jets. We also know that the forward motion will be opposed by the **drag**.

The aeroplane, therefore, can be said to be under the influence of four main forces –

1. The Lift, L, acting vertically upwards through the Centre of Pressure.

Fig 5A Long haul, large capacity
This Boeing 747–400 'Jumbo' has 350 seats, and a range of 13 528 km that could take it from Europe to Australia non-stop.

2. The Weight of the aeroplane, *W*, acting vertically downwards through the Centre of Gravity.
3. The Thrust of the engine, *T*, pulling horizontally forwards.
4. The Drag, *D*, acting horizontally backwards.

Just as for certain purposes it is convenient to consider all the weight as acting through one point, called the centre of gravity, or all the lift as acting at the centre of pressure, so we may imagine the resultant of all the drag acting at one point which, for convenience, we will call the centre of drag. Its actual position depends on the relative resistance of different parts of the aeroplane.

Conditions of equilibrium

Now, under what conditions will these four forces balance the aeroplane? That is to say, keep it travelling at a steady height at uniform velocity in a fixed direction, a state of affairs which, in the language of mechanics, is known as **equilibrium**. It is sometimes hard to convince a traveller by air that he may travel at 200 m/s and yet be in a state of equilibrium; equilibrium simply means that the existing state of affairs is remaining unchanged; in other words, that the aeroplane is obeying Newton's First Law of Motion.

In order to do this the forces acting on it must be balanced – **the lift must be equal to the weight** (this condition will keep the aeroplane at a constant height); and the **thrust must be equal to the drag** (this condition will keep the aeroplane moving at the same steady velocity).

The idea is often prevalent that the lift must be **greater** than the weight, or, as it is often expressed, the lift must 'overcome' the weight; and when it comes to the question of thrust and drag the author has known students dismiss the idea that the thrust **need only be equal** to the drag as 'contrary to common sense'.

There still remains a third condition for equilibrium. In order to maintain straight and even flight, **we must prevent the aeroplane from rotating**, and this depends not only on the magnitudes of the four forces, but also on the positions at which they act. If the centre of pressure is behind the centre of gravity, the nose will tend to drop and the tail to rise, and vice versa if the centre of pressure is in front of the centre of gravity. But we are also concerned with the lines of action of the thrust and drag, for if the line of thrust is high and the line of drag is low, these two forces also will tend to make the nose drop. Such tendencies could be prevented by the pilot using his controls, but it is the aim of the designer to make an aeroplane which will in the words of the pilot, fly 'hands off'. Therefore he must see that the forces act in the right places.

Difficulties in balancing the four forces

First, the lift. The lift will act through the centre of pressure, which will depend on the position of the wings; so the designer must be careful to place the planes in the correct position along the fuselage. But the problem is complicated by the fact that a change in the angle of attack means a movement of the lift, and usually in the unstable direction; if the angle of attack is increased the pitching moment **about the centre of gravity** will become more nose-up, and tend to increase the angle even further.

Secondly, the weight. This will act through the centre of gravity, which in turn will depend on the weight and position of every individual part of the aeroplane and the loads that it carries. Here alone is sufficient problem, but again there is a possibility of movement of the centre of gravity during flight caused, for instance, by consumption of fuel, dropping of bombs or movement of passengers. In the Concorde, fuel was actually moved from one tank to another to adjust the position of the centre of gravity.

Thirdly, the thrust. Here the problem is easier. The line of thrust is settled by the position of the propeller shaft or centre line of the jet, which in turn depend on the position of the engine or engines. In this matter the designer has little choice, but has to consider such problems as keeping the propeller clear of the ground and giving the pilot a clear view ahead; new problems arise too when the thrust can be deflected as in certain modern types.

Lastly, the drag. This is, perhaps, the most difficult of all. The total drag is composed of the drag of all the separate parts, and the designer must either estimate the drag of each part separately, and so find the total drag and its line of action, or must rely on wind tunnel experiments on a model or computed predictions; and even when the line of drag has been found it too will be liable to change at different angles of attack.

Arranging the forces

For steady flight along a straight line, whether level or not, it is **not only** necessary to balance the four forces so that they produce no resultant force; their lines of action must also be arranged so that they produce **no resultant moment,** otherwise the aircraft will rotate either nose-up or nose-down. When there is no resultant moment, the aircraft is said to be **trimmed.** Mathematically, if the sum of the moments is M, then the condition for trim is that $M=0$.

One way to achieve this would be to arrange that all of the forces act through a single point, as in Fig. 5.1. However this is not generally practical, as there are many factors that tend to alter the line of action, apart from those already stated. For example, lowering the undercarriage tends to shift the line of the resultant drag down, and on the floatplane variant of a light aircraft illustrated in Fig. 5B the position of the drag resultant would be very much lower than for the original floatless design.

It is possible to balance the moment produced by the drag and thrust being out of line, by arranging the lift and weight forces to be out of line by an amount that causes them to exactly produce the necessary balancing moment, as in Fig 5.2. However, because of the way that the lines of action of the forces tend to change according to aircraft attitude, fuel weight etc., there is no simple design solution to ensure that the resultant moment will always be zero. As described below, some active involvement of the pilot is required in order to keep the aircraft trimmed.

The tail plane and other horizontal control surfaces

The traditional method of ensuring that the aircraft can be trimmed to give no resultant moment is to provide the aircraft with an auxiliary lifting surface

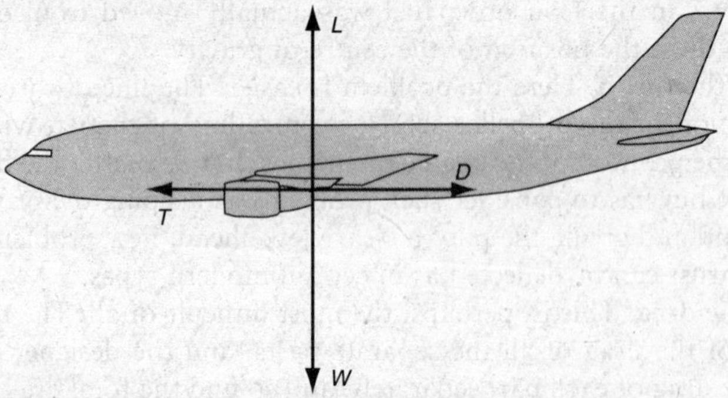

Fig 5.1 Forces acting through a single point. This arrangement is not generally practical, as the line of action of the lift tends to move around with the angle of attack

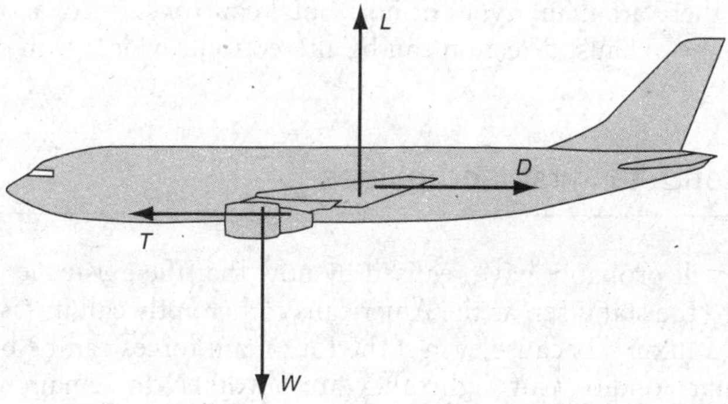

Fig 5.2 Moment due to lift and weight balanced by moment due to drag and thrust

Fig 5B Floatplane
(By courtesy of Cessna Aircraft Company, USA)

called a tail plane. The lift of the tail plane can be regulated by the pilot, and thus he can adjust the moment that it applies. Fig. 5.3 (overleaf) shows how this works. The tail plane can produce lift in either the positive (upward) or negative (downward sense) in order to produce the required moment for trim. To change the tail plane lift, either the whole surface can be pivoted, or the rear part of the surface (the elevator) can be hinged up or down. In practice, for small adjustments to the trim, it is common practice to provide a very small hinged surface or 'trim tab' in addition to the main elevator, as described later.

Nowadays, there are many types of horizontal control surface, and sometimes even the engine exhaust direction can be altered to provide a trim control.

'Tail-less' and 'tail-first' aeroplanes

The reader will probably have realised by now that the existence of this auxiliary plane – the stabiliser, as the Americans rather aptly call it – is a necessity rather than a luxury, because even if the four main forces can be balanced for one particular condition of flight, they are not likely to remain so for long. What then of the so-called **tail-less** type of aeroplane?

This type has had followers from the very early days of flying – and among birds from prehistoric times – and although the reasons for its adoption have changed somewhat, a common feature has been a large degree of sweepback, or even delta-shaped wings, so that although this type may appear to have no tail, the exact equivalent is found at its wing tips, the wings being, in fact, swept back so that the tip portion can fulfil the functions of the tail plane in the orthodox aeroplane. In fact, it is true to say that the **'tail-less' type has two tails instead of one!** (Figs 5C and 5D).

More unusual is the tail-first or canard configuration aeroplane. A most important historical example was the original Wright Flyer, which is generally accepted to have made the first controlled power-driven flight. Like many early ideas, the canard has recently made something of a come-back (Fig. 5E, overleaf), and examples are now found for many types of aircraft, but particularly for missiles and highly manoeuvrable fighters such as the Eurofighter Typhoon. A tail in front can hardly be called a tail, and this surface is commonly known now as the foreplane.

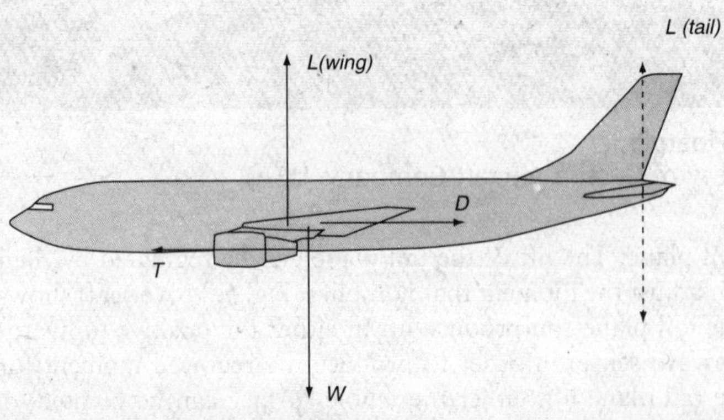

Fig 5.3 The pilot can adjust the tail lift so that the resultant moment is zero and the aircraft is trimmed. The tail lift can be either upwards or downwards

Fig 5C Tail-less – old type
(By courtesy of *Flight*)
The Westland Hill Pterodactyl.

Fig 5D Tail-less – new type
This popular form of powered microlight aircraft has been derived from hang-glider technology.

Fig 5E Tail-first
Apart from the canard layout, the Rutan Vari-Eze shows many unusual features such as a pusher propeller, composite construction, and a nosewheel that can be retracted in flight or when parked.

Loads on tail plane

But to return to the normal aeroplane. Where the four main forces can be satisfactorily balanced in themselves, the duty of the tail plane is merely to act as a 'stand-by'. Therefore, it will usually be set at such an angle, that at cruise speed it will be at zero angle of attack, thereby producing no lift. At flight speeds higher than the cruise speed, the lift coefficient must be lowered to compensate for the higher dynamic pressure otherwise the lift would be greater than the weight. This means that the aircraft must be trimmed a little more nose-down. In doing so, the centre of lift of the wing will move back, giving a nose-down pitching moment. However, if the angle of attack of the tail plane was zero for cruise, then it will now become negative, and the tail will generate a down-force (Fig. 5.4) producing a nose-up pitching moment which will tend to more than counteract the nose-down moment produced by the wing. In fact, as the flight speed increases, the pilot normally has to make a small nose-down trim adjustment. Correspondingly, at low speeds, the nose of the aircraft must be raised in order to increase the angle of attack. This means that the tail lift will now become positive (Fig.5.5). As the tail plane is equally likely to carry an upward or a downward force, it is usually of symmetrical camber, and therefore produces no lift at zero angle of attack. On a tail-first or canard aircraft, the foreplane is set at a slightly higher angle of attack than the wings for reasons of stability, and both wings produce lift in normal flight.

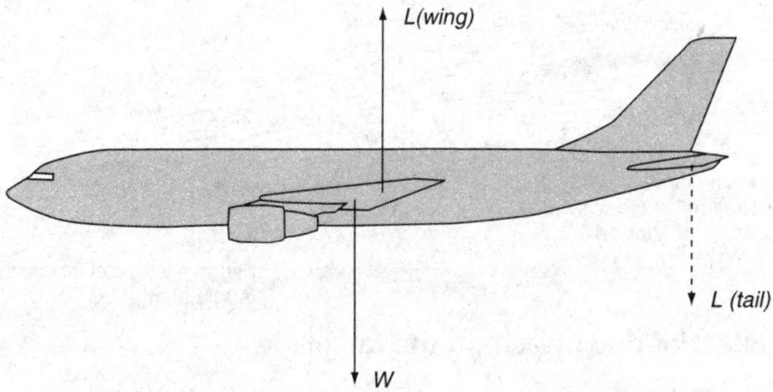

Fig 5.4 High speed: down load from tail needed to balance effect of rearward location of wing lift

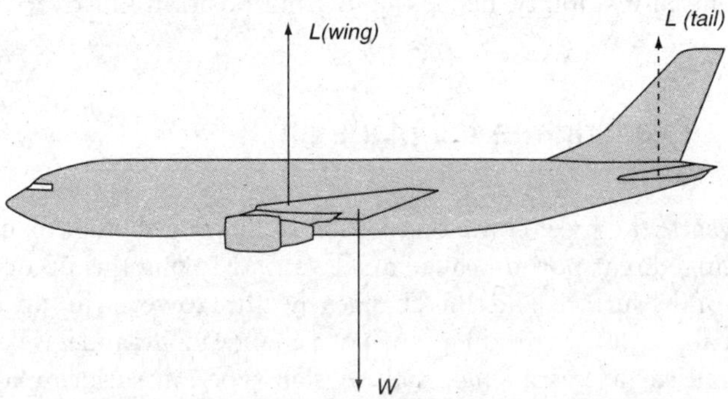

Fig 5.5 Low speed: up load from tail needed to balance effect of forward location of wing lift

Effects of downwash

In many types of aircraft the air which strikes the tail plane has already passed over the main planes, and the trailing vortices from these will cause a downwash on to the tail plane (Fig. 5.6, overleaf). The angle of this downwash may be at least half the angle of attack on the main planes, so that if the main planes strike the airflow at 4°, the air which strikes the tail plane will be descending at an angle of 2°, so that if the tail plane were given a riggers' angle of incidence of 2°, it would strike the airflow head-on and, if symmetrical, would provide no force upwards or downwards. Again, the angle of downwash will, of course, change with the angle of attack of the main planes, and it is for this reason that the angle at which the tail plane should be set is one of the difficult problems confronting the designer.

As we shall discover later, its setting also affects the stability of the aeroplane, and further difficulties arise from the fact that in a propeller-driven aircraft the tail plane is usually in the slipstream, which is a rotating mass of air and will therefore strike the two sides of the tail plane at different angles.

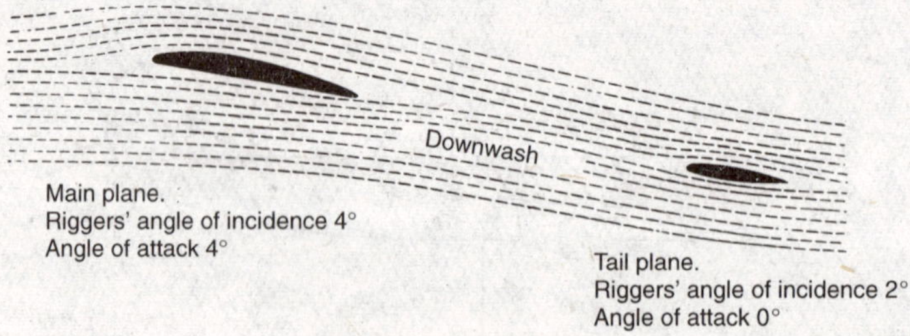

Fig 5.6 Effect of downwash on the tail plane

In jet-driven aircraft the tail plane is often set very high (Fig. 5F), to keep it clear of the hot jets, and this in turn may cause trouble since it may be shielded by the main planes at large angles of attack, resulting in what is called a deep stall and general instability, hence the low tail position illustrated in Fig. 5G.

Methods of varying the tail plane lift

For many years, the most common method of altering the tail plane lift was to provide a hinged rear portion called an elevator. Moving this up or down alters the camber of the surface, and thus changes the lift. However, in supersonic flight, changes of camber do not produce much of a change in lift, and it is better to move the whole tail surface as a single slab. A **slab** type tail surface may be seen in Figures 5G and 5H (overleaf). The moving slab tailplane is nowadays sometimes used even on low-speed aircraft as there are structural and manufacturing advantages to this type of tail. On some aircraft, particularly large transports, the two approaches may be combined. The whole tailplane can be moved, mainly to provide for trimming, and a hinged elevator is also used, mainly for control and manoeuvre.

Conditions of balance considered numerically

We have so far avoided any numerical consideration of the forces which balance the aeroplane in straight and level flight; in simple cases, however, these present no difficulty.

When there is no load on the tail plane the conditions of balance are these –

1. Lift = Weight, i.e. $L = W$.
2. Thrust = Drag, i.e. $T = D$.
3. The 'nose-down' pitching moment of L and W must balance the 'tail-down' pitching moment of T and D.

Fig 5F High tail
The tail of a Lockheed C-5 Galaxy. The elevators are split into a number of sections to provide redundancy for reasons of safety.

Fig 5G Low tail
The Alpha-Jet has a relatively low tail plane with pronounced anhedral.

The two forces, L and W, are two equal and opposite parallel forces, i.e. a couple; their moment is measured by 'one of the forces multiplied by the perpendicular distance between them'. So, if the distance between L and W is x metres, the moment is Lx or Wx newton-metres.

Similarly, T and D form a couple and, if the distance between them is y metres, their moment is Ty or Dy newton-metres.

Therefore the third condition is that –

$$Lx \text{ (or } Wx) = Ty \text{ (or } Dy)$$

To take a numerical example: Suppose the mass of an aeroplane is 2000 kg. The weight will be roughly 20 000 N. L = W. So L = 20 000 N.

Now, what will be the value of the thrust and drag? The reader must beware of falling into the ridiculous idea, which is so common among students, that the thrust will be equal to the weight! The statement is sometimes made that the 'Four forces acting on the aeroplane are equal', but nothing could be farther from the truth. $L = W$ and $T = D$, but these equations certainly do not make $T = W$. This point is emphasised for the simple reason that out of over a thousand students to whom the author has put the question, 'How do the think the thrust required to pull the aeroplane along in normal flight compares with the weight?' more than 50 per cent have suggested that the thrust must obviously equal the weight and over half the remainder have insisted that, on the other hand, the thrust must be many times greater than the weight! Such answers show that the student has not really grasped the fundamental principles of flight, for is not our object to obtain the maximum of lift with the minimum of drag, or, what amounts to much the same thing, to lift as much weight as possible with the least possible engine power? Have we not seen that the wings can produce, at their most efficient angle, a lift of 20, 25 or even more times as great as their drag?

It is true that there is a great deal of difference between an aeroplane and a wing; for whereas a wing provides us with a large amount of lift and a very much smaller amount of drag, all the other parts help to increase the drag and provide no lift in return. Actually it is not quite true to say that the wings provide all the lift, for by clever design even such parts as the fuselage may be persuaded to help. Efforts to increase the lift may be well worth while, but even total increase of lift from such sources will be small, whereas the addition of the parasite drag of fuselage, tail and undercarriage will form a large item; in old designs it might be as much, or even more than the drag of the wings, but in modern aircraft with 'clean' lines the proportion of parasite drag has

Fig 5H Slab tail planes (opposite)
(By courtesy of the Grumman Corporation)
Large slab tail planes and twin fins are evident on this photograph of the F–14.

been much reduced. None the less it may result in the reduction of the lift/drag ratio of the aeroplane, as distinct from the aerofoil, down to say 10, 12 or 15.

So it will be obvious that the **ideal aeroplane** must be one in which there is no parasite drag, i.e. in the nature of a '**flying wing**'; we should then obtain a lift some forty times the drag. At present we are a long way off this ideal except for high-performance gliders.

In our numerical example, let us assume that the lift is ten times the total drag, this being a reasonable figure for an average aeroplane. Then the drag will be 2000 N; so thrust will also be 2000 N. **This means that an aeroplane of weight 20 000 N can be lifted by a thrust of 2000 N; but the lift is not direct, the work is all being done in a forward direction – not upwards.** The aeroplane is in no sense a helicopter in which the thrust is vertical, and in which the thrust must indeed be at least equal to the weight.

To return to the problem –

$$L = W = 20\,000 \text{ N}$$

$$T = D = 2000 \text{ N}$$ (This is merely an approximation for a good type of aeroplane)

$$Lx = Ty$$

$$\text{So } 20\,000\,x = 2000\,y$$

$$\text{and } x = \frac{1}{10}\,y$$

So if T and D are 1 metre apart, L and W must be 1/10 metre apart, i.e. 100 mm. In other words, the lines of thrust and drag must be farther apart than the lines of lift and weight in the same proportion as the lift is greater than the drag (Fig. 5.7).

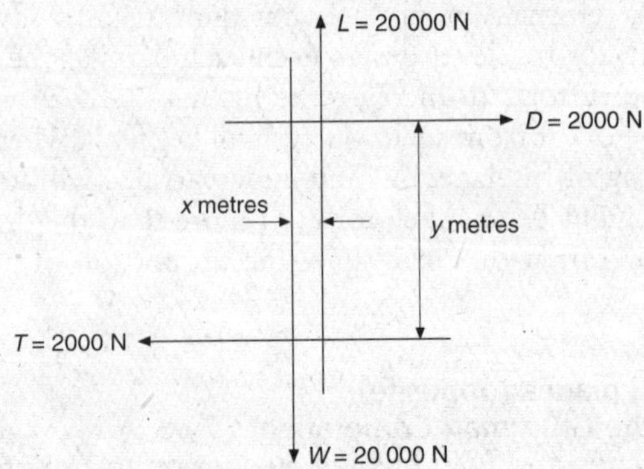

Fig 5.7 Balance of forces – no load on tail

Finding the tail load

This conclusion only applies when there is no force on the tail plane. When there is such a force the problem is slightly more complicated, but can still be solved by the principle of moments. Consider a further example –

An aeroplane weighs 10 000 N; the drag in normal horizontal flight is 1250 N; the centre of pressure is 25 mm (0.025 m) behind the centre of gravity, and the line of drag is 150 mm (0.15 m) above the line of thrust. Find what load on the tail plane, which is 6 m behind the centre of gravity, will be required to maintain balance in normal horizontal flight.

Let the lift force on the main planes = Y newtons

Let the force on the tail plane = P newtons (assumed upwards)

Then total lift $= L = Y + P$ (Fig 5.8)

$$\text{But } L = W = 10\,000 \text{ N}$$

$$\therefore Y + P = 10\,000 \text{ N} \qquad (1)$$

$$\text{Also } T = D = 1250 \text{ N} \qquad (2)$$

Take moments about any convenient point; in this case perhaps the most suitable point is 0, the intersection of the weight and thrust lines.

Nose-down moments about 0 are caused by Y and P

W and T will, of course, have no moments about 0,

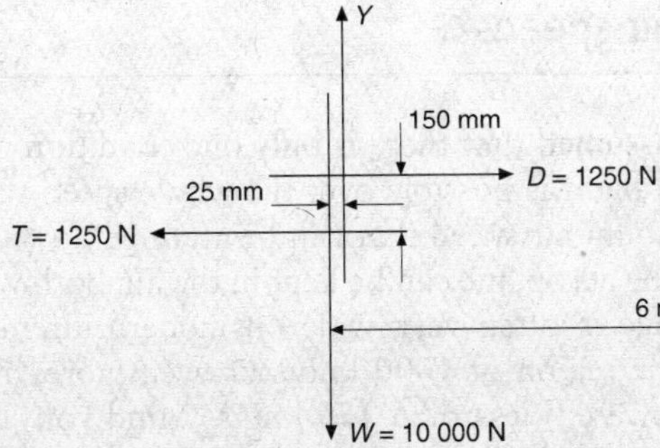

Fig 5.8 The four forces and the tail load

So total nose-down moments $= 0.025\ Y + 6\ P$

(all distances being expressed in metres)

Tail-down moment about *O* is caused by *D* only,

So total tail-down moment $= 0.15$ D

$\therefore 0.025\ Y + 6\ P = 0.15\ D$

i.e. $Y + 240\ P = 6\ D$ (3)

But from (1), $Y + P = 10\,000$

$\therefore 239\ P = 6\ D - 10\,000$

But from (2), $D = 1250$

$\therefore 239\ P = 6 \times 1250 - 10\,000$

$= 7500 - 10\,000$

$= -2500$

$\therefore P = -(2500/239) = -10.4$ N

Therefore a small **downward** force of 10.4 N is required on the tail plane, the negative sign in the answer simply indicating that the force which we assumed to be upwards should have been downwards.

The student is advised to work out similar examples which will be found in Appendix 3 at the end of the book.

Level flight at different air speeds

So far we may seem to have assumed that there is only one condition of level flight; but this is not so. Level flight is possible over the whole speed range of the aeroplane, from the maximum air speed that can be attained down to the minimum air speed at which the aeroplane can be kept in the air, both without losing height. This **speed range** is often very wide on modern aircraft; the maximum speed may be in the region of 1000 knots or even more, and the minimum speed (with flaps lowered) less than 150 knots. Mind you, though level flight is **possible** at any speed within this range, it may be very **inadvisable** to fly unduly fast when considerations of fuel economy are involved, or to fly unduly slowly if an enemy is on your tail! There is nearly always a correct speed to fly according to the circumstances.

Relation between air speed and angle of attack

An aeroplane flying in level flight at different air speeds will be flying at different angles of attack, i.e. at different attitudes to the air. Since the flight is level, this means different attitudes to the ground, and so the pilot will be able to notice these attitudes by reference to the horizon (or to the 'artificial horizon' on his instrument panel).

For every air speed – as indicated on the air speed indicator – there is a corresponding angle of attack at which level flight can be maintained (provided the weight of the aeroplane does not change).

Let us examine this important relationship more closely. It all depends on our old friend the lift formula, $L = C_L \cdot \frac{1}{2}\rho V^2 \cdot S$. To maintain level flight, the lift must be equal to the weight. Assuming for the moment that the weight remains constant, then the lift must also remain constant and equal to the weight. The wing area, S, is unalterable. Now, if we look back, or think back, to Chapter 2 we will realise that $\frac{1}{2}\rho V^2$ represents the difference between the pressure on the pitot tube and on the static tube (or static vent), and that this difference represents what is read as air speed on the air speed indicator, in other words, the **indicated air speed**. There is only one other item in the formula, i.e. C_L (the lift coefficient). Therefore if $\frac{1}{2}\rho V^2$, or the indicated air speed, goes up, C_L must be reduced, or the lift will become greater than the weight. Similarly, if $\frac{1}{2}\rho V^2$ goes down, C_L must go up or the lift will become less than the weight. Now C_L depends on the angle of attack of the wings; the greater the angle of attack (up to the stalling angle), the greater the value of C_L. **Therefore for every angle of attack there is a corresponding indicated air speed.**

This is most fortunate, since the pilot will not always have an instrument on which he can read the angle of attack, whereas the air speed indicator gives him an easy reading of air speed. That is why a pilot always talks and thinks in terms of speed, landing speed, stalling speed, best gliding speed, climbing speed, range or endurance speed, and so on. The experimenter on the ground, on the other hand, especially if he does wind tunnel work, is inclined to talk and think in terms of angle, stalling angle, angle of attack for flattest glide, longest range, and so on. This difference of approach is very natural. The pilot, after all, has little choice if he does not know the angle of attack but does know the speed. To the experimenter on the ground, speed is rather meaningless; he can alter the angle of attack and still keep the speed constant – something that the pilot **cannot** do. But, however natural the difference of outlook, it is unfortunate; and it is undoubtedly one of the causes of the gap between the two essential partners to progress, the practitioner and the theorist.

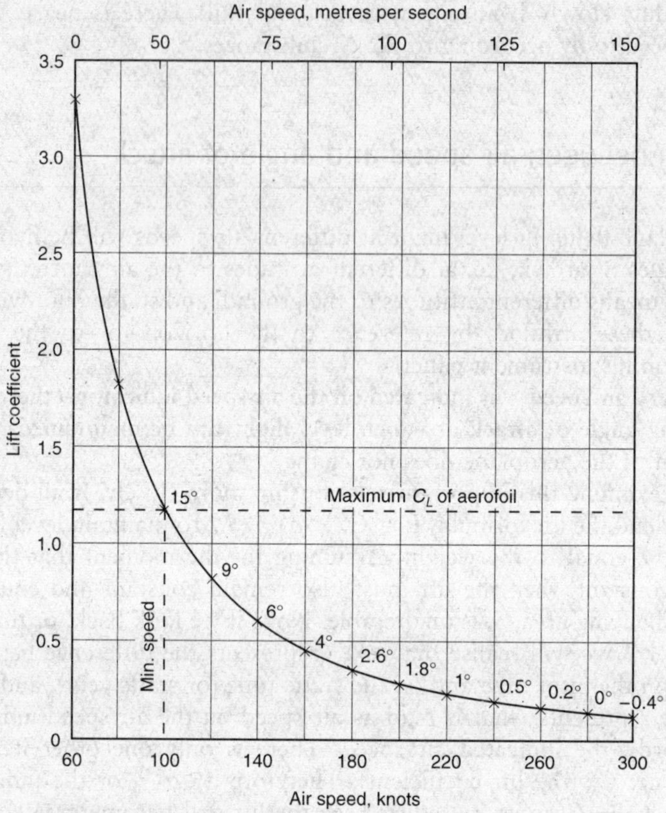

Fig 5.9 Air speed, lift coefficient and angle of attack
Aeroplane of weight 50 kN (5100 kgf); wing area 25.05 m²; wing loading 1920 N/m² (196 kgf/m²); aerofoil section characteristics Figs 3.13, 3.15, 3.16 and 3.17.

Let us examine our general statement more critically by working out some figures. Suppose the mass of an aeroplane is 5100 kg (so that its weight will be approximately 50 000 newtons and that its wing area is 26.05 m², i.e. a wing loading of about 1920 N/m². Assume that the aerofoil section has the lift characteristics shown in the lift curve in Fig. 3.13. Consider first the ground level condition, the air density being 1.225 kg/m³.

Whatever the speed, the lift must be equal to the weight, 50 kN; but the lift must also be equal to $C_L \cdot \frac{1}{2}\rho V^2 . S$, so

$$50\,000 = C_L \times \frac{1}{2} \times 1.225 \times V^2 \times 26.05$$

from which

$C_L = 3134/V^2$

Now insert values for V of 60, 80, 100 and other values up to 300 knots, converting them, of course, to m/s, and work out the corresponding values of C_L; then by referring to Fig. 3.13 read off the angle of attack for each speed. The result will be something like that shown in Table 5.1 and in Fig. 5.9. The angles given in the table are approximate since all the values are small.

Now let us see what this table and graph mean; we shall find it very interesting. In the first place at speeds below about 100 knots, the lift coefficient needed for level flight is greater than the maximum lift coefficient (1.18) provided by the aerofoil, therefore level flight is not possible below this speed. Secondly, as the speed increases to 120, 140, 160 knots, etc., the angle of attack decreases from 15° to 9°, 6°, 4°, etc.; and for each speed there is a corresponding angle of attack. We should notice, in passing, that at comparatively low speeds there is much greater change in angle of attack for each 20 knots increase in air speed than there is at the higher speeds, e.g. the angle of attack at 120 knots is 6° less than at 100 knots, whereas at 280 knots it is only 0.2° less than at 260 knots. This change in proportion is interesting, and is one of the arguments for an angle of attack indicator, which is sensitive at low speeds, which is just where the air speed indicator is most unsatisfactory.

We could, of course, continue Table 5.1 speeds higher than 300 knots, and we should find that we needed even smaller lift coefficients, and even more negative angles of attack (though never less than 21.8° since at this angle there would be no lift, whatever the speed). But at this stage we must begin to consider another factor affecting speed range, namely **the power of the engine.** What we have worked out so far is accurate enough, provided we can be sure

Table 5.1 Angle of attack derived from airspeed

Air speed		V^2	$C_L = 3134/V^2$	*Angle of attack*
knots	*m/s*			
60	30.8	949	3.31	–
80	41.2	1697	1.85	–
100	51.5	2652	1.18	15°
120	61.8	3819	0.82	9°
140	72.0	5184	0.60	6°
160	82.4	6790	0.46	4°
180	92.6	8580	0.37	2.6°
200	103.0	10 600	0.30	1.8°
220	113.2	12 800	0.24	1°
240	123.7	15 400	0.20	0.5°
260	134.0	18 000	0.17	0.2°
280	144.0	20 740	0.15	0°
300	154.0	23 700	0.13	−0.4°

of obtaining sufficient thrust. It may be that at speeds of 300 knots or above or, for that matter, at 100 knots or below, we shall not be able to maintain level flight for the simple reason that we have not sufficient engine power to overcome the drag. So the engine power will also determine the speed range, not only the top speed, but also to some extent the minimum speed.

If the reader thinks that the minimum speed of this aeroplane is rather high, we should perhaps point out, first, that we have not used flaps; secondly, that the aerofoil does not given a very high maximum lift coefficient; and, thirdly, that it has a fairly high wing loading, or ratio of weight to wing area, which, as we shall see later, has an important influence on minimum speed. All we wish to establish now is the relationship between air speed and angle of attack, and this is clearly shown by Table 5.1.

Effect of height

Table 5.1 was worked out for ground level conditions. What will be the effect of height on the relationship between air speed and angle of attack? The answer, once it is understood, is simple – but very important.

Whatever the height, the air speed indicator reading is determined by the pressure $\frac{1}{2}\rho V^2$. In this expression, V is the true air speed. As has already been explained in Chapter 2, when the air speed indicator reads 200 knots at 3000 m, it simply means that the difference in pressure between pitot and static tubes (i.e. $\frac{1}{2}\rho V^2$) is the same as when the air speed was 200 knots at ground level. Now, it is not only the pitot pressure that depends on $\frac{1}{2}\rho V^2$; so do the lift and the drag. Therefore, at the same value of $\frac{1}{2}\rho V^2$, i.e. *at the same indicated speed*, the lift and drag will be the same as at ground level, other things (such as C_L) being equal. Therefore the table remains equally true at all heights, provided the air speed referred to is the indicated speed, and not the true speed. Thus the angle of attack, or the attitude of the aeroplane to the air, is the same in level flight at all heights, provided the indicated air speed remains the same.

Effect of weight

The table was worked out for a constant weight of 50 kN. What will be the effect of changes of weight such as must occur in practical flight owing to fuel consumption, etc.? The answer to this is not quite so simple.

Suppose the weight is reduced from 50 kN to 40 kN. At the same indicated air speed, the angle of attack would be the same, and the lift would be the same as previously, i.e. 50 kN. This would be too great. Therefore, in order to reduce the lift, we must adjust the attitude, so that the wings strike the air at a smaller angle of attack, or we must reduce the speed, or both. Whatever we do, we shall get a slightly different relationship between air speed and angle of attack: the reader is advised to work out the figures for a weight of 40 kN. Although the relationship will differ from that for 50 kN weight, it will again remain constant at all heights for the same indicated speeds. To sum up the effect of weight, we can say that **the less the total weight of the aircraft, the less will be the indicated air speed corresponding to a given angle of attack.** A little calculation will show that the indicated air speed for the same angle of attack will be in proportion to the square root of the total weight.

Flying for maximum range – propeller propulsion

Whether in war or peace, we shall often wish to use an aircraft to best advantage for some particular purpose – it may be to fly as fast as possible, or as slowly as possible, or to climb at maximum rate, or to stay in the air as long as possible, or, perhaps, most important of all, to achieve the maximum distance on a given quantity of fuel. Flying for maximum range is one of the outstanding problems of practical flight, but it is also one of the best illustrations of the principles involved. To exploit his engine and aircraft to the utmost in this respect, a pilot must be not only a good flier, but also an intelligent one.

The problem concerns engine, propeller and aircraft; it also concerns the wind. In this book we are interested chiefly in the aircraft, but we cannot solve this problem, and indeed we can solve few, if any, of the problems of flight, without at least some consideration of the engine and the propeller, or jet, or rocket, or whatever it may be, and how the pilot should use them to get the best out of his aircraft. As for the wind, we shall, as usual in this subject, first consider a condition of still air.

The object in any engine regardless of type is to burn fuel so as to get energy and then to convert this energy into mechanical work. **In order to get the greatest amount of work from a given amount of fuel, we must, first of all, get the maximum amount of energy out of it, and then we must change it to mechanical work in the most efficient way.** Our success or otherwise will clearly depend to some extent on the use of the best fuel for the purpose, and on the skill of the engine designer. But the pilot, too, must play his part. To get the most heat from the fuel, it must be properly burned; this means that the mixture of air to fuel must be correct. In a piston engine, what is usually called 'weak mixture' is, in fact, not so very weak, but approximately the correct mixture to burn the fuel properly. If we use a richer mixture, some of the fuel will not be properly burned, and we shall get less energy from the same amount of fuel: we may get other advantages, but we shall not get economy. Both the manifold pressure and the revolutions per minute will affect the efficiency of the engine in its capacity of converting energy to work. The problem of the best combination of boost and rpm, though interesting, is outside the scope of this book and at this stage, too, the principles of the reciprocating and turbine engine begin to differ, while the rocket or ramjet has not got any rpm! For the reciprocating engine we can sum up the engine and propeller problem by saying that, generally speaking, the pilot will be using them to best advantage if he uses weak mixture, the highest boost permissible in weak mixture, and the lowest rpm consistent with the charging of the electrical generator and the avoidance of detonation. All this has assumed that he has control over such factors, and that the engine is supercharged and that the propeller has controllable pitch. Without such complications, the pilot's job will, of course, be easier; but the chances are that, whereas a poor pilot may get better results, the good pilot will get worse – far, far worse.

Before we leave the question of the engine and propeller, let us look at a problem which affects **all** engines in which fuel is burnt to give mechanical energy – not just piston engines.

The problem is that we cannot convert all the energy produced by burning the fuel into mechanical work, however hard we try. What is more, although in a sense we are always trying to do this (and then call the engine inefficient because we do not succeed!) we know quite well why we cannot and never will do it – it is just contrary to the laws of thermodynamics, the laws that govern the conversion of energy into mechanical work. All we can get, even in the best engines and in the hands of the best pilots, is something like 30 per cent of this figure. From each litre of fuel we can expect to get about 31 780 000 joules of thermal energy and hence only 0.3 × 31 780 000 joules i.e. about 9 500 000 joules or newton metres of mechanical energy.

This is what the engine should give to the propeller; and we may lose 20 per cent of it due to the inefficiency of the propeller, and so the aircraft will only get about 80 per cent of 9 500 000, i.e. **about 7 600 000 joules, or newton metres.**

It still seems a large figure – it is a large figure – but, as we shall see, it will not take the aircraft very far. However at this stage we are not so much concerned with the numerical value, as with the unit, and to think of it in the form of the **newton metre**. We have found that a litre of fuel, if used in the best possible way (we have said nothing about how quickly or slowly we use it) will give to the aircraft so many newton metres. **Suppose, then, that we want to move the aircraft the maximum number of metres, we must pull it with the minimum number of newtons, i.e. with minimum force. That simple principle is the essence of flying for range.**

Let us examine it more closely. It means that we must fly so that the propeller gives the least thrust with which level flight is possible. Least thrust means least drag, because drag and thrust will be equal.

Flying with minimum drag

Now, on first thoughts, we might think that flying with minimum drag meant presenting the aeroplane to the air in such an attitude that it would be most streamlined; in other words, in the attitude that would give least drag if a model of the aircraft were tested in a wind tunnel. But if we think again, we shall soon realise that such an idea is erroneous. This 'streamlined attitude' would mean high speed, and the high speed would more than make up for the effects of presenting the aeroplane to the air at a good attitude; in a sense, of course, it is the streamlined attitude that enables us to get the high speed and the high speed, in turn, causes drag. We are spending too much effort in trying to go fast.

On the other hand, we must not imagine, as we well might, that we will be flying with least drag if we fly at the minimum speed of level flight. This would mean a large angle of attack, 15° or more, and the induced drag particularly would be very high – we would be spending too much effort in keeping up in the air.

There must be some compromise between these two extremes – it would not be an aeroplane if there was not a compromise in it somewhere. Perhaps, too, it would not be an aeroplane if the solution were not rather obvious – once it has been pointed out to us! Since the lift must always equal the weight, which we have assumed to be constant at 50 kN, **the drag will be least when the lift/drag ratio is greatest**. Now, the curve of lift/drag ratio given in Fig. 3.16 refers to the aerofoil only. The values of this ratio will be less when applied to the whole aeroplane, since the lift will be little, if any greater, than that of the wing alone, whereas the drag will be considerably more, perhaps twice as much. Furthermore, the change of the ratio at different angles of attack, in other words, the shape of the curve, will not be quite the same for the whole aeroplane. None the less, there will be a maximum value of, say 12 to 1 at about the same angle of attack that gave the best value for the wing, i.e. at 3° or 4°, and the curve will fall off on each side of the maximum, so that the lift/drag ratio will be less, i.e. the drag will be greater, whether we fly at a smaller or a greater angle of attack than 4°; in other words, at a greater or less speed than that corresponding to 4°, which our table showed to be 160 knots.

A typical lift/drag curve for a complete aeroplane is shown later on in Fig. 6.3, and Table 5.2 shows the sort of figures we shall get from it,

This table is very instructive, and shows quite clearly the effect of different air speeds in level flight on the total drag that will be experienced. It shows, too, that the least total drag is at the best lift/drag ratio, which in this case is at 4° angle of attack, which, in turn, is at about 160 knots air speed (Fig. 5.10, overleaf).

The angle of attack that gives the best lift/drag ratio will be the same whatever the height and whatever the weight; it is simply a question of presenting the aeroplane to the air at the best attitude, and has nothing to do with the

Table 5.2 Airspeed and drag

Air speed		*Angle of attack*	*L/D ratio*		*Total drag*
knots	m/s		*Wing*	*Aeroplane*	newtons
100	51.5	15°	10.7	6.0	8330
120	61.8	9°	17.2	10.6	4720
140	72.0	6°	20.6	11.8	4240
160	82.4	4°	22.7	12.0	4170
180	92.6	2.6°	23.8	10.7	4670
200	103.0	1.7°	22.8	8.5	5880
220	113.2	1°	20.8	7.2	6940
240	123.7	0.5°	18.8	6.0	8330
260	134.0	0.2°	16.4	5.2	9615
280	144.0	0°	13.9	4.5	11 110
300	154.0	−0.4°	12.8	3.8	13 160

density of the air, or the loads that are carried inside the aeroplane, or even the method of propulsion.

This means that **the indicated air speed**, which is what the pilot must go by, **will be the same, whatever the height**, but will increase slightly for increased loads. The same indicated air speed means the same drag at any height, and therefore the same range.

On the other hand, the higher speed which must be used for increased weights means greater drag, because, looking at it very simply, even the same lift/drag ratio means a greater drag if the lift is greater. **So added weight** means added drag – in proportion – and therefore **less range** – also in proportion.

Let us go back to the newton metres, the 7 600 000 joules that we hope to get from 1 litre of fuel. How far can we fly on this? At 100 knots we shall be able to go 7 600 000 divided by 8330, i.e. about 912 metres; at 120 knots 1610 m; at 140 knots 1792 m; at 160 knots 1822 m; at 180 knots 1627 m; at 200 knots 1292 m; and at 220, 240, 260 and 280 knots, 1095, 912, 790, 684 metres respectively, and at 300 knots only 577 metres. These will apply at whatever height we fly. If the load is 60 kN instead of 50 kN each distance must be divided by 60/50, i.e. by 1.20; if the load is less than 50 kN each distance will be correspondingly greater.

Now to sum up this interesting argument: in order to obtain the maximum range, we must fly at a given angle of attack, i.e. at a given indicated air speed, we may fly at any height, and we should carry the minimum load; but if we must carry extra load, we must increase the air speed.

That is the whole thing in a nutshell from the aeroplane's point of view. Unfortunately, there are considerations of engine and propeller efficiency, and of wind, which may make it advisable to depart to some extent from these simple rules, and there are essential differences between jet and propeller propulsion in these respects. We cannot enter into these problems in detail, but a brief survey of the practical effects is given in the next paragraphs.

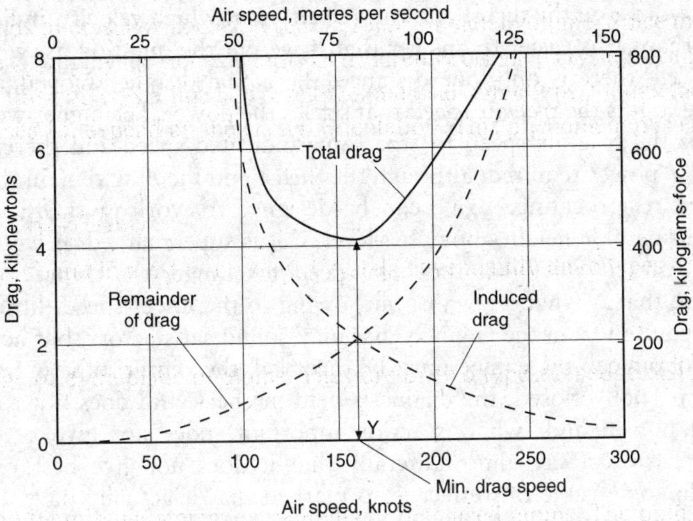

Fig 5.10 Flying for maximum range: how the total drag is made up Induced drag decreases with square of speed. Remainder of drag increases with square of speed. Total drag is the sum of the two; and is a minimum when they are equal.

Another way of thinking of the significance of flying with minimum drag is to divide the total drag into induced drag – which decreases in proportion to the square of the speed – and all-the-remainder of the drag – which increases in proportion to the square of the speed. This idea is well illustrated in Fig. 5.10 and in the numerical examples (Appendix 3).

Effects of height – propeller propulsion

So far as the aeroplane is concerned, we will get the same range and we should fly at the same indicated speed, whatever the height. Now, although the drag is the same at the same indicated speed at all heights, the power is not. This may sound strange, but it is a very important fact. If it were not so, if we needed the same power to fly at the same indicated speed at all heights, then the advantage would always be to fly high, the higher the better, because for the same power the higher we went, the greater would be our true speed. However, it can hardly be considered a proof that an idea is incorrect simply because it would be very nice if it were correct. The real explanation is quite simple. Power is **the rate of doing work**. Our fuel gives us so many newton metres, however long we take to use it. But if we want the work done quickly, if we want to pull with a certain **thrust** through a certain **distance** in a certain **time**, then the power will depend on the thrust and the distance and the time, in other words, on the **thrust** and the **velocity**. But which velocity, indicated or true? Perhaps it is easier to answer that if we put the question as, which distance? Well, there is only one distance, the actual distance moved, the true distance. So it is the true air speed that settles the power. The higher we go, the greater is the true air speed for the same indicated speed and therefore the greater the power required, although the thrust and the drag remain the same.

Now a reciprocating engine can be designed to work most efficiently at some considerable height above sea-level, if it is supercharged. If we use it at sea-level, and if we fly at the best speed for range, the thrust will be a minimum, that is what we want, but, owing to the lower speed, little power will be required from the engine. That may sound satisfactory, but actually it is not economical; the engine must be throttled, the venturi tube in its carburettor is partially closed, the engine is held in check and does not run at its designed power, and, what is more important, does not give of its best efficiency; we can say almost literally that it does not give best value for money. In some cases this effect is so marked that it actually pays us, if we **must** fly at sea-level, to fly considerably faster than our best speed and use more power, thereby using the aeroplane less efficiently but the engine more efficiently. But to obtain maximum range, both aircraft and engine should be used to the best advantage, and this can easily be done if we choose a greater height such that when we fly at the correct indicated speed from the point of view of the aeroplane, the engine is also working most efficiently, that is to say, the throttle valve is fully open, but we can still fly with a weak mixture. At this height, which may be, say, 15 000 ft (4570 m), we shall get the best out of both aeroplane and engine, and so will obtain maximum range.

Here the reader may be wondering what governs the operating height the designer chooses. This may be selected for terrain clearance, cruise above likely adverse weather conditions or the engine may be sized for take-off performance and the cruising altitude follows as a by-product to give full throttle cruise.

What happens at greater heights? At the same indicated speed we shall need more and more power; but if the throttle is fully open, we cannot get more power without using a richer mixture. Therefore we must either reduce speed and use the aircraft uneconomically, or we must enrich the mixture and use the engine uneconomically.

Thus there is a best height at which to fly, but the height is determined by the engine efficiency (and to some extent by propeller efficiency) and not by the aircraft, which would be equally good at all heights. The best height is not usually very critical, nor is there generally any great loss in range by flying below that height. It may well be that considerations of wind, such as are explained in the next paragraph, make it advisable to do so.

Range flying – effects of wind

If the flight is to be made from A towards B and back to A, then wind of any strength from any direction will adversely affect the radius of action. This fact, which at first sounds rather strange, but which is well known to all students of navigation, can easily be verified by working out a few simple examples, taking at first a head and tail wind, and then various cross-winds. But the wind usually changes in direction and increases in velocity with height, and so a skilful pilot can sometimes pick his height to best advantage and so gain more by getting the best, or the least bad, effect from the wind than he may lose by flying at a height that is slightly uneconomical from other points of view. It may pay him, too, to modify his air speed slightly according to the strength of the wind, but these are really problems of navigation rather than of the mechanics of flight.

Flying for endurance – propeller propulsion

We may sometimes want to stay in the air for the longest possible time on a given quantity of fuel. This is not the same consideration as flying for maximum range. **To get maximum endurance**, we must use **the least possible fuel in a given time**, that is to say, we must use **minimum power**. But, as already explained, power means drag × velocity, the velocity being true air speed. Let us look back to Table 5.2 showing total drag against air speed; multiply the drag by the air speed and see what happens.

Table 5.2 shows that, although the drag is least at about 160 knots, the power is least at about 125 knots (see also Fig. 5.11, overleaf). The explanation is quite simple; by flying slightly slower, we gain more (from the power point of view) by the reduced speed than we lose by the increased drag. **Therefore the speed for best endurance is less than the speed for best range** and, since we are now concerned with true speed, the lower the **height**, the better.

The endurance speed is apt to be uncomfortably low for accurate flying; even the best range speed is not always easy and, as neither is very critical, the pilot is often recommended to fly at a somewhat higher speed.

The reader who would like to consider endurance flying a little further should look back to The Ideal Aerofoil in Chapter 3. Among the desirable qualities was a high maximum value of $C_L^{3/2}/C_D$ – the quality which means minimum power, i.e. maximum endurance.

There we were considering only the aerofoil, and the aeroplane is not quite the same thing as regards values, but the idea is the same. So **for endurance flying** we must present the aeroplane to the air at the angle of attack that gives **the best value of** $C_L^{3/2}/C_D$ (for the aeroplane), and this will be a greater angle of attack and so a lower speed, than for range.

Table 5.3 Airspeed, drag and power

Air speed knots	m/s	*Drag* *newtons*	*Drag × Air speed*	*Power* *kW*
100	51.5	8330	428 995	429
120	61.8	4720	291 696	292
140	72.0	4240	305 280	305
160	82.4	4170	343 608	344
180	92.6	4670	432 442	432
200	103.0	5880	605 640	606
220	113.2	6940	785 608	786
240	123.7	8330	1 030 421	1030
260	134.0	9615	1 288 410	1288
280	144.0	11 110	1 599 840	1600
300	154.0	13 160	2 026 640	2027

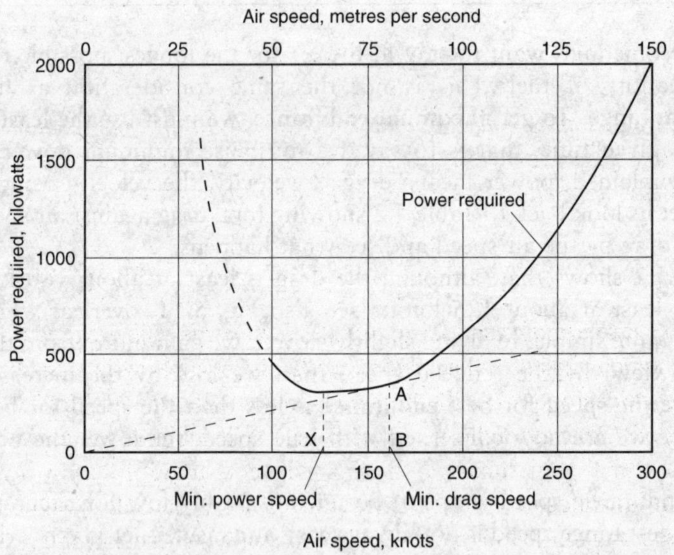

Fig 5.11 Flying for maximum endurance. Maximum endurance is at speed of minimum power (*X*). Maximum range (minimum drag) is at speed *B* since AB/OB = Power/Speed = (Drag × Speed)/Speed = Drag, and AB/OB is least when OA is a tangent to the power curve.

Flying for range – jet propulsion

In trying to get maximum range or endurance out of any aircraft we are, in effect, simply trying to get maximum value for money, the value being the range or endurance and the money being the fuel used. We shall only get the maximum overall efficiency if in turn we get the maximum efficiency at each stage of the conversion of the fuel into useful work done. The three main stages are **the engine**, the **system of propulsion**, and the **aeroplane**.

This applies to every type of aircraft – it is necessary to emphasise this point because there seems to be a growing tendency to think that jet or rocket propulsion involves completely new principles. This is not so – the principles are exactly the same, the only difference lies in the degree of importance of the various efficiencies.

From the point of view of an aeroplane, flying for maximum range means flying with minimum drag. It is in that condition that the aeroplane is most efficient no matter by what means it is driven. But if, when we fly with minimum drag, either the propulsive system, or the engine, or both, are hopelessly inefficient – then, rather obviously, it will pay us to make some compromise, probably by flying rather faster than the minimum drag speed.

From the point of view of an aeroplane, as an aeroplane, we shall obtain the same range at whatever height we fly, provided we fly in the attitude of minimum drag. But if at some heights the propulsive system, or the engine, or both, are more efficient than at other heights – then, rather obviously, it will pay us to fly at those heights so as to get the maximum overall value out of the engine-propulsion-aeroplane system.

Now an aeroplane is an aeroplane whether it is driven by propeller or jet and, as an aeroplane, the same rules for range flying will apply. But when the efficiencies of the propulsion system and the engine are included the overall effects are rather different. In the propeller-driven aeroplane we do not go far wrong if we obey the aeroplane rules although, even so, it usually pays us to fly rather faster than the minimum drag speed because, by so doing, engine and propeller efficiency is improved – and flying is more comfortable. It also definitely pays us to fly at a certain height because at that height the engine-propeller combination is more efficient. But in the main it is the aeroplane efficiency that decides the issue. Not so with the jet aircraft.

There are two important reasons for the difference –

1. Whereas the thrust of a propeller falls off as forward speed increases, the thrust of a jet is nearly constant at all speeds (at the same rpm).
2. Whereas the fuel consumption in a reciprocating engine is approximately proportional to the power developed, the fuel consumption in jet propulsion is approximately proportional to the thrust.

Both of these are really connected with the fact that the efficiency of the jet propulsion system increases with speed, and this increase in efficiency is so important that it is absolutely necessary to take it into account, as well as the efficiency of the aeroplane. When we do so we find that we shall get **greater range** if we fly a great deal faster than the **minimum drag speed**. The drag will

Table 5.4 Air speed and drag ratios

Air speed	*Drag/Speed*	*Air speed*	*Drag/Speed*
100	8330/100 = 83.3	220	6940/220 = 31.5
120	4720/120 = 39.3	240	8330/240 = 34.7
140	4240/140 = 30.3	260	9615/260 = 37.0
160	4170/160 = 26.1	280	11 110/280 = 39.7
180	4670/180 = 25.9	300	13 160/300 = 43.9
200	5880/200 = 29.4		

be slightly greater – not much, because we are on the low portion of the curve (Fig. 5.12) – the thrust, being equal to the drag, will also of course be slightly greater, and so will the fuel consumption in litres per hour. The speed, on the other hand, will be considerably greater and so we shall get more miles per hour. Everything, in fact, depends on getting the **maximum of speed compared with thrust**, or speed compared with drag. In short, we must fly at **minimum drag/speed** which as the figure shows, will always occur at a **higher speed** than that giving minimum drag (Fig. 5.12). **So to get maximum range jet aircraft must fly faster than propeller-driven aircraft** – the difference being due to the different relationship between efficiency and speed in the two systems.

As a matter of interest let us go back to the figures and work out for each speed the value of drag/speed (Table 5.4).

Since, in this instance, we are only concerned with the air speed at which minimum drag/speed occurs, there is no need to convert the knots to m/s.

Note that the minimum value for drag/speed is at about 175 knots, so the range speed for this aircraft, if driven by jets, is 175 instead of 160 knots; but, what is more important, *note the shape of the drag/speed curve* (Fig. 5.13, overleaf); whereas the other curves rise fairly steeply above the minimum value the drag/speed curve hardly rises at all between 170 and 280 knots, so with jet propulsion we can get good range anywhere between these speeds.

At what height shall we fly? That is an easy one to answer. We know that it makes no difference as far as the efficiency of the aeroplane is concerned – **but it makes all the difference to the efficiency of jet propulsion**. The aircraft will be in the same attitude, and we shall get the same drag and the same thrust, if we fly at the same indicated speed at altitude – but the true speed will be greater. Now it is the true speed that largely settles the overall efficiency so at 40 000 ft (12 200 m), where the true speed is doubled, the efficiency will be greatly increased, and, provided the fuel consumption remains proportional to thrust, the range will be similarly increased. **So to get range on jet aircraft – fly high.**

Since modern flights in jet aircraft may take place at heights such as 40 000 or 50 000 ft (12 200 m or 15 200 m) quite a large proportion of the flights may be spent in climbing and descending, and in order to obtain maximum range rather different speeds may be required say for climbing and for the level

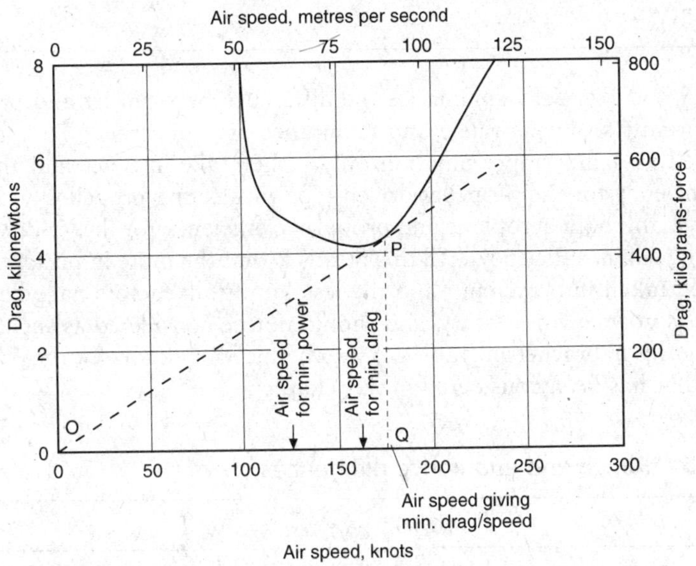

Fig 5.12 Flying for range: jet propulsion
The tangent OP to the curve of total drag gives the air speed *Q* at which drag/speed is a minimum.

portion of the flight. The best speed, for instance, for a cruising climb may be 1.3 × speed for minimum drag, and for level flight 1.2 × speed for minimum drag, i.e. for the aeroplane which we have considered, 208 and 192 knots respectively.

Flying for endurance – jet propulsion

If the argument has been followed so far, there will be no difficulty in understanding the problem of maximum endurance for jet-driven aircraft. Since fuel consumption is roughly proportional to thrust, we shall get maximum endurance by flying with minimum thrust, i.e. with minimum drag. So the endurance speed of a jet aircraft corresponds closely to the range speed of a propeller-driven aircraft, and from the comfort point of view, this makes the jet aircraft easier to fly in the condition of maximum endurance.

Since the thrust, and hence the consumption, should be the same at the same indicated speed at any height, it should not matter at what height we fly for endurance.

Summary

Table 5.5 and Fig. 5.13 summarise the difference between jet and propeller-driven aircraft so far as range and endurance are concerned. They must be considered as a first approximation only – they take into account the aeroplane efficiency for the propeller-driven type (neglecting propeller and engine efficiency), and both aeroplane and propulsive efficiency for the jet-driven type (neglecting engine efficiency). All this means is that the more important factors have been taken into account, and the less important factors have been neglected. It is **not** the whole story, and should not be considered as such.

The figures in brackets in Table 5.5 are the speeds in knots for the particular aircraft that has been considered in this chapter.

Table 5.5 Range and endurance differences

	Propeller	*Jet*
Speed for maximum range	Minimum drag (160)	Minimum drag/speed (175 up to 280)
Height for maximum range	Unimportant	High
Speed for maximum endurance	Minimum power (125)	Minimum drag (160)
Height for maximum endurance	Low	Unimportant

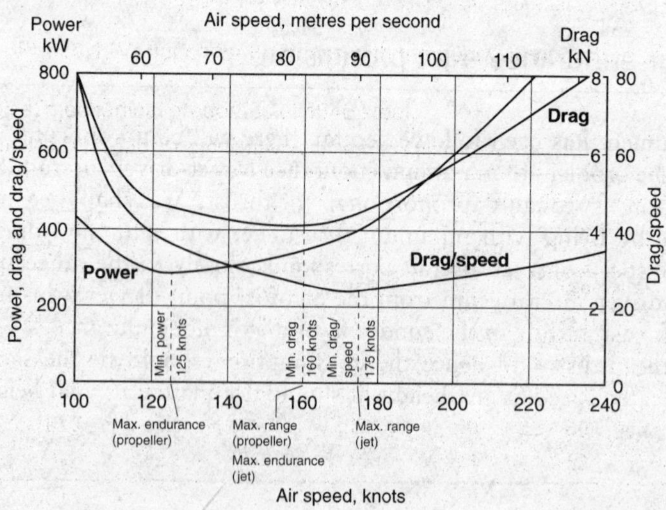

Fig 5.13 Power, drag and drag/speed curves

Before leaving this important subject it should be made clear that flying control regulations, made in the interests of safety, may sometimes make it necessary to fly at flight levels which do not exactly correspond to the best interests of either aircraft or engines.

Can you answer these?

1. What are the four most important forces which act upon an aeroplane during flight?
2. What are the conditions of equilibrium of these four forces?
3. Are these forces likely to alter in value, and to move their line of action during flight?
4. Explain how it is that an aeroplane can fly level at a wide range of air speeds.
5. Is the relationship between air speed and angle of attack the same at height as at sea-level?
6. What is the effect of weight on the relationship between air speed and angle of attack?
7. On a propeller-driven aircraft –
   (a) Why will we get less range if we fly too high?
   (b) At what height should we fly for best endurance?
   (c) Why is the air speed for best endurance different from the air speed for best range?
8. On a jet-driven aircraft –
   (a) Under what conditions should we fly for maximum range?
   (b) At what height should we fly for maximum range?
   (c) At what speed and height should we fly for maximum endurance?

For solutions see Appendix 5.
In Appendix 3 you will find some simple numerical examples on the problems of level flight.

CHAPTER 6

Gliding and landing

Gliding

Let us next consider the flight of an aeroplane while **gliding** under the influence of the force of gravity and without the use of the engine.

Of the four forces, we are now deprived of the thrust, and therefore when the aeroplane is travelling in a steady glide it must be kept in a state of equilibrium by the lift, drag, and weight only. This means that the total aerodynamic force – that is to say, the resultant of the lift and drag – must be exactly equal and opposite to the weight (see Fig. 6.1). But the lift is now at right angles to the path of the glide, while the drag acts directly backwards parallel to the gliding path.

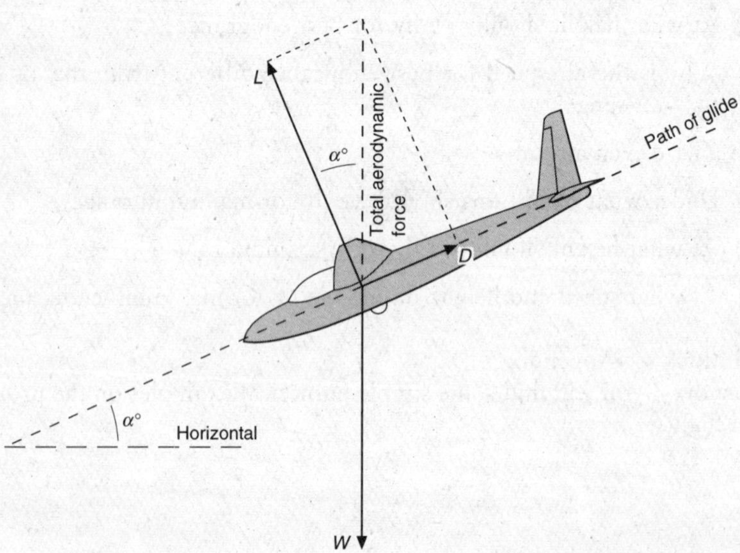

Fig 6.1 Forces acting on an aeroplane during a glide

Gliding angle

By a process of simple geometry, it is easy to see that the angle formed between the lift and the total aerodynamic force is the same as the angle α between the path of the glide and the horizontal, which is called the gliding angle. Therefore

$$D/L = \tan \alpha$$

This means that the less lower value of D/L – i.e. the greater the value of L/D – the flatter will be the gliding angle.

From this simple fact we can very easily come to some important conclusions; for instance –

1. **The tangent of the gliding angle is directly dependent on the *L/D*,** which is really the 'efficiency' of the design of the aeroplane, and therefore the more 'efficient' the aeroplane, the farther it will glide, or, expressing it the other way round, the measurement of the angle of glide will give a simple estimate of the efficiency of the aeroplane.

   The word 'efficiency' is apt to have a rather vague meaning, and we are using it here in a particular sense. We are concerned only with the success or otherwise of the designer in obtaining the maximum amount of lift with the minimum of drag, or what might be called the 'aerodynamic' merit of the aeroplane. For instance, our conclusion shows that any improvement which reduces the drag will result in a flatter gliding angle.

   It will be noticed that this is the same criterion as for maximum range, so that an aeroplane that has a flat gliding angle should also be efficient at flying for range, neglecting the influence of the propulsion efficiency.

2. **If an aeroplane is to glide as far as possible, the angle of attack during the glide must be such that the lift/drag is a maximum.**

   The aeroplane is so constructed that the riggers' angle of incidence is a small angle of, say, 2° or 3°. This particular angle is chosen because it is the most suitable for level flight. As was explained when considering the characteristics of aerofoils, the modern tendency is to make this angle rather less than the angle of maximum L/D (because we are out for speed), but, even so, it will be within a degree or so of that angle, so it is true to say that the angle of attack during a flat glide will be very nearly the same as that during straight and level flight, and almost exactly the same as when flying for maximum range with piston engines.

   The pilot finds it fairly easy to maintain ordinary horizontal flight at the most efficient angle because the fuselage is then in a more or less horizontal position. When gliding however, the task is not always so easy. Sophisticated modern aircraft may be fitted with an angle of attack indicator, but on older and simpler types this is not normally the case. Fortunately, as in level flight, there is a direct connection between the air speed and the angle of attack, and therefore the air speed can be found which gives the best gliding angle, and this acts as a guide to the pilot. The fact remains, however, that it requires considerable skill, instinct, or whatever one likes to call it, on the part of a pilot to glide at the flattest possible angle. This is the type of skill which is especially needed by the pilot of a motorless glider or sailplane.

It should be noted that, although there is a relationship between air speed and angle of attack on the glide just as there is in level flight, the relationship is not exactly the same, and the speed that gives the flattest gliding angle is usually rather less than the speed that gives maximum range. The difference, however, is small and the principle is the same.

3. If the pilot attempts to glide at an angle of attack either greater or less than that which gives the best *L/D*, then in each case the path of descent will be steeper.

Perhaps this conclusion may be considered redundant because it is simply another way of expressing the preceding one. It is purposely repeated in this form because there seems to be such a strong natural instinct on the part of pilots to think that if the aeroplane is put in a more horizontal attitude it will glide farther. Even if one has never flown it is not difficult to imagine the feelings of a pilot whose engine has failed, and who is trying to reach a certain field in which to make a forced landing. It gradually dawns on him that in the way in which he is gliding he will not reach that field. What, then, could be more natural than that he should pull up the nose of his aeroplane in his efforts to reach it? What happens? In answer to this question the student often says that he will stall the aeroplane. Not necessarily. He should in the first place have been gliding nowhere near the stalling angle, but at an angle of attack of only about 3° or 4°, so that he has many degrees through which to increase the angle before stalling. But what will most certainly happen is that the increase in angle will decrease the value of *L/D* and so **increase the gliding angle**, and although the aeroplane will lie flatter to the horizontal, it will glide towards the earth at a steeper angle and will not reach even so far as it would otherwise have done. The air speed during such a glide will be less than that which gives the best gliding angle.

Suppose, on the other hand, that, when a pilot is gliding at the angle of attack which gives him the greatest value of *L/D*, he puts the nose of the aeroplane down, this will decrease the angle of attack which, as before, will decrease the value of *L/D* and therefore **increase the steepness of the gliding path**, the air speed this time being greater than that which gives the best gliding angle.

It is not easy to visualise the angle of attack during a glide, and the reader, like the pilot, must be careful not to be confused between the direction in which the aeroplane is pointing and the direction in which it is travelling. It is hoped that the figures may help to make this important point clear (Figs 6.2 and 6.3, overleaf).

In the previous chapter we discovered that the ratio of lift to drag of complete aeroplanes may be in the neighbourhood of 8, 10 or 12 to 1. These values correspond to gliding angles of which the tangents are 1/8, 1/10 and 1/12, i.e. approximately 7°, 6° and 5° respectively. Thus, neglecting the effect of wind, a pilot will usually be in error on the right side if he assumes that he can glide a kilometre for every 200 metres of height, i.e. if he reckons on a gliding angle of which the tangent is 1/5.

Real and apparent angles of glide

Let us remember once again that gliding must be considered **as relative to the air.** To an observer on the ground an aeroplane gliding into the wind may appear to remain still or, in some cases, even to ascend. In such instances there must be a wind blowing which has both a horizontal and an upward velocity, and to an observer travelling on this wind in a balloon the aeroplane would appear to be travelling forwards and descending. When viewed from the ground an aeroplane gliding against the wind will appear to glide more steeply, and will, **in fact**, glide more steeply relative to the ground (Fig. 6.4); and when gliding with the wind it will glide less steeply than the real angle measured relative to the air – the angle as it would appear to an observer in a free balloon.

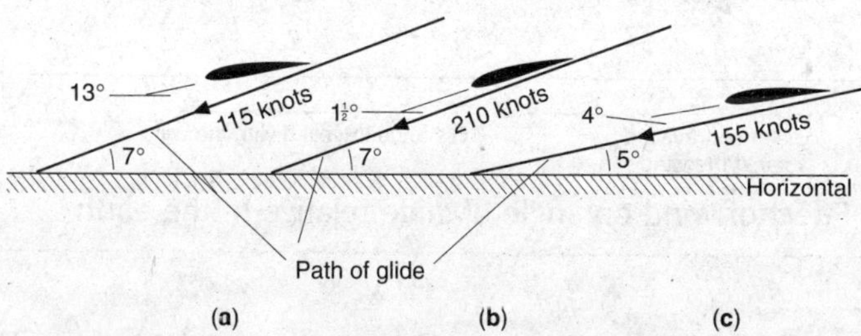

Fig. 6.2 How the angle of attack affects the gliding angle

(*a*) Slow glide at slope of 1 in 8 (7°). Angle of attack 13°. Speed 115 knots.

(*b*) Fast glide at slope of 1 in 8 (7°). Angle of attack $1\frac{1}{2}$°. Speed 210 knots.

(*c*) Flattest glide at slope of 1 in 12 (5°). Angle of attack 4°. Speed 155 knots.

Note. In the diagram the gliding angles, and the differences between them, have been exaggerated so as to bring out the principles.

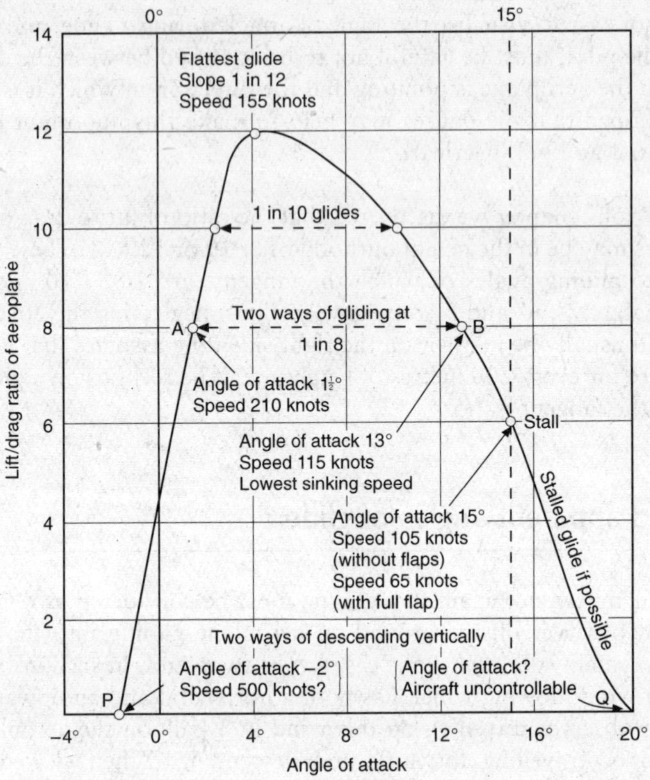

Fig 6.3 Lift/drag curve and gliding angles of aeroplane

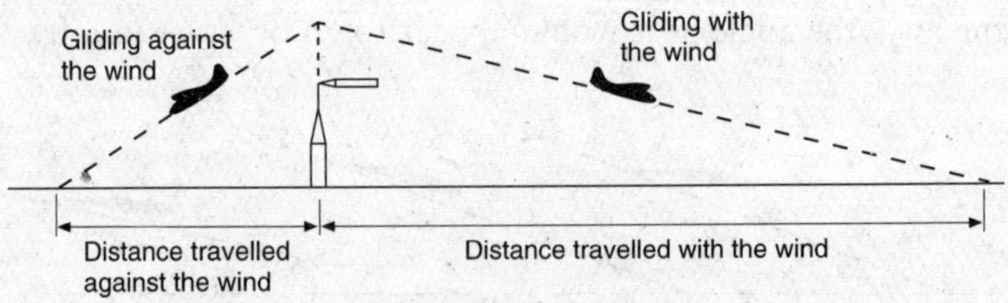

Fig 6.4 Effect of wind on angle of glide relative to the earth

Effect of weight on gliding

It is commonly thought that heavy aeroplanes should glide more steeply that light aeroplanes, but a moment's reflection will make one realise that this is not so, since the gliding angle depends on the ratio of lift to drag, which is quite independent of the weight. Neither in principle nor in fact does weight have an appreciable influence on the gliding angle, but what it does affect is the air speed during the glide.

Look back for a moment at Fig. 6.1. Imagine an increase in the line representing the weight; there will need to be a corresponding increase in the total aerodynamic force, and a greater lift and a greater drag. But the proportions will all remain exactly the same, the same lift/drag ratio, the same gliding angle. But the greater lift, and greater drag, can only be got by greater speed. If we now think back to flying for range, it will be remembered that the condition was the same: greater weight meant greater speed. But there is an interesting and important difference in this case. In flying for range, greater speed meant greater drag, greater thrust, and so less range. In gliding without engine power, greater speed means greater drag, but now the 'thrust' is provided by the component of the weight which acts along the gliding path and this, of course, is automatically greater because the weight is greater. **So greater weight does not affect the gliding angle and does not affect the range, on a pure glide – but it does affect the speed.**

Endurance on the glide

The conclusion of the previous paragraph might perhaps lead one to ask whether, in that case, there is any need for a sailplane to be built of light construction. The answer is definitely – Yes. A sailplane (Fig. 6A, overleaf) must have a flat gliding angle if it is to get any distance, any range from its starting point; but, even more important, it must have a **low rate of vertical descent or sinking speed**; it must be able to stay a long time in the air and be able to take advantage of every breath of rising air, however slight. Sailplane pilots do sometimes add ballast so as to increase the flight speed as this can be useful under certain circumstances. However, a description of such advanced sailplane techniques is best left to books devoted specifically to that subject. It is easy to see that the rate of vertical descent depends both on the angle of glide and on the air speed during the glide. Therefore to get a low rate of descent we need a good lift/drag ratio, i.e. good aerodynamic design, and a low air speed, i.e. low weight.

Actually we shall get a lower rate of descent by reducing speed below that which gives the flattest glide; this is because we gain more by the lower air speed than we lose by the steeper glide. Thus there is an 'endurance' speed for gliding just as for level flight and, as before, it is lower than the range speed, and corresponds to the speed for minimum power requirement.

Disadvantages of flat gliding angle

It should not be thought that a flat gliding angle is always an advantage; when approaching a small airfield near the edge of which are high obstacles, it is

Fig 6A Gliding
(By courtesy of Slingsby Sailplanes Ltd)
The Skylark 4 with large aspect ratio and good value of lift/drag; needing spoilers to increase gliding angle when necessary.

advisable to reach the ground as soon as possible after passing over such obstacles. In these circumstances a flat gliding angle is a definite disadvantage, and even if the aeroplane is dived steeply it will pick up speed and will tend to float across the airfield before touching the ground.

The gliding angle can be steepened by reducing the ratio of lift to drag; this can be done by decreasing the angle of attack (resulting in too high a speed), or by increasing the angle of attack (resulting in an air speed which may be too low for safety), or by using an **air brake** (Fig. 6.5). The last is by far the most satisfactory means, and the air brake may take the form of some kind of flap, such as was described in the chapter on aerofoils; but the modern tendency is to use the various types of flap when lift is required, and separate air brakes or spoilers when drag is required.

Landing

The art of landing an aeroplane consists of bringing it in contact with the ground at the lowest possible vertical velocity and, at the same time, somewhere near the lowest possible horizontal velocity relative to the ground. It is true that in certain circumstances a fast landing may be permissible, and that some modern aircraft are **flown onto the ground** in a definitely unstalled con-

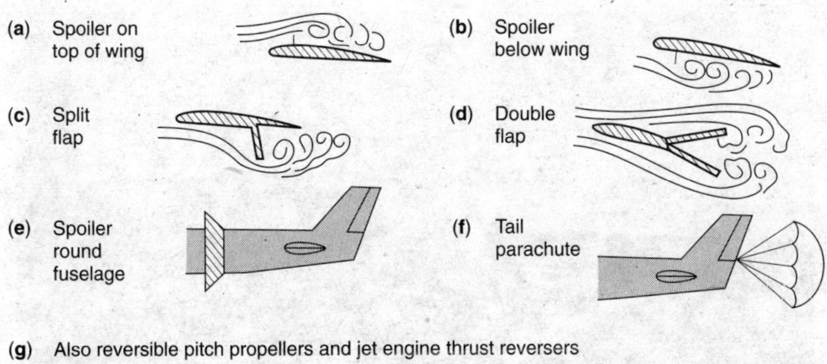

Fig 6.5 Air brakes

dition, but the general rule applies to the landings of many slower and lighter types, and especially to forced landings in which everything usually depends on the minimum horizontal velocity being achieved.

The reader will have noticed that it is the horizontal velocity **relative to the ground** which must be reasonably low. Now, the first step in this direction is to land against the wind and so reduce the ground speed. This, however, is entirely up to the pilot; in our present problem we are only concerned with a low air speed. Given this low air speed, the pilot can, by landing into the wind obtain a low ground speed. In the case of landing on the decks of ships (Fig. 6B, overleaf), if the ship herself steams into the wind, the ground speed will be still further reduced. Supposing, for instance, the minimum air speed of an aeroplane is 80 knots, the wind speed is 20 knots, and the ship is steaming at 30 knots into the wind, then the 'ground' speed of the aeroplane when landing will be only 30 knots; while if the wind speed had been 50 knots, the 'ground' speed would have been reduced to nil – a perfectly possible state of affairs.

In an early chapter it was mentioned that the wind speed is apt to be irregular near the ground, and it is when landing that such irregularity may be important. If the wind speed suddenly decreases, the aircraft, owing to its inertia, will tend to continue at the same ground speed and will therefore lose air speed, and, if already flying near the critical speed, may stall. Similarly, if the wind speed suddenly increases, the aircraft will temporarily gain air speed and will 'balloon' upwards, making it difficult to make contact with the ground at the right moment. Such instances may occur in changeable and gusty winds, in up-currents caused by heating of parts of the earth's surface, in cases of turbulence caused by the wind flowing over obstructions such as hills and hangars, and due to wind gradient. Of these, wind gradient is probably the most important, and the most easily allowed for. An aeroplane, when landing against a high wind, will encounter a decreasing wind speed as it descends through the last few feet and will be in danger of stalling unless it has speed in hand to compensate for the air speed lost. If landing up a slope or towards a hangar, one may suddenly run into air which is blanketed by the

Fig 6B Deck landing
(By courtesy of the former British Aircraft Corporation, Preston)
The Jaguar, designed by Breguet and BAC, landing on a deck.

obstruction, or a head wind may even become a following wind blowing up the hill or towards the hangar. In a really high wind, and when flying a small light aircraft, these conditions may be dangerous, and the obvious moral is to allow for them by approaching to land at a higher speed than usual.

The vertical velocity of landing can be reduced to practically nothing provided the forward velocity is sufficient to keep the aeroplane in horizontal flight – that is to say, provided the lift of the wings is sufficient to balance the weight of the aeroplane.

We have already seen that there is a definite relationship between the indicated air speed and the angle of attack. Fig. 6.6 illustrates the attitudes of an aeroplane at various speeds and the corresponding angles of attack required to maintain level flight: (*a*) shows the attitude of maximum speed; (*b*) that of normal cruising flight; (*c*) that for an ordinary landing; and (*d*) the attitude when fitted with flaps and slots and flying as slowly as possible.

Now, since lift must equal weight, and must also equal $C_L \cdot \frac{1}{2}\rho V^2 \cdot S$, it is quite obvious that if V is to be as small as possible, C_L must be as large as possible. The pilot may never have heard of a lift coefficient, and he may be none the worse a pilot for that; but, consciously or unconsciously, he will increase C_L by increasing the angle of attack until he decides (it matters not whether his decision is based on scientific knowledge, instinct or bitter experience!) that

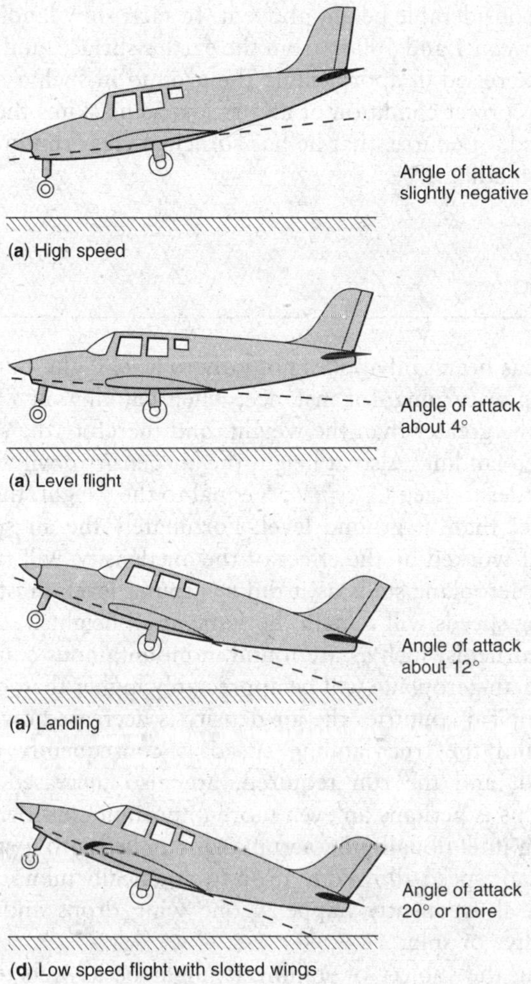

Fig 6.6 Attitudes for level flight

any further increase in the angle of attack will decrease rather than increase the lift; in other words, until he has come near to that stalling angle which we considered so fully when dealing with aerofoils. At this angle (about 15° to 20° in the case of an ordinary aerofoil), C_L is at its maximum, and therefore V is a minimum.

If the pilot, through lack of any of the three qualities mentioned above, exceeds this angle, then both C_L and V will decrease; therefore $C_L \cdot \frac{1}{2}\rho V^2 \cdot S$ can no longer equal the weight and the aeroplane will commence to drop. For 20 m, 50 m, or more, the vertical component of velocity will increase and the nose of the aeroplane will drop, therefore the pilot must beware that, when he does this experiment of flying as slowly as possible, he is either very near the ground or at a considerable height above it. In fact, slow landings should not be practised between 1 and 500 m from the earth's surface, and the whole skill of the pilot is exercised in approaching the ground in such a manner that he has reached the correct condition of affairs just as he skims the surface of the runway, provided, of course, that he has sufficient clear run in front in which to pull up after landing.

Stalling speed

Much of what has been said applies not only to level flight, but to stalls when gliding, climbing or turning; for instance, when banking on a turn the lift on the wings must be greater than the weight, and therefore the stalling speed is higher than when landing. Also at height the air density ρ will be less, and this means that in order to keep $C_L \cdot \frac{1}{2}\rho V \cdot S$ equal to the weight, the stalling speed V will be greater than at ground level. Fortunately **the air speed indicator,** which is in itself worked by the effect of the air density, will record the same speed when the aeroplane stalls as it did at ground level; in other words, the indicated stalling speeds will remain the same at all heights.

But on high airfields, such as are found in mountainous countries, the true landing speed of an aeroplane will be appreciably higher than on sea-level airfields; and in tropical countries the air density is decreased owing to the high temperatures, and the true landing speed is consequently increased. The taking-off speed, and the run required, are also increased in both these instances, and this is perhaps an even more important consideration.

When stalling intentionally the aeroplane is pulled into a steeply climbing attitude and the air speed allowed to drop to practically nil until the nose suddenly drops or, as frequently happens, one wing drops and the aeroplane commences to dive or spin.

Before leaving the subject of stalling it might be as well to mention that there has always been difficulty in deciding upon an exact definition of stalling or stalling speed. The stall occurs because the smooth airflow over the wing becomes separated – but this is a gradual process. At quite small angles of attack there is some turbulence near the trailing edge; as the angle increases, the turbulence spreads forward. What is even more important is that it also spreads spanwise, usually from tip to root on highly-tapered wings, and from root to tip on rectangular wings. If we define the stall as being the break up of the airflow, when did it occur? There may be buffeting of the tail plane or main planes, but this too may be slight and unimportant, or it may be violent. As a result of the change from smooth to turbulent airflow the curve of lift coefficient reaches a maximum and then starts to fall. We defined the stalling angle in Chapter 3 as the **angle at which the lift coefficient is a maximum**. But how does the pilot know that it is at its maximum value? All the pilot knows is that if he tried to fly below a certain speed he gets into difficulties. How great the difficulties depends on the type of aircraft, and the extent to which the pilot can overcome them depends on a lot of things, but particularly on his own skill.

In fact, there are different definitions of stalling according to the point of view of the person who wishes to define it – the pilot looks at it one way, the aerodynamicist another, and so on. What is important is that each should realise that it is his own definition, and that all these things do not necessarily occur at the same time.

Possibilities of lower minimum speeds

What are the possibilities of reducing this minimum velocity of flight?

In all forms of transport, with the exception of flying, the maximum speed attainable is a major consideration. But in the exceptional case of flight it is equally important to obtain a low minimum speed as it is to obtain a high maximum speed. This low speed is of such importance that it is apt to be exaggerated at the expense of the maximum speed. Whatever we say about obtaining a low landing speed, we must never forget that the chief advantage of flight over other means of transport depends on the high speed obtainable. But, provided we bear this in mind, everything must be done to reduce the landing (and taking off) speed, because only in this way can flying be made a popular and safe means of transport. The minimum speed of most light aeroplanes is as much as 50 or 60 knots, and of some more than 100 knots.

High lift aerofoils

What, then, has been done, and what can yet be done to decrease minimum speed? If V is to be small, C_L must be as large as possible. In other words, we must have a larger lift coefficient. So the aerofoils which give the largest maximum lift coefficient will give the lowest minimum speeds. Unfortunately, however, these aerofoils are usually those with a large drag, and so they seriously affect the high speed end of the range. Therefore we must turn to some device by which the shape of the aerofoil can be altered during flight, and so we naturally think of flaps and slots.

In an earlier chapter (Fig. 3.32) we noticed the effect of various kinds of flaps and slots on maximum lift and speed range. The idea of variable camber is an old one, but it is only in recent years, when maximum speeds have increased so much, that the problem has become really urgent and these devices have come into their own. In this respect necessity has proved to be the mother of invention, and many and varied have been the devices which have been tried. It is not easy to compare the respective merits of all these types of slots and flaps, or combinations of slot and flap, because so many conflicting qualities are required. If a low speed was our only aim, the problem would be comparatively simple, the device giving the highest maximum lift coefficient being the most suitable. But what we really need is a **low minimum speed and a high maximum speed, i.e. a good speed range**. This condition means that the device must be such that it can be altered, or will alter automatically, from the position giving maximum lift (e.g. slot open or flap down), to the position of minimum drag (e.g. slot closed or flap neutral). Even that is not the end of our requirements for, having landed as slowly as possible, we must pull up quickly after landing. The former (slow landing) needs high lift, the latter (quick pull-up) needs much drag, lift being of no consequence at all. For a quick pull-up we really need a definite **air brake** which will assist the wheel brakes. Notice, however, that an air brake cannot reduce actual landing speed, it can only improve the pull-up after landing. Once we are on the ground we want to get rid of lift as quickly as possible to achieve minimum wheel brake effectiveness; hence the use of lift dumpers after touchdown. These are usually devices which disrupt the flow over the top of the wing, increasing drag, and decreasing lift.

Yet another aspect of the problem, so far as landing is concerned, is the question of **attitude**, and in this respect some of the otherwise most effective types of slots and flaps are at a disadvantage, for they attain their maximum lift coefficient at a greater angle of attack than the ordinary aerofoil; this means that in order to make full use of them the angle of attack when landing may need to be 25°, or even more. But when an aeroplane with a tail-wheel type of undercarriage rests with its main and tail wheels on the ground, the angle of inclination of the wings is only about 15°. With a nose-wheel type of undercarriage the problem is if anything, worse – as the reader will no doubt realise – for, in order to land at an angle of 25°, we are faced with four possibilities all of which have serious drawbacks –

1. **To allow the tail to touch the ground before the main wheels** (Fig. 6.7*a*). This is hardly a practical proposition.
2. **To have a much higher undercarriage** (Fig. 6.7*b*). This will cause extra drag and generally do more harm than good.
3. **To provide the main planes with a variable incidence gear** similar to that which is sometimes used for tail planes (Fig. 6.7*c*). This involves considerable mechanical difficulties.
4. **To set the wings at a much greater angle to the fuselage** (Fig. 6.7*d*). This means that in normal flight the rear portion of the fuselage sticks up into the air at an angle which not only looks ridiculous, but which is inefficient from the point of view of drag. It gives the appearance of a 'broken back', but has sometimes been used for aircraft designed for deck landing since it not only gives a low landing speed, but a quick pull-up after landing.

   The only real answer lies in the design of flaps and slots which must be such that the effective camber of the wing can be altered so as to give

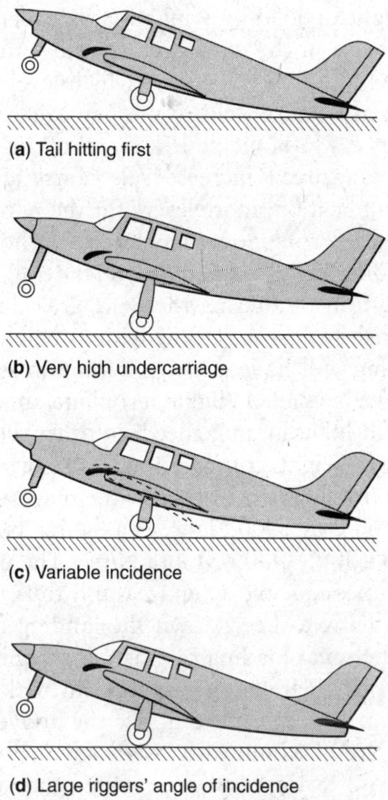

Fig 6.7 Difficulties of landing at large angles of attack

maximum lift and still maintain a reasonable attitude for landing. Improvements in design on these lines have resulted in a real, and by no means negligible, reduction in landing speed (or perhaps, more correctly, has halted further increases in landing speed) but at the expense of ever more sophistication – and ever more complication.

Wing loading

We assumed at an earlier stage, that the area of the wings was bound to remain constant, but inventors have from time to time investigated the problem of providing wings with **variable area**.

Since $W = C_L \cdot \frac{1}{2}\rho V^2 \cdot S$

$W/S = C_L \cdot \frac{1}{2}\rho V^2$

The fraction *W/S*, weight divided by wing area, is called the wing loading of the aeroplane. An increase in the wing area should reduce the value of *W/S*, and so also reduce the minimum velocity at which level flight is possible.

The objections to variable area are chiefly mechanical; the operating gear means extra weight and so, since *W* will increase, it by no means follows that *W/S* will actually decrease if the wing area is increased; also, once again, the extra sophistication means more complication, more levers for the pilot to fiddle with, more chances of something going wrong. Some flap systems do, however, give an increase in wing area during landing as they extend beyond the trailing edge of the wing.

Apart from the question of altering the wing area during flight, the equation $W/S = C_L \cdot \frac{1}{2}\rho V^2$ shows us that, other things being equal, the aeroplane with a low wing loading will have a lower minimum speed than one with a high wing loading. The so-called 'light aeroplane' may have a high wing loading, and therefore a high landing speed; in other words, it is not a question of weight, but of weight compared with wing area, that settles the minimum speed. The wing loading of a sailplane may be less than 100 N/m^2, of a training aeroplane 300 to 1000 N/m^2, of a fighter, bomber or airliner anything from 1500 up to 2000, 3000 or more N/m^2. The modern tendency is to increase wing loading by reducing wing area and thus raising the maximum speed, and then using flaps to keep down the landing speed. The student is advised to work out the wing loading of existing aeroplanes and to compare the figures obtained with their landing speeds; in making this comparison, however, the student must be careful to notice the above phrase 'other things being equal', because the maximum lift coefficient of the aerofoil used also affects the result. An old example of high wing loading, very high for that time, was the 1977 N/m^2 of the S.6b Schneider Trophy racing seaplane; the corresponding figure for fighters like the 'Spitfire' and 'Hurricane' at the beginning of the Second World War was 1187 N/m^2, and for the German fighter, Messerschmitt 109, 1522 N/m^2. Modern figures may be considerably higher than these; the Concorde was about 4800 N/m^2.

Method of finding minimum landing speed

It is easy to work out simple problems on minimum or landing speeds by using the now familiar formula –

$$\text{Weight} = \text{Lift} = C_L \cdot \frac{1}{2}\rho V^2 \cdot S$$

If we denote the maximum value of the lift coefficient by C_L max, and the landing speed by V_L, then our formula becomes

$$W = C_L \cdot \max \frac{1}{2}\rho V^2 \cdot S$$

Consider this problem –

Find the wing area required for an aeroplane of mass 1500 kg, if the minimum landing speed is to be 35 knots (65 km/h) and the maximum value of the lift coefficient for the aerofoil used is 1.2 (assume the air density to be 1.225 kg/m³).

Data: $W = 14\,715$ N

$\rho = 1.22=$ kg/m³

$V_L = 65$ km/h $= 18$ m/s

C_L max $= 1.2$

$S = ?$

$$\text{So } 14\,715 = 1.2 \times \frac{1}{2} \times 1.225 \times 18 \times 18 \times S$$

$$\therefore S = \mathbf{62\ m^2\ approx}$$

This is rather a large wing area for an aeroplane of this weight, and it is doubtful whether the structure involved would not make the total weight greater than 14 715 N, in which case, of course, the landing speed would be above 35 knots.

Suppose we could use a flapped wing with a maximum lift coefficient of 1.8 instead of 1.2, then, neglecting any small increase in weight, the necessary wing area to produce the same landing speed would be –

$(1.2/1.8) \times 62 =$ **41 m² approx**

It would certainly be much easier to design a wing structure of this size so as to conform to a total weight of 14 715 N and, further, the reduced wing area would enable a much greater maximum speed to be obtained.

As another problem, let us compare the minimum landing speeds of the following –

(*a*) A sailplane of wing loading 100 N/m².

(*b*) A training machine of wing loading 400 N/m².

(*c*) A fighter of wing loading 1500 N/m².

(*d*) The S.6b of wing loading about 2000 N/m².

(*e*) An airliner of wing loading 3000 N/m².

Supposing other things to be equal, e.g. taking ρ as 1.225 kg/m^3, and assuming each machine is fitted with an aerofoil section having a maximum lift coefficient of 1.12, then –

(*a*) Wing loading $= W/S = 100 = C_L \max \frac{1}{2}\rho V_L{}^2$

$$= 1.12 \times \tfrac{1}{2} \times 1.225 \times V_L{}^2$$

$$\therefore V_L{}^2 = (100 \times 2)/(1.12 \times 1.225) = 146$$

$\therefore V_L =$ **12 m/s, i.e. 23 knots or 43 km/h**

Similarly for

(*b*) Landing speed = 47 knots (87 km/h)

(*c*) 91 knots (169 km/h)

(*d*) 106 knots (195 km/h)

(*e*) ? knots (? km/h). Reader, work it out.

Such is the type of problem which confronts the designer of an aeroplane in the very early stages, when, by a process of calculations, he has to decide such important items as the wing area, the type of aerofoil, and the landing speed. It will now be obvious that in order to settle these he must know the weight – the weight of an aeroplane which he has not yet commenced to design! Here comes the first great guess; but it is a guess based on experience, and often proves remarkably accurate. A decision as to landing speed and as to the type of aerofoil will then decide the wing area, on which the whole lay-out of the aeroplane depends, so it will be seen how important is this question of landing speed and its influence on the whole design of the finished aeroplane.

Landing speeds and the future

So far as landing speed is concerned, we are reaching an interesting stage in the history of aviation. Wing loadings are still going up; they went up slowly but surely for the first thirty years of flight, and rather less slowly, but more surely, during the Second World War – and since. There is at the moment no sign of any halt in this progress – for progress it certainly is. We must assume, therefore, that wing loadings will go still farther.

During this time, slots and flaps, and then better flaps, have been invented, and the maximum value of C_L has gone up from just over 1 to about 3, or even 4, for a good modern aerofoil section with slotted flaps and slots extending along 60 per cent of the wing span. When the maximum C_L was 1.22 (RAF

15), a wing loading of 500 N/m^2 was considered high; but with a C_L max of 3, even 5000 N/m^2 has already been exceeded. Now a C_L max of 1.22 and *W/S* of 500 gives a landing speed of about 50 knots (93 km/h) whereas a C_L max of 3 and *W/S* of 5000 gives a landing speed of about 102 knots (188 km/h).

So we have accepted a considerably higher landing speed – but how long can this go on? The increase in wing loading has already had a greater effect than the increase in maximum lift coefficient, but so far we have discovered better and better flaps. Now, however, it would seem that there is not much hope of any further great improvement in flaps – so what of the future? The first thing we must do is clear enough – flaps, and slots too, must extend along the whole span of the wing, perhaps also under the fuselage; this has already been done in some types of aircraft, sometimes by arranging that the ailerons act also as flaps, therefore known as '**flaperons**' or by dispensing with the ailerons altogether and adopting an alternative form of lateral control, such as spoilers – this will be discussed later, or by differential movement of the tail surfaces, thus known as '**tailerons**'.

Such methods might give us another 40 per cent increase in maximum lift coefficient and a landing speed of about 86 knots (158 km/h) for a wing loading of 5000 N/m^2. So what are the prospects for the future? Well, there has been no shortage of ideas, and the patent offices of the world contain hundreds of examples, ranging from the good to the impractical or ill-conceived. Amongst the good ideas is that of using high pressure air bled from the engine compressor to help induce attached flow over the flaps (**blown flaps**) at large deflections, or even to produce a downward curtain of air which can turn the main airflow through extremely large angles: the so-called **jet flap**. Such devices, though effective, are however complex and heavy, and have generally only been used for specialised aircraft, usually military. An example of an apparently promising idea that never went into general use is the Custer channel wing shown in Fig. 6.8. The idea was quite simple; the propellers drew air over the curved wing sections, thereby allowing the flow to remain attached at high angles of attack. However, there are two main problems with most of such ideas. Firstly the lifting surface also needs to have a good lift to drag ratio at high speed, and secondly, as the minimum speed of the aircraft decreases, it becomes more and more difficult to provide adequate stability and control by conventional means. Other approaches to providing lift at low or even zero speed are dealt with below.

Fig 6.8 The idea of the 'Custer' channel wing
The engines – with pusher propellers – were suspended in the channels with the ailerons just outboard of the channels.

Short and vertical landing and take-off

The idea of low landing speeds or even vertical landing was a much sought-after goal even in the earliest days of flying. In fact, until the twentieth century, the only manned flights were made in vertical take-off devices: balloons and airships. Over the years, there has been considerable research into both vertical and short take-off aircraft, and one firm outcome has been the coining of two acronyms, **VTOL** for Vertical Take-off and Landing, and **STOL** for Short Take-off and Landing. To this has been added some hybrids like **STOVL**, Short Take-off and Vertical Landing. In the following sections, some of the more successful or promising types are described.

The gyroplane

Perhaps the first really practical achievement in this direction was the Cierva Autogiro in 1923 The Autogiro was not a helicopter, but a gyroplane, and a gyroplane differs from a helicopter in one vital particular in that, whereas in a helicopter the wings or blades are rotated by the power of the main engine, in a gyroplane the rotating wings are not driven except by the action of the air upon them, and this in turn is caused by the forward or downward air speed of the aircraft. Thus forward speed is necessary in a gyroplane, just as it is in a conventional aeroplane and, as in the latter, it is provided by the thrust of an ordinary engine and propeller. A modern gyroplane is shown in Fig. 6C (top).

Space does not permit a full consideration of the principles underlying this very interesting flying machine, and it must suffice to say that the secret lies in the fact that by a suitable inclination backwards of the axis the wings rotate automatically in such a way that even when the forward speed of the aircraft is far lower than the stalling speed of a conventional aeroplane the rotating wings are still striking the air at a considerable velocity, and can thus provide sufficient lift to keep the aircraft in the air. In this way the forward speed can be reduced to 5 or 10 knots which, in a slight head wind, means a ground speed of practically nil.

When the Cierva Autogiro first appeared it succeeded in achieving, in one fell swoop, a landing speed such as the experimenters of all nations had considered to be a dream of the distant future. Moreover later developments resulted, in 1935, in a 'jump start', or true vertical take-off, achieved by rotating the wings on the ground (by engine power), and then suddenly increasing the pitch, or angle of attack of the blades.

The gyroplane, sometimes called an autogyro (the spelling Autogiro was a trade name), still has its advocates, and still exists in various forms, including very light and simple single-seaters (Fig. 6C).

Fig 6C STOL
Top: The autogyro after almost disappearing as an historical curiosity has made a comeback in the form of a popular ultralight.
Bottom: The de Havilland Canada Dash-7, original mainstay of the London City (STOL) Airport. The four propellers provided a' high-energy airflow over much of the wing.
(Photo by courtesy of Terry Shwetz, de Havilland Canada.)

The helicopter

In a true helicopter in normal flight, the upward thrust of the revolving blades must be equal to the weight; forward motion is produced by inclining the effective axis of the rotor forward which normally entails tilting the nose of the helicopter down. The actual means by which this is achieved is rather complicated and involves altering the cyclic variations in rotor blade incidence. Owing to the reaction of the torque of the lifting blades the whole aircraft will try to rotate in the opposite direction, this resulting in a tendency to yaw which corresponds to the rolling tendency due to the propeller on a fixed-wing aircraft. The yawing tendency can be counteracted either by an auxiliary propeller or by a jet reaction system at the tail; this can also be used to provide directional control and to save the necessity of having a rudder (Fig. 6D).

The development of successful helicopters has involved the solution of many other practical problems. The blade going into the wind (the wind produced by the motion of the aircraft) gets more lift, and drag, than the blade going down wind. This happens whenever the machine moves in any direction, forwards, backwards or sideways; but it becomes an even more serious problem when moving at high speeds because the tip portions of the blade going into the wind meet compressibility problems before the aircraft itself is moving anywhere near the speed of sound; this fact has so far limited the speed of helicopters to something like 200 knots (370 km/h). The helicopter has other failings too, vibration is apt to be excessive, as is the noise when compared with aircraft of similar power.

Attempts to solve these and other problems have resulted in wings that must not only rotate, but also have cyclically and collectively variable incidence, hinges to give variable dihedral (resulting in a kind of flapping motion of the blades), drag hinges to allow the blades to bend backwards, and sometimes, a tilting axes of rotation. Then there are the auxiliary propellers at the tail, or jets, some have had jets too at the wing tips to rotate the blades, some have had more than one set of rotating wings (like contra-rotating propellers), and there have been many other ingenious devices, all helping to some extent to solve the problems, but all contributing to the complication and weight of an already complicated type of aircraft.

Recent developments have included the use of so-called 'rigid' rotors, where the flapping and drag hinges are replaced by flexures. This has reduced the mechanical complexity, and also the weight of the rotor head. Another development has been the shrouding of the tail rotor in a kind of fin. This results in less drag and reduces the possibility of accidental contact with the tail rotor. A further development has been the replacement of the tail rotor with a variable air jet at the rear to perform the same function. This type, known as a **NOTAR**, (standing for No Tail Rotor) is much quieter than the conventional helicopter, and also avoids the possibility of damage to or from the tail rotor.

The main limitation of a conventional helicopter remains its low maximum speed. One of the reasons for this is that in order to generate lift, a significant

Fig 6D VTOL – helicopters
Top: The tiny but popular Robinson R-22, with piston engine and rubber belt drive. Flying it feels like trying to ride a monocycle at first.
Bottom: At the other extreme, the massive Russian Mil Mi-26 TM capable of lifting a load of 196 000 newtons (20 tonnes).

part of the retreating blades must be going faster than the relative air speed past the helicopter. This means that the advancing ones will be going more than twice as fast as the aircraft, and will therefore go supersonic when the aircraft is still flying at significantly less that half the speed of sound. Several solutions have been proposed and even developed. One of these is the so-called **'compound'** helicopter. This has the normal features of a conventional helicopter but also has small wings and a separate engine or engines to provide forward thrust directly. At high speed, the wings provide most of the lift, allowing the rotor to be rotated relatively slowly or even be feathered. A number of examples have been built, but so far it remains something of a rarity. A possible alternative is provided by development of the coaxial rotor helicopter which has two contra-rotating rotors mounted on the same shaft or sometimes two close shafts with intermeshing rotors. On this arrangement, it is possible to arrange for the lift to be generated primarily by the advancing blades of the rotors, the retreating blades providing little or no lift. Since they do not have to provide lift, the retreating blades can be feathered and do not need to go faster than the relative flow speed past the aircraft. This of course means that that the advancing blade can be going correspondingly slowly. Many practical examples of coaxial rotor helicopters have been manufactured and they are not uncommon, particularly in military applications. They have the added advantage that no tail rotor is required. Their disadvantage lies in the extreme complexity of the rotor head, and the problems caused by intersecting rotor wakes. As yet, the potential of the coaxial rotor for very high speed flight has not yet been realised on production examples.

Jet lift

Until recently, the only serious contenders to the helicopter have been **jet lift** and **thrust vectoring** aircraft, and the only really successful one was the Harrier, (known as the AV-8 in the USA). The Harrier (Fig. 6G, later) uses downward deflection of the engine to produce a lifting thrust until sufficient forward speed has been obtained for the wings to take over the job of lifting. The engine outlet nozzles, two from the hot exhaust and two from the compressor, can be 'vectored' (varied in outlet angle) from vertical or even slightly forward, to horizontal in order to obtain the desired balance between lift and forward thrust. The thorny problem of providing stability and control at low speed is solved by using four 'puffer' jets of compressed air bled from the engine. These are situated at the nose, the tail and both wing tips. Like the helicopter, this aircraft can be flown backwards forwards and sideways. It can also hover and land and take off vertically. Although it is capable of vertical take-off, it is more economical in terms of fuel burn for the aircraft to make a short horizontal take-off, with the engines set at an intermediate angle so as to provide thrust and some lift. It may therefore be classified as V/STOL, meaning that it can use vertical and short take-off.

During the cold war, the Soviet Union developed a rival to the Harrier, the Yakovlev Yak-36. This aircraft used an alternative method of jet lift in that it employed separate engines for lift instead of just relying on thrust vectoring. For many years, however, the Harrier was the only really practical jet lift aircraft in service anywhere, but after much political indecision and argument, the Lockheed Martin F-35 JSF (Joint Strike Fighter) finally emerged. Some variants of this have similar V/STOL capabilities to the Harrier, and a much higher top speed. Thrust vectoring is used, but in addition, a fan driven by the engine provides part of the lift.

Other short and vertical take-off aircraft

As with high-lift devices, there have been very many ideas for vertical or short take-off aircraft. Ignoring the untried or less promising examples, we are left with three main approaches to the problem. These are –

(*a*) **Tilting the whole aircraft**, as shown in Fig. 6E (overleaf). A small number of successful experimental aircraft were built, but the practical problems (such as seeing where you are going on landing) remained too daunting for further development.

(*b*) **Tilting** the engines and wings for take off and landing, whilst leaving the main body of the aircraft horizontal. Several experimental aircraft have been flown, but no production types have been forthcoming.

(*c*) **Tilting the rotors/propellers or the engines and rotors/propellers.** Again, many experimental aircraft of this type have been flown, but it was only after much political indecision and a number of serious technical set-backs that the go-ahead was finally given for series production of the Boeing-Bell V 22 Osprey tilt-rotor aircraft (Fig. 6F, overleaf). This aircraft combines the features of conventional aircraft and helicopters. In high speed flight, the rotors are tilted forward and become large diameter propellers, so the aircraft does not suffer the same speed limitations as a conventional helicopter. The cost and complexity are however high. Another aircraft using this principle is the smaller Bell/Augusta BA609.

Surface skimming vehicles

The hovercraft can reasonably be counted as a type of aeroplane, as it does fly, albeit just above the ground. The hovercraft is lifted above the ground or water surface by means of a fan, which produces a raised pressure under the vehicle. Leakage of the air from the underside is restricted by a peripheral

Fig 6E VTOL
(By courtesy of the General Dynamics Corporation, USA)
Pogo aircraft built for US Navy; after vertical take-off it could level off and fly horizontally; for landing the pilot pointed the nose straight up, and backed down tail first onto small wheels at trailing edges of wings and fins.

Fig 6F Convertiplane
(By courtesy of Bell Helicopter Textron)
The Boeing-Bell V-22 Osprey. Capable of VTOL and VSTOL. The engines can be rotated to provide conventional forward flight after take-off.

'curtain' of high speed air, normally supplemented by a flexible 'skirt'. At one time, the hovercraft was seen as a major new type of transport vehicle, and for many years hovercraft operated a ferry service across the English Channel. In fine weather this provided a much faster alternative to the conventional ferries. However, the hovercraft were expensive to run and maintain. Also they could not be operated in very heavy seas, and were prone to generate nausea amongst the passengers even in moderate seas. Nowadays hovercraft are mostly used for very specialised operations, for example in swampy areas, and for recreational and competition use. For ferry duties, they have been largely superseded by various types of catamaran and hydrofoil.

Another type of vehicle that flies just above the surface is the surface skimmer or ekranoplane. These aircraft resemble a conventional aircraft, but by flying just above the surface, take advantage of the 'ground effect' which greatly reduces the wing downwash, and consequently reduces the accompanying trailing vortex drag. Some extremely large and very fast aircraft of this type were under development in the old Soviet Union just before its collapse, but subsequent economic difficulties brought further development to a virtual standstill. The main disadvantages of the ekranoplane are the problems of ensuring the correct ground clearance at all times, and the very large amount of engine power needed to get the aircraft to lift off from the water. Several large turbojets were needed on the larger experimental aircraft, and once lift-off had been achieved, most of them became redundant for the rest of the journey.

The complete approach and landing

During the few years preceding the Second World War, a new technique in flying was developed. In this book we are not concerned with the art of flying as such, but we are very much concerned with the alteration of technique because it was brought about by progress in the science of flight accompanied by the corresponding changes in aeroplane design. To the outward eye the main change was that the monoplane, rather suddenly, took precedence over the biplane. Less obvious, perhaps but more important, was the increase in wing loadings – which incidentally was itself the reason behind the ascendancy of the monoplane. With increase in wing loading came higher landing speed,

Fig 6G V/STOL
The Harrier is capable of vertical take-off, very short take-off, and conventional take-off using its vectored-thrust Pegasus turbofan engine; it can also hover and fly sideways or backwards. Control at low speed is by reaction jet nozzles situated at the wing-tips and either end of the fuselage.
(Photo courtesy of Nigel Cogger)

Fig 6H VTOL – another way
After a century of neglect, the hot-air balloon has made a major comeback. This one is tethered and is being used for pilot training. The wicker basket provides good impact absorption on landing.

higher landing speed meant that flaps – once a luxury – became a necessity, and flaps, in their turn, were largely responsible for the new technique, especially in so far as it affects the approach and landing.

A pilot, when approaching a landing ground, may find that he is either undershooting or overshooting. If he is undershooting there is little that he can do (assuming, of course, that he is already gliding at the best angle) except use his engine power to flatten his glide. In the old days such a method was considered bad flying; if the engine was functioning satisfactorily, it showed lack of judgment; if the engine was out of action – that is to say in the case of a forced landing – it could not be done. Nowadays the engine-assisted approach, as it is called, is a standard method of approaching to land. It might almost be said that on the most modern machines the glide approach is only used as practice for a forced landing.

In the anxiety to avoid undershooting there is a natural tendency to overshoot, especially since it would seem to be easy to lose any unnecessary height. In practice it is not quite so easy. In older types of aircraft the following methods were available as means of losing height in the event of overshooting –

(*a*) Sideslipping.

(*b*) Prolonging the glide by S-turns.

(*c*) Putting the nose down and gliding fast.

(*d*) Holding the nose up and gliding slowly.

The objections to the last two have already been explained; the first two methods, on the other hand, were successfully employed for very many years.

Then came the modern type of aircraft – its superior streamlining gave it a very flat gliding angle, so flat that even a slight degree of overshooting caused it to float much too far before landing. But that was not all – it did not like sideslipping (reasons for this will be given later), and with its very flat angle of glide the S-turn did not result in sufficient loss of height.

Necessity may be the mother of invention, but in this little bit of aviation history the invention existed before the necessity arose; flaps had been in use for many years, but they had not really been fully applied to their modern purposes. These purposes can best be described by considering the process of approaching and landing a light private aircraft by the old-fashioned glide approach technique. This typically consists of five separate phases (Fig. 6.9) – the glide, the flattening-out, the float or hold-off, the landing, and the pull-up. In each and all of these flaps have their part to play. Let us consider them in turn.

First, then, let it be understood that the last 150 m or so of the glide should be straight, without any slipping or turning to one side or the other. This can only be done if the pilot has means of controlling the gliding angle relative to the earth without unduly raising or lowering the air speed. Flaps can give him the means to do this, at any rate over a limited range of gliding angles. As the flaps are lowered both lift and drag are increased. The increase in lift tends to flatten the gliding angle and to make it possible to glide at a slower air speed without approaching dangerously near to the stall. The increase in drag tends to steepen the gliding angle, and gliding attitude, of the aeroplane for the same air speed. The net effect depends on whether lift or drag has the greater proportional increase, i.e. on whether the L/D ratio is raised or lowered, and that in turn depends on how much the flaps are lowered and on the type of flap.

So much then for the actual glide – we can sum it up by saying that flaps give us at least some control over the gliding angle (Fig. 6I, overleaf).

Next comes the process of **flattening out**; this involves a change of direction, and so an acceleration, and force, towards the centre of the curved path. This force must be provided by the wings which must therefore have more speed and more angle, so in effect the stalling speed is higher. Now the steeper the original glide, the greater the change in flight path involved, and so the more speed must there be in hand for flattening out. All this is very annoying – it means that the more steeply we glide, the faster must we glide; just what we were trying to avoid. The solution is to use engine power, as we shall consider later.

After flattening out, we must lose any excess speed – this may be called **the float or hold-off**. In this the drag of flaps or air brakes play their part, as do the wings themselves as they are brought to the angle for actual landing.

After the float comes **the landing** (Fig. 6J, overleaf). This, in a sense, is momentary only but the landing speed is of the utmost importance because it settles both the gliding speed and the distance to pull up after landing. The problem of landing speed has already been fully discussed, but it must be emphasised –

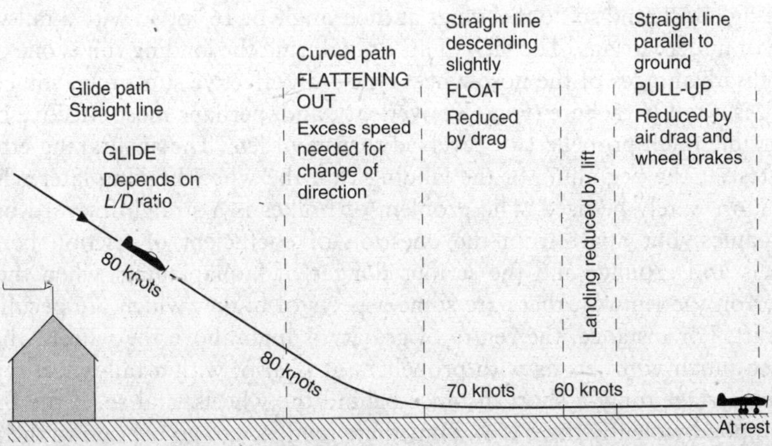

Fig 6.9 The approach and landing
Air speeds shown are approximate only, but indicate at what parts of the approach and landing there is a loss of air speed.

Fig 6I Ready for landing
The huge Antonov AN-124 with Fowler flaps and leading edge slats extended, and undercarriage lowered, showing the ten pairs of main wheels.

because it is so often misunderstood – that drag, whether caused by flaps or anything else, cannot reduce landing speed; that is entirely a question of lift.

After the landing, **the pull-up**. This at least is easy to understand; what we need for a quick pull-up is drag – wheel brakes and air brakes – the more the better, provided the aircraft can stand it and does not tip on its nose. In addition to actual air brakes, some types of flap, when fully lowered, give good braking effect, and so do the wings at their angle of 16° or so with a tail-wheel type of undercarriage. The lack of air drag during the landing run is one of the few disadvantages of the nose-wheel type – an effective substitute, sometimes used, is a tail parachute (Fig. 6K, overleaf), and, perhaps most effective of all, reversible pitch propellers or reversed thrusts of jets. The air-braking effect is greatest at the beginning of the landing run, the wheel-braking later when it can more safely be used. The problem of brakes is a straightforward one of mechanics, but apart from the question of coefficient of friction between wheels and ground, and the serious danger of 'aquaplaning' when there is water on the runway, there are some aspects of brakes which are peculiar to aircraft. For instance, the centre of gravity is high above the wheels, though not so much with jets as with propellers; it is also, with a tail-wheel type of undercarriage, only a short distance behind the wheels, and so, if the brakes are applied violently there is an immediate tendency to go over on to the nose. Another difficulty is that if a tail-wheel aeroplane starts to swing, the centre of gravity, being behind the wheels, will cause the swing to increase. This may be checked by the differential action of the brakes, but it is interesting to note that the tricycle or nose-wheel undercarriage (Figs 6J, 6K and others) can remove the cause of this and other troubles. When this type of undercarriage is used, as it is almost universally on modern aircraft, the centre of gravity is in front of the main wheels and there is no tendency to swing, and at the same time the aircraft is prevented from going on to its nose by the front wheel. The effect on braking, and consequent shorter pull-up has to be seen – or, better still, tried – to be believed. And there are other advantages too.

So much for the glide approach and landing; for forced landings it is the only way. For approaching over high obstacles at the edge of an airfield it may also come in useful. In all other circumstances, with modern types of aircraft, the **engine-assisted approach** is better, so let us sum up the reasons as to why it is preferred –

1. By slight adjustments of the throttle the path of glide can be flattened or steepened at will.
2. The gliding path is flatter, so there is less change of path in flattening out, less acceleration, less extra lift required, less increase in stalling speed, and thus less excess speed is needed, and the glide can safely be made more slowly.
3. In propeller-driven aircraft the extra speed of the slipstream over elevators and rudder makes these controls more effective – their effectiveness enables us to counteract wind gradient and turbulence effects near the ground.
4. Since there is less excess speed to be lost, the float is reduced.

Fig 6J Landing
A BAe 146 about to touch down, Fowler flaps extended and rear air-brake doors open.

Fig 6K Pulling up after landing
Eurofighter Typhoon using an arrester parachute to reduce the landing run.

5. An engine, already running, will respond more readily to the throttle if it is found necessary to make another circuit before landing.
6. For all the reasons already given less judgment is required – the whole process is easier.

Well, all that sounds pretty convincing, and not only is it perfectly sound reasoning but it is amply confirmed by experience.

As was mentioned in Chapter 3 many modern types of aircraft are literally **'flown onto the ground'** at speeds well above the stalling speed; this is where real **lift dumpers** come into their own – these are larger than normal air brakes or spoilers, and cannot be operated unless the undercarriage is locked down, and the aircraft is bearing on the wheels and the engines throttled back – they 'kill' the lift, keep the aircraft on the ground and make the wheel brakes more effective. They are essentially for use after landing, and are a common feature on gliders.

Effect of flaps on trim

The lowering or raising of flaps affects the airflow not only over the lower surface of the wing, but also over the upper surface – probably the more important effect – and in front of the wing and behind the wing (Fig. 6.10). The airflow in turn affects the pressure distribution and the forces and moments on the wing and on the tail plane. It is hardly surprising, therefore, that the trim may be affected, but it may seem curious that the lowering of the flap sometimes tends towards nose-heaviness, sometimes tail-heaviness.

Consider the top surface of the wing. When the flap is lowered, the air flows faster over the top, especially near the leading edge. There will be greater suction here and the chances are that the centre of pressure on the top surface will move forward, thus tending towards tail-heaviness.

The downwash behind the wing will be large; and if the tail plane is so situated as to receive the full benefit of this downwash, there will be a downward force on the tail plane, tending towards tail-heaviness.

In a low-wing aircraft the low position of the drag on the flap, especially when fully lowered, will tend towards nose-heaviness. On a high wing aircraft the drag, being high, may tend towards tail-heaviness.

The net effect on the pitching moment depends entirely on the type of flap or slots used, on how much they are lowered, and on the situation of the tail plane. Slotted flaps, and flaps that move backwards so increasing the rear portion of the wing area, will nearly always cause a nose-down moment which sometimes has to be counteracted by leading edge slots and flaps.

Sometimes, too, the change of trim is in one direction for the first part of the lowering of the flap, usually tail-heavy; and in the other direction, nose-heavy, when full flap is lowered. In some aircraft the effects, whether by design or good luck, so cancel each other that there is little or no change of trim, and no one is more pleased than the pilot.

It should be noted that the technique of landing a modern high performance aircraft, be it military or civilian, is nothing like the straightforward seat-of-

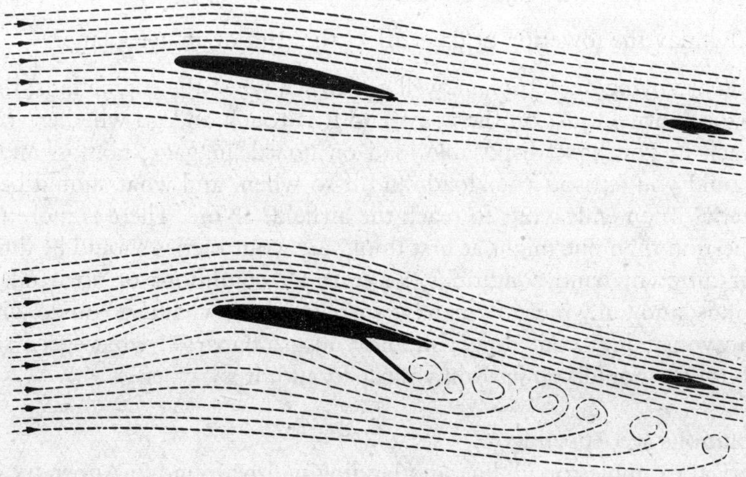

Fig 6.10 Flow over wing and tail – flaps up and flaps down

the-pants procedure described earlier for the case of a light private aircraft. For such modern aircraft, the speed, height, angle of descent and a host of other factors such as the flap and power setting must be correct within very close limits as the aircraft crosses the airfield threshold. If not, the pilot must abort the approach and try again. Landing a modern airliner manually requires a great deal of skill and concentration, and the majority of landings are nowadays made under automatic control with the pilot merely keeping a watchful eye, and being ready to take over at any instant in the event of a system malfunction. In order to maintain their skill (and their licence) pilots are, however, required to make a certain proportion of landings under manual control. It would be virtually impossible to land machines such as the Space Shuttle without some form of computer assistance.

Can you answer these?

1. If, when an aeroplane is gliding at its minimum angle of glide, the pilot attempts to glide farther by holding the nose of the aeroplane up, what will be the result, and why?
2. Discuss the effect of flaps on the gliding angle.
3. How does the load carried in an aeroplane affect the gliding angle and gliding speed?
4. Does the flattest glide give the longest time in the air? If not, why not?
5. Does (a) the stalling speed, (b) the stalling angle, change with height?
6. What are the advantages of the engine-assisted approach?
7. Why may the lowering of flaps affect the trim of an aeroplane?
8. You are flying an aeroplane well out to sea when the engine fails; there is a good airfield just on the coast and it is touch and go whether you can reach it; you have disposable load on board, luggage, bombs, and fuel; should you jettison your load, and if so when, and what should be your tactics in an endeavour to reach the airfield? (Note. There is more in this question than one might at first think, e.g. your tactics should be different for different wind conditions, so consider conditions of no wind, head winds, and tail winds; if you have surplus speed, what can you do with it?; can you reduce your drag?; should you use flaps?; at what speed should you fly?; should you jettison your load and, if so, when?)

For solutions see Appendix 5.
Numerical examples on gliding and landing will be found in Appendix 3.

CHAPTER 7

Performance

It may seem rather illogical that we should first consider level flight, then gliding and landing, and now the take-off, climb and general performance of the aeroplane. But there is method in our madness. Level flight is, as it were, the standard condition of flight with which all other manoeuvres are compared. Gliding, too, involves simple fundamental principles, in some ways more elementary than those of level flight. Landing we have used to illustrate the principles of flight at low speeds. All these have followed quite naturally one on the other, but the take-off is a problem on its own.

Taking-off

The pilot needs skill and practice before he can be sure of making a good take-off, one of the main problems being to keep the aircraft on a straight and narrow path. This difficulty applies mainly to propeller-driven aircraft, and has already been discussed in Chapter 4. In general, it may be said that the object during the take-off is to obtain sufficient lift to support the weight with the least possible run along the ground. In order to obtain this result the angle of attack is kept small during the first part of the run so as to reduce the drag; then, when the speed has reached the minimum speed of flight, if the tail is lowered and the wings brought to about 15° angle of attack, the aeroplane will be capable of flight. Although by this method the aeroplane probably leaves the ground with the least possible run, it is apt to be dangerous because, once having left the ground, any attempt to climb by further increase of angle will result in stalling and dropping back on to the ground. Therefore it is necessary to allow the speed to increase beyond the stalling speed before 'pulling-off', and sometimes the aeroplane is allowed to continue to run in the tail-up position until it takes off of its own accord (Fig. 7A, overleaf).

The process of taking-off is largely influenced by such things as the runway surface, and although of extreme interest, the subject is too complex to be

Fig 7A Taking-off
The Myasischev M-55 high-altitude aircraft for surveillance and geophysics research.

within the scope of this book. In order to reduce the length of run, and increase the angle of climb after leaving the ground – so as to clear obstacles on the outskirts of the airfield – the take-off will, when possible, be made against the wind. Other aids to taking-off are slots, flaps or any other devices which increase the lift without unduly increasing the drag, and, essential in propeller-driven high-speed aircraft, the variable-pitch propeller.

The question as to whether or not flaps should be used for taking-off depends upon whether the increased lift of the flap, with the resulting decrease in taking-off speed, makes up for the lower acceleration caused by the increased drag of the flap. But the problem is a little more complicated than that, because while we wish to avoid drag throughout all the take-off run, we only really need the extra lift at the end, when we are ready to take off. No doubt we could get off most quickly by a sudden application of flap at this stage, but such a method would certainly be dangerous. The lift type of flap helps the take-off considerably, other types may have some beneficial effect if used at a moderate angle, and in practice some degree of flap is nearly always used for take-off in modern high-speed types of aircraft if only because it reduces the otherwise very high take-off speed with consequent wear of tyres.

Some interesting problems arise in connection with the take-off. Modern undercarriages may tuck away nicely during flight, but when lowered they are less streamlined than a fixed undercarriage and their drag may hamper the take-off quite considerably; the lower undercarriage that can be used with jets is a great advantage in this respect. Again, just as landing speeds go up with high wing loading, so do take-off speeds, and the length of run needed to attain such speeds is liable to become excessive. The idea of catapulting is an old one but its use has mostly been restricted to carrier-borne aircraft. The assistance of rockets gives much the same effect as catapulting, but has rarely been used. Refuelling in the air does not sound like a form of assisted take-off, but it does present possibilities in that an aircraft can be taken off lightly loaded as regards fuel.

Finally, although we have considered STOL and VTOL in their effect on landing, we must not forget that they are at least as important for take-off.

Climbing

During level flight the power of the engine must produce, via the propeller, jet or rocket, a thrust equal to the drag of the aeroplane at that particular speed of flight. If now the engine has some reserve of power in hand, and if the throttle is further opened, **either** –

(a) The pilot can put the nose down slightly, and maintain level flight at an increased speed and decreased angle of attack, or

(b) The aeroplane will commence to climb (Fig. 7B).

A consideration of the forces which act upon an aeroplane during a climb is interesting, but slightly more complicated than the other cases which we have considered.

Fig 7B Climbing
An Airbus A-330 using flaps to demonstrate its steep climb capability.

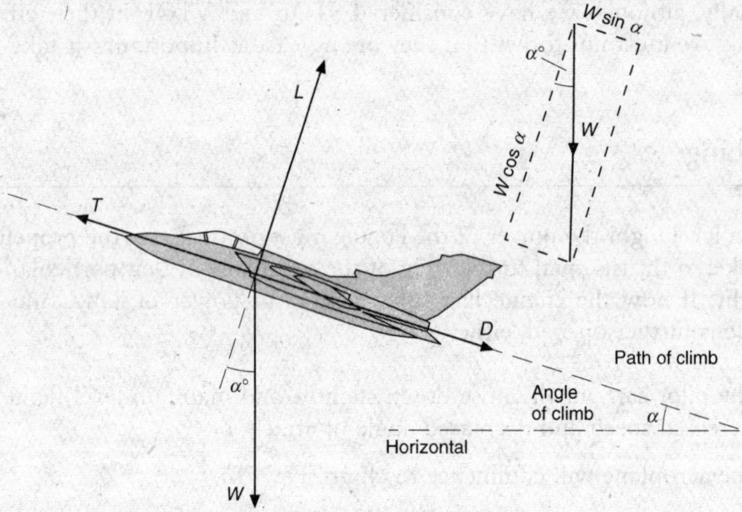

Fig 7.1 Forces acting on an aeroplane during a climb

Assuming that the path actually travelled by the aeroplane is in the same direction as the thrust, then the forces will be as shown in Fig. 7.1. If α is the angle of climb, and if we resolve the forces parallel and at right angles to the direction of flight, we obtain two equations –

(1) $T = D + W \sin \alpha$ (2) $L = W \cos \alpha$

Translated into non-mathematical language, the first of these equations tells us that during a climb the thrust needed is greater than the drag and increases with the steepness of the climb. This is what we would expect. If a vertical climb were possible, α would be 90° and therefore sin a would be 1, so the first equation would become $T = D + W$, which is obviously true because in such an extreme case the thrust would have the opposition of both the weight and the drag. Similarly if $\alpha = 0°$ (i.e. if there is no climb), $\sin \alpha = 0$. Therefore $W \sin \alpha = 0$. Therefore $T = D$, the condition which we have already established for straight and level flight.

Power curves – propeller propulsion

An interesting and more practical way of approaching the climbing problem is by means of what are called **performance curves.** By estimating the power available from the engine and the power required for level flight at various speeds, we can arrive at many interesting deductions. It is largely by this method that forecasts are made of the probable performance of an aeroplane, and it is remarkable how accurate these forecasts usually prove to be.

The procedure for jet and rocket systems of propulsion is rather different because, as already mentioned, we must think in terms of thrust rather than power. They will therefore be dealt with separately, and the following discussion relates primarily to piston-engined aircraft.

The deduction of the curve which gives the power output of the engine is outside the scope of this book, as it depends on a knowledge of the characteristics of the piston (or gas turbine) engine for a propeller-driven aircraft. From this curve must be subtracted the power which is lost through the inefficiency of the propeller (the efficiency of a good propeller at reasonable speed, but falling off on both sides depending on rpm, is about 80 per cent). The resulting curve (Fig. 7.2) shows the power which is **available** at various forward speeds of the aeroplane.

The power which will be **required** is found by estimating the drag. For this purpose the wing drag and the parasite drag are usually found separately, the former from the characteristics of the aerofoils and the latter by estimating the drag of all the various parts and summing them up. Another method of finding the total drag is by measuring the drag of a complete model in a wind tunnel and scaling up to full size. After the total drag has been found at any speed, the power is obtained by multiplying the drag by the speed, as in flying for endurance in Chapter 5, e.g. if the total drag is 4170 N at 82.4 m/s –

Power required = 4170 × 82.4 = **344 kW**

And in a similar way the power required is found at other speeds, the lower curve in Fig. 7.2 illustrating a typical result. The reader may be puzzled as to

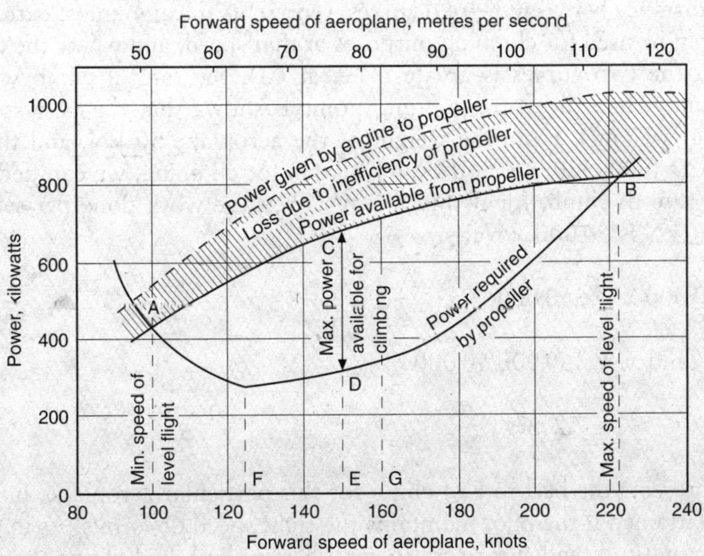

Fig 7.2 Power available and power required

why the power required increases so rapidly at low speeds; the explanation is that in order to maintain level flights at these low speeds, a very large angle of attack is required, and this results in an increase of drag in spite of the reduction in speed. If the argument sounds familiar, it is simply because we are returning to the same argument as when discussing range and endurance. The figures we have just quoted are taken from that argument, and the curve of power required in Fig. 7.2 is based on the aeroplane of Chapter 5. This follow-up of the same aeroplane will make the power curves more interesting and instructive.

It should be noted that there will be no fundamental difference in the shape of the power required curve for jet and propeller-driven aircraft. It is in the power available that the difference lies.

Maximum and minimum speeds of horizontal flight

From the combination of the two curves (Fig. 7.2) some interesting deductions can be made. Wherever the power available curve is above the power required curve, level flight is possible, whereas both to the left and right of the two intersections it becomes impossible for the rather obvious reason that we would require more power than we have available! Therefore the intersection A shows the least possible speed (51 m/s, as we had discovered before for other reasons), and the intersection B the greatest possible speed (115 m/s), at which level flight can be maintained. Between the points A and B the difference between the power available and the power required at any particular speed, i.e. the distance between the two curves, represents the amount of extra power which can be used for climbing purposes at that speed, and where the distance between the two curves is greatest, i.e. at CD, the rate of climb will be a maximum, while the corresponding point E shows that the best speed for climbing is 77 m/s. From the weight of the aeroplane 50 kN, and the extra power CD (680–320, i.e. 360 kW) available for climbing, we can deduce the vertical rate of climb, for if this is x m/s, then the work done per second in lifting 50 kN is 50 000 x watts,

$$\text{So } 50\,000\,x = 360\,000$$

$$\text{and } x = 360\,000/50\,000$$

$$= \mathbf{7.2\ m/s}$$

This represents the best rate of climb for this particular aeroplane, but it will only be attained if the pilot maintains the right speed of 77 m/s. As in gliding, there is a natural tendency to try to get a better climb by holding the nose up higher but, as will be seen from the curves, if the speed is reduced to 62 m/s only about 250 kN will be available for climbing, and this will reduce the rate of climb to 250 000/50 000, i.e. 5 m/s. Similarly, at speeds above 77 m/s the rate of climb will decrease, although it will be noticed that between certain speeds the curves run roughly parallel to each other and there is very little change in the rate of climb between 72 and 88 m/s; obviously at 51 m/s and again at 115 m/s, the rate of climb is reduced to nil, while below 51 and above 115 m/s the aeroplane will lose height.

As a matter of interest, the speeds for maximum endurance (F), 64 m/s, and maximum range (G), 82 m/s, have also been marked. As was explained in Chapter 5, these are the best speeds from the point of view of the aeroplane, but they may have to be modified to suit engine conditions. Note that the speed for maximum endurance (F) could be deduced from the curve, since it is the lowest point on the curve, i.e. the point of minimum power required for level flight. The speed for maximum range (G), however, must be obtained from Table 5.2, which showed the drag at various speeds.

Effect of changes of engine power

We have so far assumed that for a certain forward speed of the aeroplane the power available is a fixed quantity. This, of course, is not so, since the power of the engine can be varied considerably by manipulating the engine controls. If the curve shown in Fig. 7.2 represents the power available at some reasonably economical conditions and in weak mixture, then we shall be able to get more power by using rich mixture, and the absolute maximum power by opening the throttle to the maximum permissible boost and using the maximum permissible rpm – with fixed-pitch propellers this will simply be a case of full throttle. From this we shall get a curve of emergency full power (Fig. 7.3, overleaf). It will be noticed that the minimum speed of level flight is now slightly lower – very slightly, so slightly as to be unimportant. The maximum speed is, as we might expect, higher – perhaps not so much higher as we might expect (118 instead of 115 m/s). The most important change is in the rate of climb: 460 kW surplus power is now available for climbing, and the rate of climb is 9.2 m/s instead of 7.2 m/s.

Except in special circumstances, it is inadvisable to fly with the engine 'flat out', and, even so, full power must be used only for a limited time or there will be a risk of damage to the engine. The effects of decreasing the power are also shown in Fig. 7.3. From the point of view of the aeroplane, it makes no difference whether the power is decreased by reducing boost, or lowering the rpm, or both; but for fuel economy it is generally advisable to lower the rpm. It will be noticed that as the power is reduced, the minimum speed of level flight becomes slightly greater, the maximum speed becomes considerably less, and the possible rate of climb decreases at all speeds.

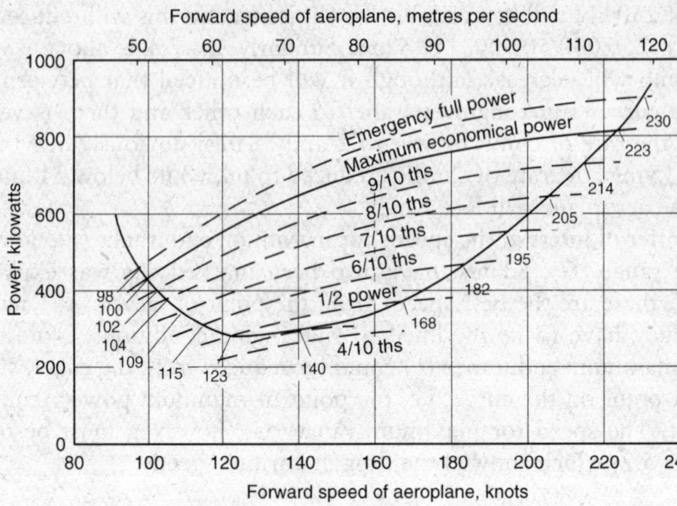

Fig 7.3 Effects of engine power

All this is what we might expect, with the possible exception of the fact, which pilots often do not realise, that the lowest speeds can be obtained with the engine running at full throttle. However, this flight condition cannot easily be sustained in practice because a small, inadvertent, decrease in speed would mean an increase in required power and a simultaneous decrease in available power. The speed reduction would then 'run away'; a condition called speed instability.

Eventually, as the engine is throttled down, we reach a state of affairs at which there is only one possible speed of flight. This is the speed at which least engine power will be used, and at which we shall therefore obtain maximum endurance. It is rather puzzling to find that this speed (72 m/s) is different from the speed (64 m/s) at the lowest point of the power required curve. This is because the engine and propeller efficiency is slightly better at 72 than at 64 m/s.

Effect of altitude on power curves

We have not yet exhausted the information which can be obtained from these performance curves, for if we can estimate the corresponding curves for various heights above sea-level we shall be able to see how performance is affected at different altitudes. There is much to be said, and much has been said, on the subject of whether it is preferable to fly high or to fly low when travelling from one place to another. It is one of those many interesting problems about flight to which no direct answer can be given, chiefly because there are so many conflicting considerations which have to be taken into account. Some of them, such as the question of temperature, wind and the quantity of oxygen in the air, have already been mentioned when dealing with the atmosphere, but the most important problem is that of performance.

How will the performance be affected as the altitude of flight is increased? Well, as the altitude increases, the air density decreases. Therefore, to support the same aircraft weight whilst maintaining the same lift coefficient and attitude, it is necessary to increase the speed so as to maintain the same dynamic pressure. What this means is that we need to fly at **the same indicated air speed.** This also means that the drag will remain the same as before. Thus, from the point of view of the airframe, it appears at first as though there is no disadvantage in flying high. Indeed we can actually fly faster for the same amount of drag and thrust. The main trouble with flying high though comes from the engine and propulsion system. The propulsion problems may be summarised as follows.

1. The power required is the product of the speed and the drag, so although the drag and thrust can remain the same when the height is increased, the increase in the speed means that the power required will increase.
2. The power output of piston engines falls as the air density reduces, and although we can compensate for this by adding a supercharger or turbocharger, there is a limit to the amount of supercharging that it is practical to use.
3. Since we need to fly faster to maintain the same lift and drag, the propeller will also have to go faster, and we will run into problems associated with compressible flow over the propeller.

   In addition to these propulsion problems there are some other important airframe and operational considerations.
4. The cockpit and cabin need to be pressurised, which adds to the complexity and weight.
5. As the height and speed are progressively increased, we will eventually encounter the problems associated with compressibility, especially in respect of the propeller.

Whatever attempts are made to mitigate the difficulties of flight at high altitudes, in propeller-driven aircraft the general tendency remains for the power available to decrease and the power required to increase with the altitude (Fig. 7.4, overleaf). This will cause the curves to close in towards each other, resulting in a gradual **increase in the minimum speed** and a **decrease in the maximum speed,** while the distance between the curves, and therefore the rate of climb, will also become less. Any pilot will confirm that this is what actually happens in practice, although, as previously mentioned, he may be

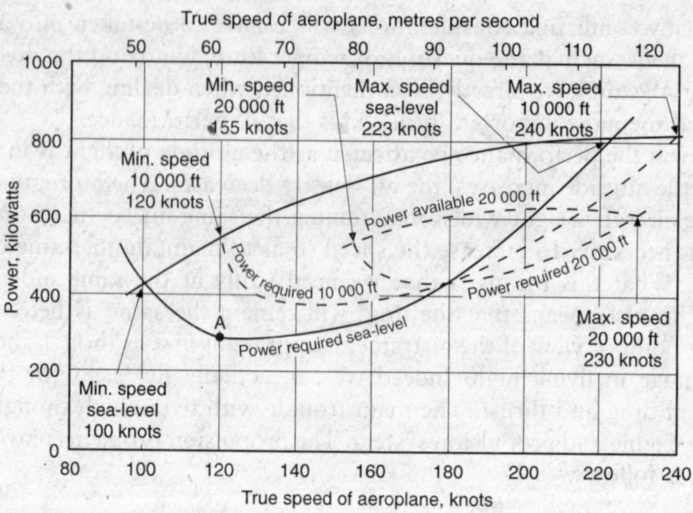

Fig 7.4 Effects of altitude

somewhat deceived by the fact that the air speed indicator is also affected by the change in density and consequently reads lower than the true air speed. This is really what accounts for the curve of power required moving over to the right as the altitude increases; if the curves were plotted against indicated speed, the curves for 3000 m and 6000 m would simply be displaced upwards, compared with that for sea-level. The difference between true and indicated speed also accounts for another apparent discrepancy in that the curves as plotted (against **true** speed) suggest that the air speed to give the best rate of climb **increases** with height. This is so, but the indicated speed for best rate of climb **falls** with height.

For certain purposes good performance at high altitudes may be of such importance that it becomes worth while to design the engine, propeller and aeroplane to give their best efficiencies at some specified height, such as 10 000 metres. It may then happen that performance at sea-level is inferior to that at the height for which the machine was designed. Even so, above a certain critical height, 6000, 9000 m or whatever it may be, performance will inevitably fall off and so the performance curves will be very similar, except that the highest curve of power available will correspond to the critical height. In such aircraft it may well be that the advantages of flying high outweigh the disadvantages.

Ceiling

This process of improving performance at altitude cannot be continued indefinitely and we shall eventually reach such a height that there is only one possible speed for level flight and the rate climb is nil. This is called the **ceiling.**

It requires extreme patience and time to reach such a ceiling, and, owing to the hopeless performance of the aeroplane when flying at this height, it is of little use for practical purposes, and therefore the idea of a **service ceiling** is introduced, this being defined as that height at which the rate of climb becomes less than 0.5 m/s, or some other specified rate.

Effect of weight on performance

It is sometimes important to be able to calculate what will be the effect on performance of increasing the total weight of an aeroplane by carrying extra load. Here again the performance curves will help us.

If the weight is increased, the lift will also have to be increased. So we must either fly at a larger angle of attack or, if we keep the same angle of attack, at a higher speed. This speed can easily be calculated thus –

Let old weight = W, new weight = W_1.

Let V be the old speed, and V_1 be the new speed at the same angle of attack.

Since angle of attack is the same, C_L will be the same.

$$\therefore \mathrm{W} = C_L \cdot \frac{1}{2}\rho \mathrm{V}^2 \,.\, S$$

$$\text{and } \mathrm{W}_1 = C_L \cdot \frac{1}{2}\rho V_1^{\,2} \,.\, S$$

$$\text{so } V_1/V = \sqrt{(\mathrm{W}_1/\mathrm{W})}$$

Such problems always become more interesting if we consider actual figures; so suppose that we wish to carry an extra load of 10 000 N on our aeroplane which already weighs 50 000 N; then –

$$V_1/V = \sqrt{(60\,000/50\,000)} = \sqrt{1.2} = 1.095$$

Since the angle of attack is the same, the lift/drag ratio remains constant, and the corresponding drag, D, will be

$$D_1 = D\frac{L_1}{L} = D \times \frac{65\,000}{50\,000} = 1.2D$$

The corresponding power, P, is thus:

$$P_1 = D_1 V_1 = 1.2 \times 1.095DV = 1.2P$$

So, for example, if we take the point on the power-required curve (Fig. 7.4) marked A, the corresponding point A_1 will be at

$$V_1 = 1.095V, P_1 = 1.2P$$

$$\text{i.e. } V_1 = 1.095 \times 60 = 65.7 \text{ m/s}$$

$$\text{and } P_1 = 1.2 \times 310 = 372 \text{ kW}$$

In a similar way, for each angle of attack, new speed and new power can be calculated, and thus a new curve of power required can be drawn for the new weight of the aeroplane.

It is interesting to note that the net effects of the additional weight are exactly the same as the effects of an increase of altitude, i.e. –

1. Slight reduction in maximum speed.
2. Large reduction in rate of climb.
3. Increase in minimum speed.

In short, the curve of power required is again displaced upwards. It will be noticed that there is too a slight increase in the best speed to use for climbing. (This must not be confused with rate of climb.)

In spite of the similarity in effect of increase of weight and increase of altitude it should be noted that the increase of weight does not affect the reading of the air speed indicator, and so the results apply equally well whether we consider true or indicated air speed.

The problems of operating piston engines at increasing altitude are such that their use is nowadays normally limited to flight at relatively low altitudes. This means that the piston engine is now relegated to use for light general aviation and specialist purposes. For propeller propulsion at higher altitudes and speeds, gas-turbine based **turboprop propulsion** is more appropriate, because the efficiency of the engine is less affected by altitude. The problem of propeller efficiency at high speed has been addressed with some success in recent years, and with very advanced propellers, it is indeed possible to fly even at transonic speeds. High-speed propellers are characterised by swept tips, giving a scimitar shape. The efficiency of propeller propulsion is theoretically greater than that of a pure jet, and for this reason, the turboprop is widely used for applications such as small regional airliners, transport aircraft and military trainers, where flight at very high altitude and speed is not necessary. Despite the advantages in terms of fuel consumption, the use of propellers does add to the weight, cost, complexity and servicing requirements.

Influence of jet propulsion on performance

In this chapter, we have so far looked at the performance of propeller driven aircraft, and it is now time to consider the effects of using turbojet propulsion. In Chapter 5 we saw that for the jet engine, thrust and fuel consumption do not change much with speed, and therefore the speeds for optimum range and endurance at any fixed altitude are faster than for a propeller driven aircraft with the same airframe. Also, as explained earlier in this chapter, by flying higher we will fly faster for a given amount of drag and thrust. This means that the optimum speeds for range and endurance also increase with altitude. Fortunately, the efficiency of the jet also rises with altitude, and we find that with turbojet propulsion, not only do the optimum speeds increase with height, but the number of kilometres that we can fly for each kilogram of fuel increases. Thus, for a jet aircraft, it is advantageous in terms of range, endurance and fuel cost to fly high and fast (Fig. 7C, overleaf). The limitation on maximum speed is normally provided by the onset of serious compressibility effects: transonic flow, which is dealt with in Chapter 11. The feature of obtaining the best economy at high speed is one reason for the popularity of jet propulsion for airliners, where passengers naturally want to get to their destinations as quickly and cheaply as possible. Other advantages are the low noise in the cabin, and the lack of vibration.

When considering the performance of jet aircraft, it is important to appreciate some important differences between the way that jet and piston-engined aircraft are controlled. On a piston-engined aircraft, the pilot controls the **engine power** by means of the throttle control, so called, because it traditionally controlled the **air flow** into the engine by throttling it. On a jet aircraft, the pilot controls the engine by adjusting the **fuel flow**, and it is the **thrust** that is controlled directly, rather than the power. Despite the fact that this control does not actually throttle anything, it still tends to be called the throttle, for historic reasons.

Another difference between piston and jet propelled aircraft is that the higher speed of the latter means that the speed relative to the speed of sound becomes very important. For jet aircraft performance estimates, therefore, we need to display the data in a different form from that used for piston-engined aircraft. It is better to work in terms of **thrust** and **drag** rather than power required and power available, and we need to know how the thrust and drag vary **with Mach number** and altitude. This means that the simple performance calculations that we are able to use for practical purposes for low speed piston-engined aircraft are no longer appropriate for **high-speed** jet-engined aircraft. The proper method of dealing with their performance is rather complicated and beyond the intended level of this book, but in the numerical questions associated with this chapter, we have included some simple calculations for jet aircraft at low speeds where compressibility effects can be ignored.

747

Can you answer these?

1. 'When an aeroplane is climbing, the lift is less than the weight.' Explain why this statement is not so inconsistent as it sounds.
2. What is the effect of altitude on the maximum and minimum speeds of an aeroplane?
3. Distinguish between 'ceiling' and 'service ceiling'.
4. In attempting to climb to the ceiling, should the air speed be kept constant during the climb?
5. If the load carried by an aeroplane is increased, what will be the effects on performance?

For solutions see Appendix 5.
Numerical examples on Performance will be found in Appendix 3.

Fig 7C Cruise performance (opposite)
(By courtesy of the Boeing Company)
Two models of the Boeing 747. The nearer aircraft is a 747–400 which cruises at around 965 km/h (600 mph) at 9150 m, and has a cruise range of 13 528 km with 412 passengers.

CHAPTER 8

Manoeuvres

In a sense, any motion of an aeroplane may be considered as a manoeuvre. In no other form of transport is there such freedom of movement. An aeroplane may be said to have six degrees of freedom which are best described in relation to its three axes, defined as follows –

The **longitudinal axis** (Fig. 8.1) is a straight line running fore and aft through the centre of gravity and is horizontal when the aeroplane is in 'rigging position'.

The aeroplane may travel backwards or forwards along this axis. Backward motion – such as a tail-slide – is one of the most rare of all manoeuvres, but forward movement along this axis is the most common of all, and is the main feature of **straight and level flight.**

Any rotary motion about this axis is called **rolling**.

The **normal axis** (Fig. 8.1) is a straight line through the centre of gravity, and is vertical when the aeroplane is in rigging position. It is therefore at right angles to the longitudinal axis as defined above.

The aeroplane may travel upwards or downwards along this axis, as in **climbing** or **descending**, but in fact such movement is not very common, the climb or descent being obtained chiefly by the inclination of the longitudinal axis to the horizontal, followed by a straightforward movement along that axis.

Rotary motion of the aeroplane about the normal axis is called **yawing.**

The **lateral axis** (Fig. 8.1) is a straight line through the centre of gravity at right angles to both the longitudinal and the normal axes. It is horizontal when the aeroplane is in rigging position and parallel to the line joining the wing tips.

The aeroplane may travel to right or left along the lateral axis; such motion is called **sideslipping or skidding.**

Rotary motion of the aeroplane about the lateral axis is called **pitching.**

These axes must be considered as moving with the aeroplane and always remaining **fixed relative** to the aeroplane, e.g. the lateral axis will remain parallel to the line joining the wing tips in whatever attitude the aeroplane may

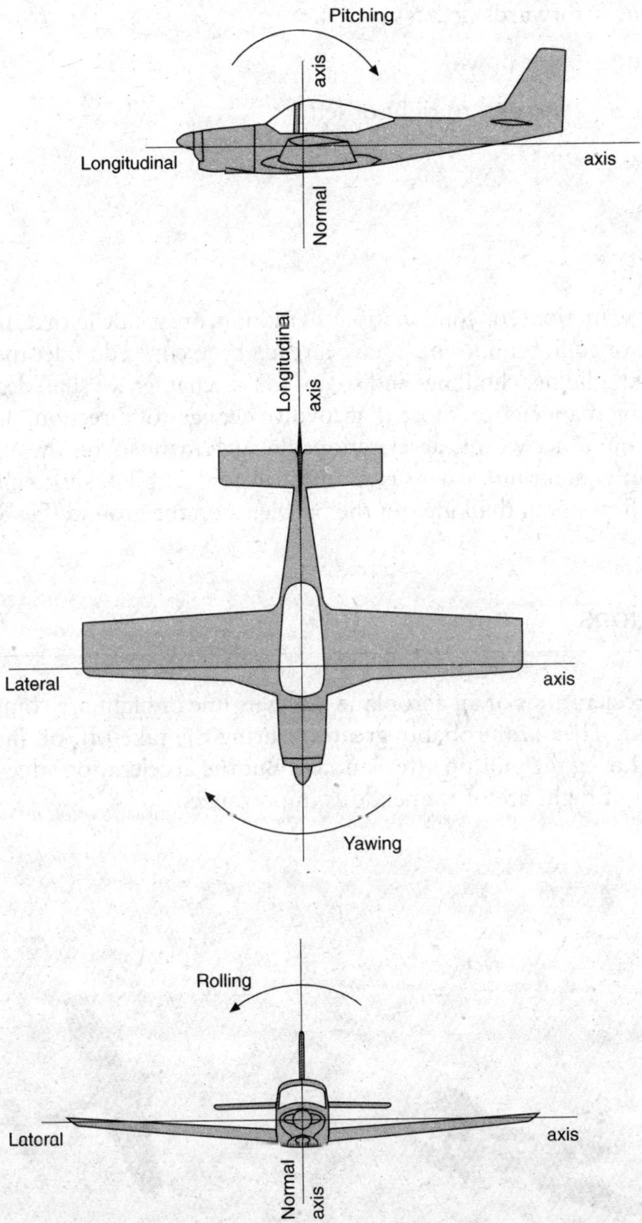

Fig 8.1 The three axes

be, or, to take another example, during a vertical nose-dive the longitudinal axis will be vertical and the lateral and normal axes horizontal.

So the manoeuvres of an aeroplane are made up of one or more, or even of all the following –

1. Movement forwards or backwards.
2. Movement up or down.
3. Movement sideways, to right or left.
4. Rolling.
5. Yawing.
6. Pitching.

Some of these motions, or combinations of motion, are gentle in that they involve only a state of equilibrium. These have already been covered under the headings of level flight, gliding, climbing, and so on. In this chapter we shall deal with the more thrilling manoeuvres, those that involve changes of direction, or of speed, or of both – in other words, **accelerations.** In such manoeuvres the aeroplane is no longer in equilibrium. There is more thrill for the pilot; more interest, but more complication, in thinking out the problems on the ground (Fig 8A).

Accelerations

Now the accelerations of an aeroplane along its line of flight are comparatively unimportant. They are probably greatest during the take-off, or, in the negative sense, during the pull-up after landing. But the accelerations due to change in direction of flight are of tremendous importance.

Fig 8A Manoeuvres
A clipped-wing Spitfire in mock combat manoeuvres with a Chance Vought Corsair.

As we have already discovered, when a body is compelled to move on a curved path, it is necessary to supply a force towards the centre, this force being directly proportional to the acceleration required. Such a force is called the centripetal force. The body will cause a reaction, that is to say an outward force, on whatever makes it travel on a curved path. This reaction is called by some people the centrifugal force.

If an aeroplane is travelling at a velocity of V metres per second on the circumference of a circle of radius r metres, then the acceleration towards the centre of the circle is V^2/r metres per second per second.

Therefore the centripetal (or centrifugal) force is $m \times$ acceleration, where m is the mass of the aeroplane in kilograms,

$$= mV^2/r \text{ newtons}$$

In practice aeroplanes very rarely travel for any length of time on the arc of a circle; but that does not alter the principle, since any small arc of a curve is, for all practical purposes, an arc of some circle with some radius, so all it means is that the centre and the radius of the circle keep changing as the aeroplane manoeuvres.

The acceleration being V^2/r shows that the two factors which decide the acceleration, and therefore the necessary force, are velocity and radius, the velocity being squared having the greatest effect. **Thus curves at high speed, tight turns at small radius, need large forces towards the centre of the curve.**

We can easily work out the acceleration, V^2/r. For instance, for an aeroplane travelling at 82 m/s on a radius of 200 metres, the acceleration is $(82 \times 82)/200 = 34\,\text{m/s}^2$, which is a little less than $4 \times$ the acceleration due to gravity. For convenience, this is often written as $4g$. However, before we go any further, we need to try to clear up some misunderstanding about the use of the symbol g. This letter is unfortunately used to represent several different things. Firstly, it represents the gravitational constant. By multiplying the mass by the gravitational constant g, we obtain the weight (written in mathematical terms $W=m\,g$). The gravitational constant g has a value of $9.81\,\text{m/s}^2$ on the earth's surface, and this makes it look like an acceleration. In this context it is not an acceleration. As we noted in Chapter 1, a book resting on a fixed table has no acceleration, but it does have a weight equal to $m \times g$ newtons. The trouble is that g is also used to represent the acceleration due to gravity. If the book falls off the table, it will accelerate at $9.81\,\text{m/s}^2$; the acceleration *will* then be equal to g ($9.81\,\text{m/s}^2$). Fortunately, although g is commonly used to represent two different things, it always has a value of $9.81\,\text{m/s}^2$, so we can still use it for both purposes, as long as we make quite clear whether we are talking about an acceleration or a weight. Finally, to make things even more confusing, pilots (and racing car drivers) often talk about pulling a certain number of '**g**'s. Unfortunately, this usage is so common that we cannot ignore it. This quantity is not an acceleration, it is just a number. It has no units, and it simply represents **a factor which, when multiplied by the weight, gives the total force**

that must be applied to a body to balance the combined effects of gravity and centripetal acceleration. It is really a load factor, because it tells us how the loads and stresses in the airframe increase during a manoeuvre. In this book we will try to make things easier by using a bold letter **'g'** in inverted commas for this quantity.

Let us try to show how the difference in usage works, by taking a simple example of an aircraft with a **mass** of 1000 kg, which therefore has a **weight** of 1000*g* newtons, (since $W=m\,g$). We will take as an example, the case of this aircraft pulling out of a dive, where it is subjected to a centripetal acceleration of 3*g* at the bottom of the manoeuvre. The **centripetal force** required to provide this acceleration will be $1000 \times 3g$ newtons (since force = mass × acceleration). Now $1000 \times 3g$ newtons is a force equal to 3 times the weight. This centripetal force will be provided by the lift from the wings. However, the wing has to support the weight of the aircraft ($1000 \times g$ newtons) as well as providing the centripetal acceleration, so the **total lift force** must be $1000 \times (3g + 1g)$ newtons, which is **4 × the weight.** The pilot therefore refers to this as a **4 'g'** manoeuvre. Not only will the lift and the stresses on the airframe by 4 times their normal level flight value, but the pilot will feel as though he weighs 4 times as much as usual. From the above, you can begin to appreciate the problem. A 4 **'g'** pullout involves a 3*g* centripetal acceleration!

Now consider what happens if the aircraft is at the top of a loop at the same speed, and with the same radius of curvature. At the top of the loop, the centripetal acceleration required will still be 3*g*, and the total force required to produce this will still be $1000 \times 3g$ newtons. However the weight of the aircraft will provide part the centripetal force required ($1000 \times g$ newtons), leaving the wing lift (now pulling downward) to provide the extra $1000 \times 2g$ newtons. Notice that although the centripetal acceleration is 3*g*, the wings only have to provide twice as much lift force as in level flight: that is **2 × the weight.** Also, the pilot will not be squashed so firmly into his seat as at the bottom of the pull-out manoeuvre, as his weight is trying to pull him out of the seat. Because the lift required is only **2 × the weight,** the load factor is only 2, and the pilot would call this a 2 **'g'** manouevre.

If the loop were performed at the same speed, but with three times the radius, then the centripetal acceleration would be 1*g*, so the required centripetal force would be 1 × the weight, and the weight alone could provide all the centripetal force required. The wings would need to produce no lift at all. This is therefore called a zero **'g'** manoeuvre. The pilot would temporarily feel weightless. Indeed, performing an outside loop or 'bunt', that is with the aircraft the right way up (see later in this chapter), with a 1*g* centripetal acceleration is used as a way of providing trainee astronauts with an experience of the weightless conditions that they will encounter in space flight.

Pulling out of a dive

We have already looked at one example involving a pull-out from a dive. Let us now look at some of the effects of this manoeuvre. Firstly, we have seen that the wings will have to generate more lift, in order to provide the necessary centripetal acceleration. This means that the bending stresses in the wing will increase. In fact, in a 3 '**g**' pull out, they will be three times as high as for normal level flight. Design and safety requirements and regulations will specify the maximum load factor that the aircraft must be able to withstand, and all the stresses in the aircraft have to be determined for this condition. The required load factor will depend on the usage. Fully aerobatic types have a high specified maximum load factor, whereas civil transports have a lower requirement. You are not allowed to roll or loop airliners, although a roll has been performed on at least one occasion (without any passengers present). The pilot will be made aware of the maximum number of '**g**'s that the aircraft is permitted to be subjected to.

In order to provide the necessary increase in lift, the lift coefficient C_L must be increased by increasing the angle of attack. This in fact goes up as the square root of the '**g**' factor. Clearly, there can come a point where the maximum C_L is reached, and any attempt to further increase it will result in

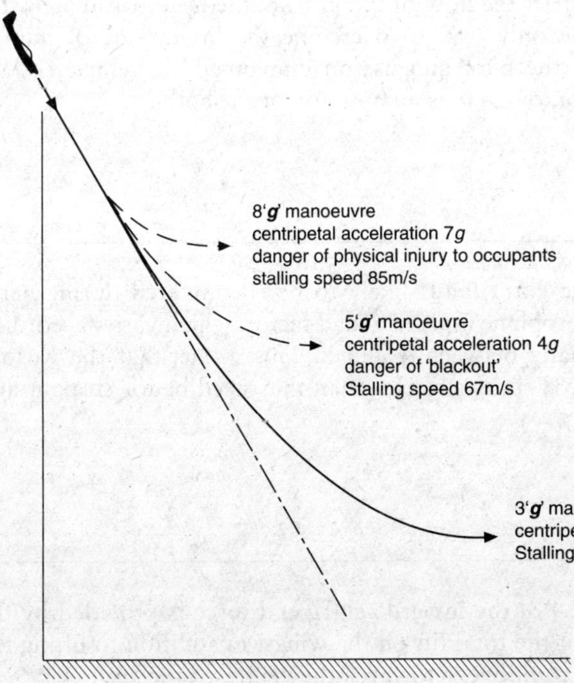

Fig 8.2 Pulling out of a dive

the aircraft stalling. The greater the centripetal acceleration required, the higher will be the stalling speed. Stalling whilst attempting to pull out too steeply is a condition that pilots must avoid at all times. The consequences can be disastrous, as many dive-bomber pilots have found to their cost. Figure 8.2 illustrates the case of an aircraft, having a level flight stalling speed of 30 m/s, pulling out of a steep dive. The load factor or '*g*' level, the stalling speed and the centripetal accelerations are given for various cases.

Apart from the loads on the airframe, any manoeuvre involving large centripetal accelerations will have a physical effect on the pilot. The pilot's head will feel heavy, and he will experience difficulty in moving his arms, which now feel several times heavier than normal. Even at 1.5 '*g*'s, writing becomes difficult. Worse than this, however, are the effects that the centripetal acceleration can have on his blood circulation. At around 4 to 5 '*g*', his heart, which is a pressure pump, will start to have difficulty pumping blood to his head, and if this is too severe, everything will appear to turn grey at first, and then he will be in danger of 'blacking out' and losing consciousness.

Apart from the problem of stalling, this physiological factor also imposes a limit on the severity of the manoeuvre that can be performed. In fighter aircraft, several means have been employed to increase the amount of centripetal acceleration that can be tolerated. One simple approach is to have the pilot lying as near to horizontal as possible whilst still being able to see where he is going. Another involves the use of special '*g*' suits which inflate at strategic points to temporarily restrict the flow of blood from the head. Ultimately, for extreme manoeuvres, the only way to overcome the limitations of human physiology is to remove the pilot and use an unmanned air vehicle (UAV), either remotely controlled, or even as an autonomous robot.

The load factor

In order to allow for the extra loads likely to be encountered during aerobatics, every part of an aeroplane is given a **load factor**, which varies according to conditions, being usually between 4 and 8. This means that the various parts are made from 4 to 8 times stronger than they need be for straight and level flight.

Turning

In an ordinary turn (Fig. 8B) the inward centripetal force is provided by the aeroplane **banking** so that the total lift on the wings, in addition to lifting the aeroplane, can supply a component towards the centre of the turn (Fig. 8.3, overleaf).

Suppose an aeroplane of weight W newtons to be travelling at a velocity of V metres per second on the circumference of a circle of radius r metres, then the acceleration towards the centre of the circle is V^2/r metres per second per second.

Therefore the force required towards the centre

$$= WV^2/gr \text{ newtons}$$

If the wings of the aeroplane are banked at an angle of θ to the horizontal, and if this angle is such that the aeroplane has no tendency to slip either inwards or outwards, then the lift L newtons will act at right angles to the wings, and it must provide a **vertical component**, equal to W newtons, to balance the **weight**, and an **inward component**, of WV^2/gr newtons, **to provide the acceleration towards the centre.**

This being so, it will be seen that –

$$\tan \theta = (WV^2/gr) \div W = V^2/gr$$

Fig 8B Turning
A Mustang making a tight turn near the ground.

This simple formula shows that there is a correct angle of bank, θ, for any turn of radius r metres at a velocity of V m/s, and that this angle of bank is quite independent of the weight of the aeroplane.

Consider a numerical example –

Find the correct angle of bank for an aeroplane travelling on a circle of radius 120 m at a velocity of 53 m/s (take the value of g as 9.81 m/s^2).

$$V = 53 \text{ m/s}$$

$$r = 120 \text{ m}$$

$$\tan\theta = V^2/gr = (53 \times 53)/(9.81 \times 120) = 2.38$$

$$\therefore \theta = 67^\circ \textbf{ approx}$$

What would be the effect if the velocity were doubled, i.e. 106 m/s?

$$\tan\theta \text{ would be } 4 \times 2.38 = 9.52$$

$$\therefore \theta = 83^\circ \textbf{ approx}$$

What would be the effect if the velocity were 53 m/s as in the first example, but the radius was doubled to 240 m instead of 120 m?

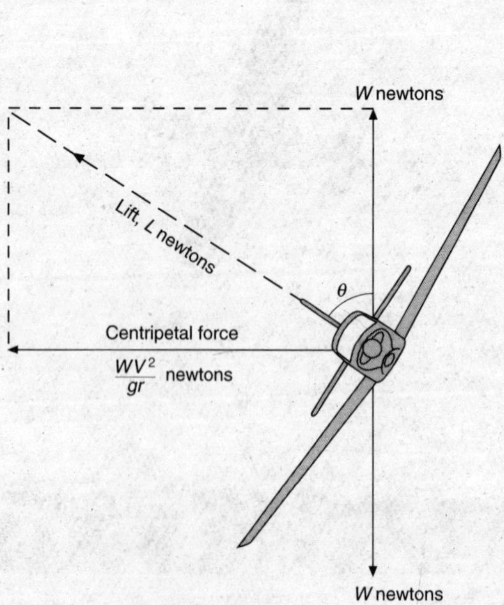

Fig 8.3 Forces acting on an aeroplane during a turn

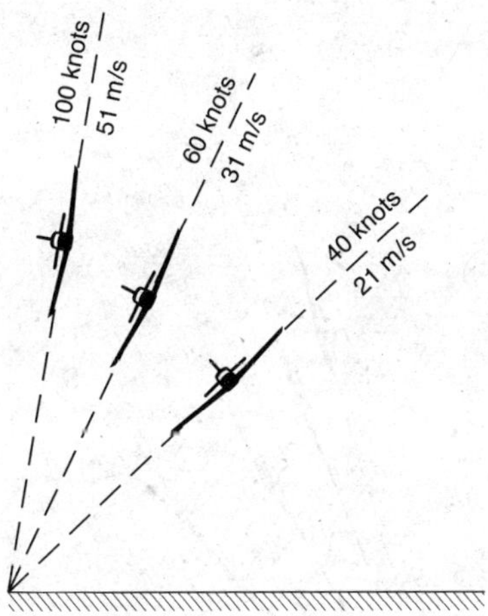

Fig 8.4 Correct angles of bank
Radius of turn 50 metres

tan θ would be 2.38/2 = 1.19

$\therefore \theta = 49°$ **approx**

Thus we see that an increase in velocity needs an increase in the angle of bank, whereas if the radius of the turn is increased the angle of bank may be reduced, all of which is what we might expect from experiences of cornering by other means of transport. Figures 8.4 and 8.5 (overleaf) show the correct angle of bank for varying speeds and radii; notice again how the speed has more effect on the angle than does the radius of turn.

Loads during a turn

It will be clear from the figures that the lift on the wings during the turn is greater than during straight flight; it is also very noticeable that the lift increases considerably with the angle of bank. This means that structural components, such as the wing spars, will have to carry loads considerably greater than those of straight flight.

Mathematically, $W/L = \cos\theta$, or $L = W/\cos\theta$

i.e. at 60° angle of bank, lift = $2W$, stalling speed, 85 knots (44 m/s)

at 70° angle of bank, lift = $3W$, stalling speed, 104 knots (53 m/s)

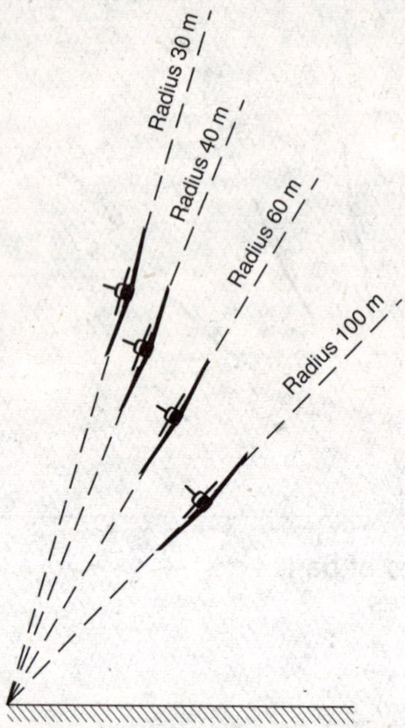

Fig 8.5 Correct angles of bank
Air speed 60 knots (31 m/s)

at 75° angle of bank, lift = 4W, stalling speed, 120 knots (62 m/s)

at 84° angle of bank, lift = 10W, stalling speed, 190 knots (98 m/s)

These figures mean that at these angles of bank, which are given to the nearest degree, the loads on the wing structure are 2, 3, 4, and 10 times respectively the loads of normal flight. This is simply our old friend g again, but in this instance it is certainly better to talk in terms of load than of g because the accelerations, and the corresponding loads, are in a horizontal plane while the initial weight is vertical; it is no longer a question of adding by simple arithmetic.

Whatever the angle of bank, the lift on the wings must be provided by $C_L \cdot \frac{1}{2}\rho V^2 \cdot S$. It follows, therefore, that the value of $C_L \cdot \frac{1}{2}\rho V^2 \cdot S$ must be greater during a turn than during normal flight, and this must be achieved either by increasing the velocity or increasing the value of C_L. Thus it follows that the stalling speed, which means the speed at the maximum value of C_L, must go up in a turn; as before it will go up in proportion of the square root of the wing loading, and the stalling speeds corresponding to the various angles of bank are shown in the table assuming, as for the pull-out of a dive, a stalling speed in level flight of 60 knots (31 m/s). These are all fairly steep banks; for banks up to 45° or so the loads are not serious, there is no danger of blacking out, and the increase of stalling speed is quite small – even so, it needs watching if one is already flying or gliding anywhere near the normal stalling speed, and suddenly decides to turn. At steep angles of bank we have to contend not only with the considerable increase of stalling speeds but with all the same problems as arose with the pull-out, i.e. blacking out, injury to pilot and crew, and the possibility of structural failure in the aircraft. It may seem curious that the angle of bank should be the deciding factor, but it must be remembered that the angle of bank (provided it is the correct angle of bank) is itself dependent on the velocity and radius of the turn, and these are the factors that really matter. In the history of fighting aircraft the ability to out-turn an opponent has probably counted more than any other feature, and from this point of view the question of steeply banked turns is one of paramount importance. An aspect of this question which must not be forgotten is that of engine power; steep turns can only be accomplished if the engine is powerful enough to keep the aeroplane travelling at high speed and at large angles of attack, perhaps even at the stalling angle. The normal duties of the engine are to propel the aeroplane at high speed at small angles of attack, or low speed at large angles of attack, but not both at the same time. The need for extra power in steeply banked tight turns has resulted in a technique in which the pilot embarking on such a manoeuvre suddenly applies all the power available.

Correct and incorrect angles of bank

We have so far assumed that the aeroplane is banked at **the correct angle** for the given turn. Fortunately the pilot has several means of telling whether the bank is correct or not (Fig. 8.6, overleaf), and since the methods help us to understand the mechanics of the turn, it may be as well to mention them here.

A good indicator is **the wind itself, or a vane, like a weather cock, mounted in some exposed position.** In normal flight and in a correct bank the wind will come from straight ahead (neglecting any local effects from the slipstream); if the bank is too much, the aeroplane will sideslip inwards and the aeroplane, and pilot if he is in an open cockpit, will feel the wind coming from the inside of the turn, whereas if the bank is too small, the wind will come from the outside of the turn, due to an outward skid on the part of the aeroplane.

Another indication would be a plumb-bob hung in the cockpit out of contact with the wind. In normal flight this would, of course, hang vertically; during a correct bank it would not hang vertically, but in exactly the same position relative to the aeroplane as it would in normal flight, i.e. it would bank with the aeroplane. If over-banked the plumb-line would be inclined inwards; if under-banked, outwards from the above position. This plumb-bob idea, in the form of a pendulum, forms the basis of the sideslip indicator which is provided by the top pointer of the so-called **turn and bank indicator.** The pointer is geared so as to move in such a way that the pilot must move the control

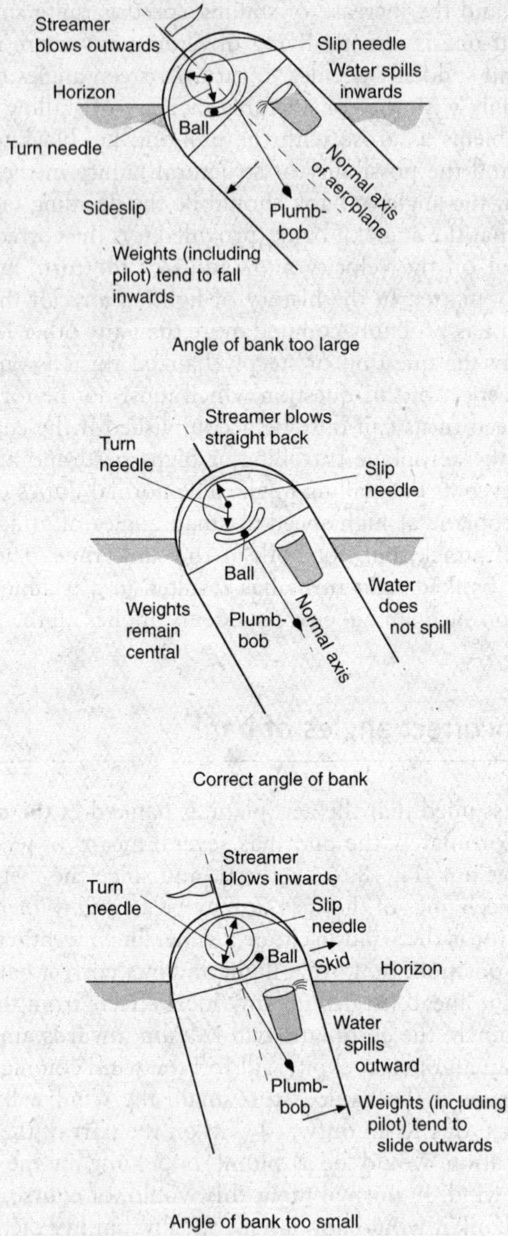

Fig 8.6 **Effects of correct and incorrect angles of bank**

column **away from** the direction of the pointer, this being the instinctive reaction. Sometimes a curved transparent tube containing a metal ball is used, and again the control column must be moved away from the indication given on the instrument. It is interesting to note that in early aeroplanes the slip indicator was, in effect, a spirit level, the tube being curved the opposite way and with a bubble (in liquid) instead of the ball; the pilot was then told to 'follow the bubble' – **not** the instinctive reaction. Nowadays such simple mechanical devices are being replaced by electronic or digital displays which neverthless often mimic the apearance of the older instruments. Figure 8.6 shows how a tumbler full of water would not spill even when tilted at 80° in a correct bank; if the bank were too small it would spill outwards over the top lip of the tumbler!

Lastly, during a correct bank the pilot will sit on his seat without any feeling of sliding either inwards or outwards; in fact, he will be sitting tighter on his seat than ever, his effective weight being magnified in the same proportions as the lift so that if he weighs 800 N in normal flight he will feel that he weighs 8000 N when banking at 84°! If he over-banks he will tend to slide inwards, but outwards if the bank is insufficient.

Other problems of turning

In order to get into a turn the pilot puts on bank by means of the ailerons, but once the turn has commenced the outer wing will be travelling faster than the inner wing and will therefore obtain more lift, so he may find that not only is it necessary to take off the aileron control but actually to apply opposite aileron by moving the control column against the direction of bank – this is called **holding off bank**.

An interesting point is that this effect is different in turns on a glide and on a climb. On a **gliding turn** the whole aircraft will move the same distance downwards during one complete turn, but the inner wing, because it is turning on a smaller radius, will have descended on a steeper spiral than the outer wing; therefore the air will have come up to meet it at a steeper angle, in other words the inner wing will have a larger angle of attack and so obtain more lift than the outer wing. The extra lift obtained in this way may compensate, or more than compensate, the lift obtained by the outer wing due to increase in velocity. Thus in a gliding turn there may be little or no need to hold off bank.

In a **climbing turn**, on the other hand, the inner wing still describes a steeper spiral, but this time it is an upward spiral, so the air comes down to meet the inner wing more than the outer wing, thus **reducing the angle of attack on the inner wing**. So, in this case, the outer wing has more lift **both** because of velocity **and** because of increased angle, and there is even more necessity for holding off bank than during a normal turn.

Another interesting way of looking at the problem of gliding and climbing turns is to analyse the motion of an aircraft around its three axes during such turns. In a flat turn, i.e. a level turn without any bank, the aircraft is yawing only. In a banked level turn, the aircraft is yawing and pitching – in the extreme of a vertically banked turn it would be pitching only. But in a gliding or climbing turn the aircraft is pitching, yawing and rolling. In a **gliding** turn it is rolling

inwards; in a **climbing** turn, **outwards**. The inward roll of the gliding turn causes the extra angle of attack on the inner wing, the outward roll of the climbing turn on the outer wing. Many people find it difficult to believe this. If the reader is in such difficulty conviction may come from one of two methods; which will suit best will depend upon the reader's temperament. The mathematically-minded may like to analyse the motion in terms of the following (Fig. 8.7) –

The rate of turn of the complete aeroplane (about the vertical), Ω.

The angle of bank of the aeroplane, θ.

The angle of pitch of the aeroplane, ϕ.

A little thought will reveal the fact that the

Rate of yaw $= \Omega \,.\, \cos\phi \,.\, \cos\theta$.

Rate of pitch $= \Omega \,.\, \cos\phi \,.\, \sin\theta$.

Rate of roll $= \Omega \,.\, \sin\phi$.

Translating this back into English, and taking one of the extreme examples, when $\theta = 0$, i.e. no bank, and $\phi = 0$, i.e. no pitch, $\cos\theta$ and $\cos\phi$ will be 1, $\sin\theta$ and $\sin\phi$ will be 0.

$\therefore$ rate of yaw $= \Omega =$ rate of turn of complete aeroplane.

Rate of pitch and rate of roll are zero. All of which we had previously decided for the flat turn.

The reader (mathematically-minded) may like to work out the other extremes such as the vertical bank ($\theta = 90°$) or vertical pitch ($\phi = 90°$), or better still the more real cases with reasonable values of θ and ϕ.

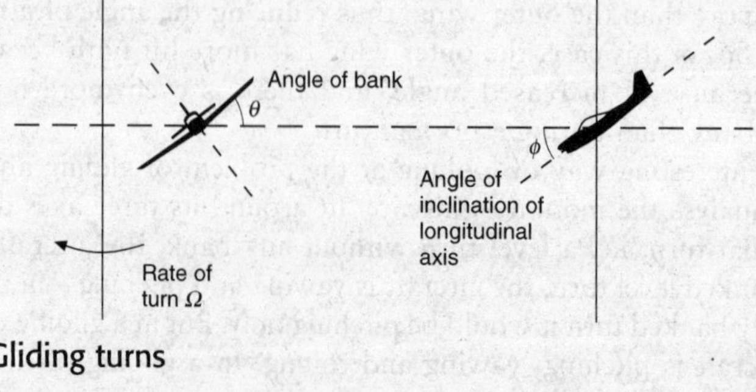

Fig 8.7 Gliding turns

Notice that the rate of roll depends entirely on the angle of pitch, i.e. the inclination of the longitudinal axis to the vertical – if this is zero, there is no rate of roll even though the aircraft may be descending or climbing.

What about the reader who does not like mathematics? Get hold of a model aeroplane, or, failing this, a waste-paper basket and spend a few minutes making it do upward and downward spirals; some people are convinced by doing gliding and climbing turns with their hand and wrist – and their friends may be amused in watching!

At large angles of bank there is less difference in velocity, and in angle, between inner and outer wings, and so the question of holding off bank becomes less important; but much more difficult problems arise to take its place.

First, though, let us go back to the other extreme and consider what is called a '**flat turn**', i.e. one that is all yaw and without any bank at all.

Very slight turns of this kind have sometimes been useful when approaching a target for bombing purposes, but otherwise they are in the nature of 'crazy flying', in other words, incorrect flying, and good pilots always try to keep their sideslip indicator in the central position. Actually flat turns are rather difficult to execute for several reasons. First, the extra velocity of the other wing tends to bank the aeroplane automatically; secondly, the lateral stability (explained later) acts in such a way as to try to prevent the outward skid by banking the aeroplane; thirdly, the side area is often insufficient to provide enough inward force to cause a turn except on a very large radius; fourthly, the directional stability (also explained later) opposes the action of the rudder and tends to put the nose of the aircraft back so that it will continue on a straight path. Taking these four reasons together, it will be realised that an aeroplane has a strong objection to a flat turn!

Modern aircraft have a small side surface and if this is coupled with **good directional stability**, for the last two reasons particularly, a flat turn becomes virtually impossible. So much is this so that it is very little use applying rudder to start a turn, the correct technique being to put on bank only.

Controls on steep banks

The turning of an aeroplane is also interesting from the control point of view because as the bank becomes steeper **the rudder gradually takes the place of the elevators**, and vice versa. This idea, however, needs treating with a certain amount of caution because, in a vertical bank for instance, the rudder is nothing like so powerful in raising or lowering the nose as are the elevators in normal horizontal flight. Incidentally, the reader may have realised that a vertical bank, without sideslip, is theoretically impossible, since in such a bank the lift will be horizontal and will provide no contribution towards lifting the weight. If it is claimed that such a bank can, in practice, be executed, the explanation must be that a slight upward inclination of the fuselage together with the propeller thrust provides sufficient lift.

This only applies to a **continuous** vertical bank in which no height is to be lost; it is perfectly possible, both theoretically and practically, to execute a turn in which, for a few moments, the bank is vertical, or even over the vertical. In the latter case the manoeuvre is really a combination of a loop and a turn.

Generally speaking, the radius of turn can be reduced as the angle of bank is increased, but even with a vertical bank there is a limit to the smallness of the radius because, quite apart from the question of side-slipping, the lift on the wings (represented by $C_L \cdot \frac{1}{2}\rho V^2 . S$) must provide all the force towards the centre, i.e. m . V^2/r or WV^2/gr.

$$\text{Thus} \quad WV^2/gr = C_L \cdot \frac{1}{2}\rho V^2 . S$$

$$\text{or} \quad r = 2W/(C_L \cdot \rho S \cdot g)$$

Now, in straight and level flight the stalling speed (V_s) is given by the equation

$$W = L = C_L \max \cdot C_L \cdot \frac{1}{2}\rho V_s^2 . S$$

If we substitute this value of W into our formula for the radius we get

$$r = (2 \cdot C_L \max \cdot C_L \cdot \frac{1}{2}\rho V_s^2 . S)/(C_L \cdot \rho S \cdot g)$$

$$\text{i.e.} \quad r = (V_s^2/g) \times (C_L \max/C_L)$$

This shows that the radius of turn will be least when C_L is equal to C_L max, i.e. when the angle of attack is the stalling angle, and radius of turn = V_s^2/g. It is rather interesting to note that the minimum radius of turn is quite independent of the actual speed during the vertical banks; it is settled only by the stalling speed of the particular aeroplane. Thus, to turn at minimum radius, one must fly at the stalling angle, but any speed may be employed provided the engine power is sufficient to maintain it. In actual practice, the engine power is the deciding factor in settling the minimum radius of turn whether in a vertical bank or any other bank, and it must be admitted that it is not usually possible to turn on such a small radius as the above formula would indicate.

This formula applies to some extent to all steep turns and shows that the aeroplane with the lower stalling speed can make a tighter turn than one with a higher stalling speed. (We are referring, as explained above, to the stalling speed in straight and level flight.) But in order to take advantage of this we must be able to stand the *g*'s involved in the steep banks, and we must have engine power sufficient to maintain turns at such angles of bank.

Aerobatics

The usual aerobatics are loops, spins, rolls, sideslips, and nose-dives, to which may be added upside-down flight, the inverted spin, and the inverted loop. The manoeuvres may also be combined in various ways, e.g. a half loop followed by a half roll, or a half roll followed by the second half of a loop.

There are many reasons why aerobatics should be performed in those types of aircraft which are suitable for them. They provide excellent training for accuracy and precision in manoeuvre, and give a feeling of complete mastery of the aircraft, which is invaluable in all combat flying. They may also be used for exhibition purposes, but modern aircraft are so fast and the radius on which they can turn or manoeuvre is so large that, in many ways, they provide less of a spectacle than older types. Not least, aerobatics increase the joy and sensations of flight to the pilot himself – not quite so much to other occupants of the aircraft!

The movements of the aeroplane during these aerobatics are so complicated that they baffle any attempt to reduce them to the terms of simple mechanics and, indeed, to more advanced theoretical considerations, unless assumptions are made which are not true to the facts.

Figs 8.8 and 8.9 (overleaf) show the approximate path travelled by a slow type of aircraft during a loop and the corresponding 'accelerometer' diagram which shows how the force on the wings varies during the manoeuvre. From this it will be seen that, as in many manoeuvres, the greatest loads occur at the moment of entry. Notice also that even at the top of the loop the load is very little less than normal – that is to say, that the pilot is sitting firmly on his seat in the upward direction, the loads will still be in the same direction **relative to the aircraft** as in normal flight, and our plumb bob will be hanging upwards!

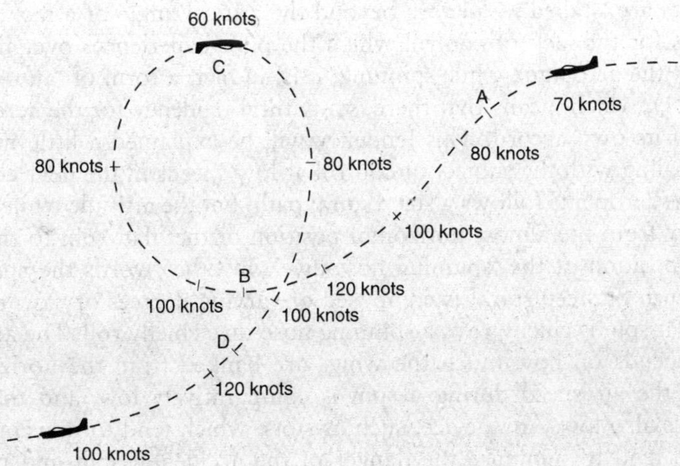

Fig 8.8 A loop

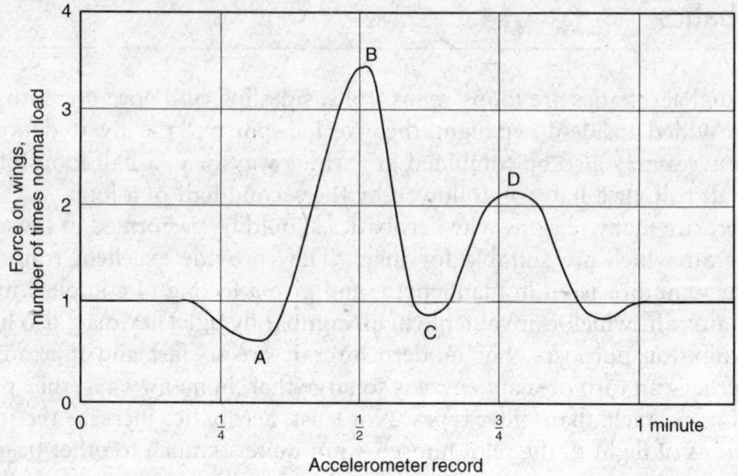

Fig 8.9 Accelerometer diagram for a loop

Only in a bad loop will the loads at the top become negative, causing the loads on the aircraft structure to be reversed and the pilot to rely on his straps to prevent him from falling out.

Simple theoretical problems can be worked out on such assumptions as that a loop is in the form of a circle or that the velocity remains constant during the loop, but the error in these assumptions is so great that very little practical information can be obtained by attempting to solve such problems, and it is much better to rely on the results of practical experiments.

A Spin (Fig. 8.10) is an interesting manoeuvre, if only for the reason that at one time there stood to its discredit a large proportion of all aeroplane accidents that had ever occurred. It differs from other manoeuvres in the fact that the wings are 'stalled', – i.e. are beyond the critical angle of attack, and this accounts for the lack of control which the pilot experiences over the movements of the aeroplane while spinning; it is, in fact, a form of 'auto-rotation' (Fig. 8.11), which means that there is a natural tendency for the aeroplane to rotate of its own accord. This tendency will be explained a little more fully when dealing with the subject of control at low speeds in the next chapter. In a spin the aeroplane follows a steep spiral path, but the attitude while spinning may vary from the almost horizontal position of the 'flat' spin to the almost vertical position of the 'spinning nose-dive'. In other words the spin, like a gliding turn or steep spiral is composed of varying degrees of yaw, pitch and roll. A flat spin is chiefly yaw, a spinning nose-dive chiefly roll. The amount of pitch depends on how much the wings are banked from the horizontal. In general, the air speed during a spin is comparatively low, and the rate of descent is also low. Any device, such as slots, which tend to prevent stalling, will also tend to minimise the danger of the accidental spin and may even make it impossible to carry out deliberately. The area and disposition of the

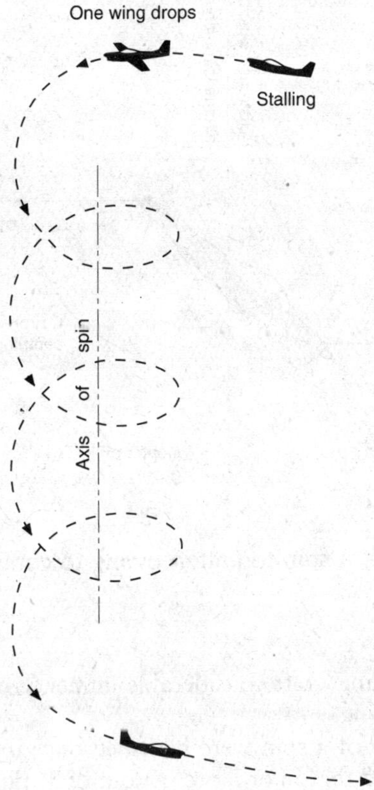

Fig 8.10 A spin

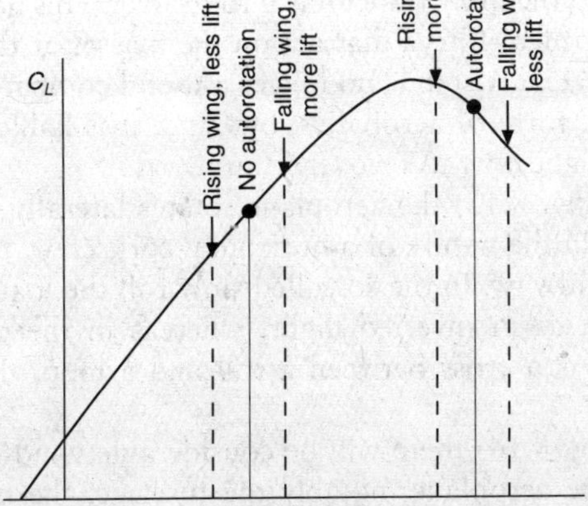

Fig 8.11 Cause of auto-rotation

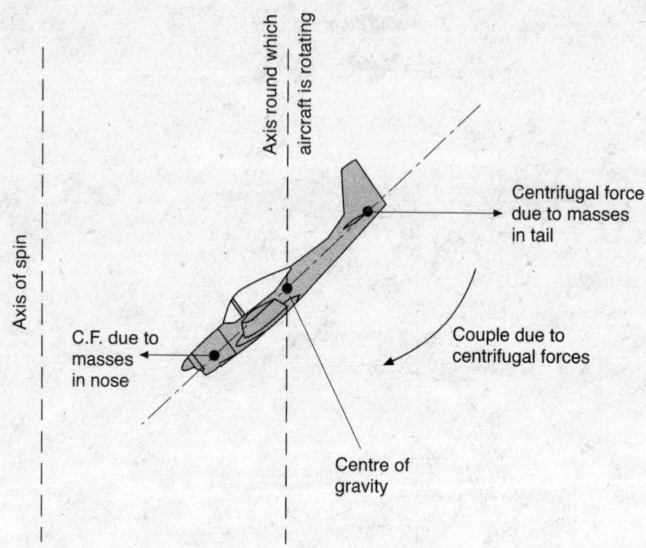

Fig 8.12 Tendency of a spin to flatten owing to centrifugal force

fin, rudder, and tail plane exert considerable influence on the susceptibility of the aeroplane to spinning.

Many of the terrors of a spin were banished once it was known just what it was. We then realised that in order to get out of a spin we must get it out of the stalled state by putting the nose down, and we must stop it rotating by applying 'opposite rudder'. In practice, the latter is usually done first, because it is found that the elevators are not really effective until the rotation is stopped. The farther back the centre of gravity, and the more masses that are distributed along the length of the fuselage, the flatter and faster does the spin tend to become and the more difficult is it to recover. This flattening of the spin is due to the centrifugal forces that act on the masses at the various parts of the aircraft (Fig. 8.12). A spin is no longer a useful combat manoeuvre, nor is it really a pleasant form of aerobatics, but since it is liable to occur accidentally, pilots are taught how to recover from it.

During a **roll** (Fig. 8.13) the aeroplane rotates laterally through 360°, but the actual path is in the nature of a horizontal corkscrew, there being varying degrees of pitch and yaw. In the so-called slow roll the loads in the 180° position are reversed, as in inverted flight, whereas in the other extreme, the barrel roll, which is a cross between a roll and a loop, the loads are never reversed.

In a **sideslip** (Fig. 8.14) there will be considerable wind pressure on all the side surfaces of the aeroplane, notably the fuselage, the fin and the rudder, while if the planes have a dihedral angle the pressure on the wings will tend to bring the machine on to an even keel. The sideslip is a useful manoeuvre for

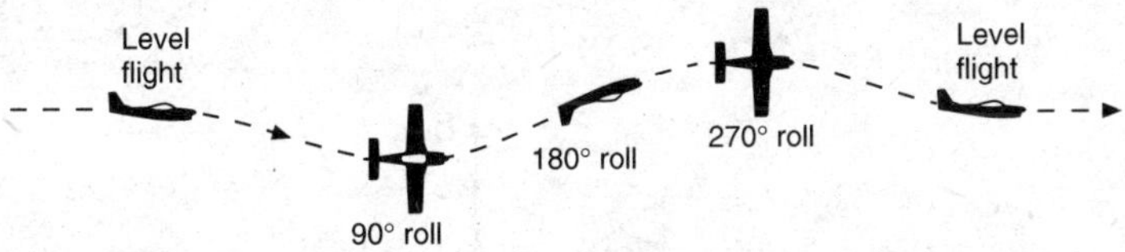

Fig 8.13 A roll

losing height or for compensating a sideways drift just prior to landing, but, as already mentioned, modern types of aircraft do not take very kindly to sideslipping. The small side area means that they drop very quickly if the sideslip is at all steep, and the directional stability is so strong that it may be impossible to hold the nose of the machine up (by means of the rudder), and the dropping of the nose causes even more increase of speed.

A **nose-dive** is really an exaggerated form of gliding; the gliding angle may be as great as 90° – i.e. vertical descent – although such a steep dive is rarely performed in practice. If an aeroplane is dived vertically it will eventually reach a steady velocity called the terminal velocity. In such a dive the weight is entirely balanced by the drag, while the lift has disappeared. The angle of attack is very small or even negative, there is a large positive pressure near the leading edge on the top surface of the aerofoil, tending to turn the aeroplane on to its back, and this is balanced by a considerable 'down' load on the tail plane (Fig. 8.15, overleaf). In such extreme conditions the terms used are apt to be misleading; for instance, the 'down' load referred to is horizontal, while the lift, if any such exists, will also be horizontal. The terminal velocity of modern aeroplanes is very high, and it makes little difference whether the engine is running or not. They lose so much height in attaining the terminal velocity that, in practice, it is doubtful whether it can ever be reached. As was only to be expected, the problems which accompany the attainment of a speed near to the speed of sound first made themselves felt in connection with the nose-dive, especially at high altitudes. At that time these compressibility effects were a special feature of nose-diving, but there was one consoling feature – as one got nearer the earth, terminal velocities were lower (owing to the greater

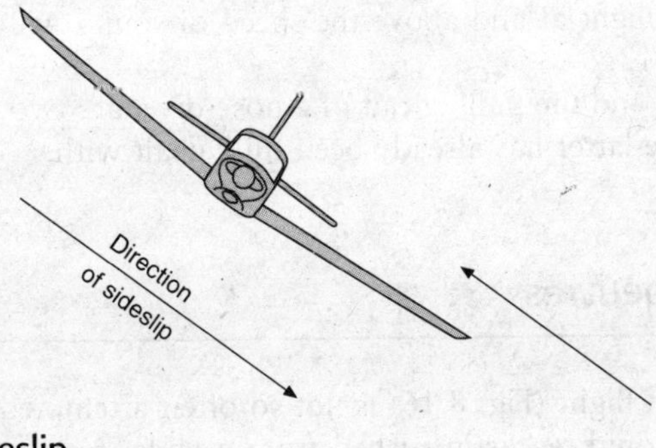

Fig 8.14 A sideslip

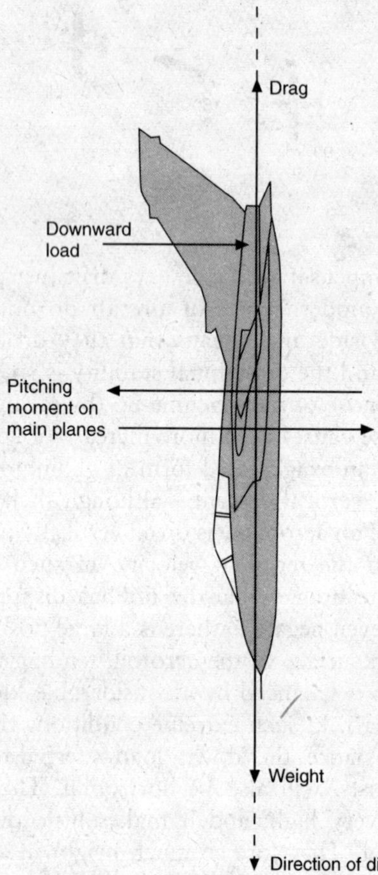

Fig 8.15 Nose-dive

density), and the speed of sound was higher (owing to the higher temperature). So if one got into trouble high up, there was always a chance of getting out of it lower down. But nowadays, when more and more aircraft can exceed the speed of sound in level flight, and when compressibility troubles are no longer associated with fears of the unknown, these ideas are out of date, and the whole subject of flight at and above the speed of sound will be considered in Chapters 11 and 12.

The nose-dive, and the pulling-out of a nose-dive, are two entirely different problems, and the latter has already been fully dealt with.

Inverted manoeuvres

Real upside-down flight (Fig. 8.16) is not so often attempted as is commonly supposed, and should be distinguished from a glide in the inverted position, which does not involve problems affecting the engine. If height is to be maintained during inverted flight, the engine must, of course, continue to run and this necessitates precautions being taken to ensure a supply of fuel and, with a piston engine, the proper functioning of the carburettor. The aerofoil will be inverted, and therefore, unless of the symmetrical type, will certainly be inefficient; while in order to produce an angle of attack, the fuselage will have to be in a very much 'tail-down' attitude. The stability will be affected, although some aircraft have been more stable when upside-down than the right way up, and considerable difficulty has been experienced in restoring them to normal flight. In spite of all the disabilities involved, some aeroplanes are capable of maintaining height in the inverted position (Fig. 8C, overleaf).

The **inverted spin** is in most of its characteristics similar to the normal spin; in fact, in some instances pilots report that the motion is more steady and therefore more comfortable. As in inverted flight, however, the loads on the aeroplane structure are reversed and the pilot must rely on his straps to hold him in the machine.

The **inverted loop**, or 'double bunt', in which the pilot is on the outside of the loop (Fig. 8.17, overleaf), is a manoeuvre of extreme difficulty and danger. The difficulty arises from the fact that whereas in the normal loop the climb to the top of the loop is completed while there is speed and power in hand and engines and aerofoils are functioning in the normal fashion, in the inverted loop the climb to the top is required during the second portion of the loop, when the aerofoils are in the inefficient inverted position. The danger is incurred because of the large reversed loads and also because of the physiological effects of the pilot's blood being forced into his head. It was a long time before this manoeuvre was successfully accomplished, and once it had been, so many foolhardy pilots began to attempt it – often with fatal results – that it had to be forbidden, except under very strict precautions and regulations.

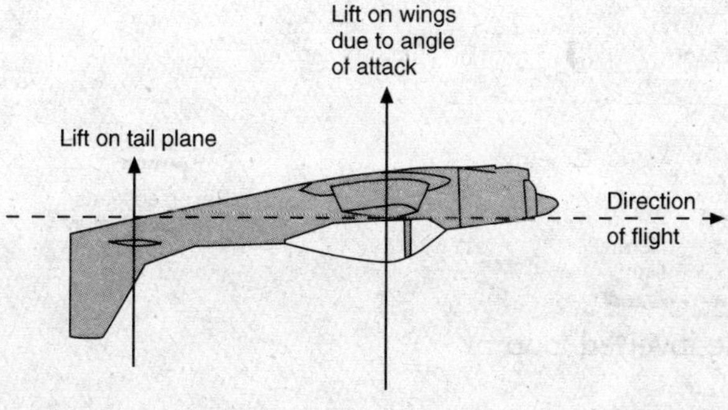

Fig 8.16 Upside-down flight

Fig 8C Inverted flight
(By courtesy of '*Flight*')
And to illustrate upside-down flight, the Hurricane of the Second World War; note the tail-down attitude usually associated with inverted flight.

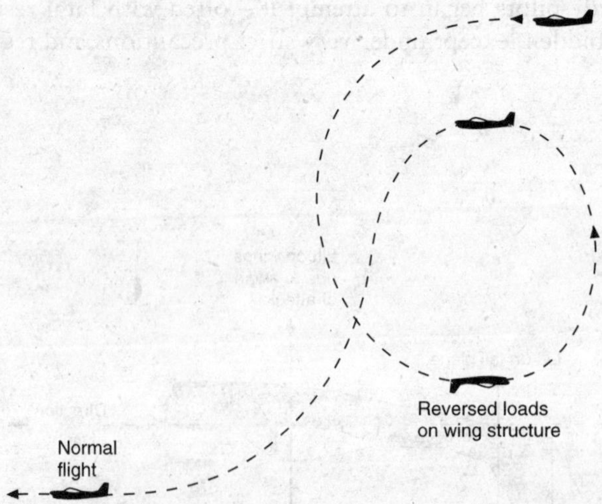

Fig 8.17 The inverted loop

'Bumpy' weather

In addition to the loads incurred during definite aerobatics, all aircraft are required to face the effects of unsteady weather conditions. Accelerometer records show that these may be quite considerable, and they must certainly be reckoned with when designing commercial aircraft. Where aeroplanes are, in any case, required to perform aerobatics, they will probably be amply strong enough to withstand any loads due to adverse weather. The conditions which are likely to inflict the most severe loads consist of strong gusty winds, hot sun, intermittent clouds, especially thunder clouds in which there is often considerable turbulence, and uneven ground conditions; a combination of all these factors will, almost certainly, spell a 'rough passage'. The turbulence and upcurrents that may be encountered in severe thunderstorms and in cumulo-nimbus clouds can sometimes be such as to tax the strength of the aeroplane and the flying skill of the pilot.

Manoeuvrability

Before leaving the subject of manoeuvres we ought to mention that the inertia of an aeroplane – or, to be more correct, **the moment of inertia of the various parts** – will largely determine the ease or otherwise of handling the machine during manoeuvres. Without entering into the mathematical meaning of moment of inertia, we can say that, in effect, it means the natural resistance of the machine to any form of rotation about its centre of gravity. Any heavy masses which are a long distance away from a particular axis of rotation will make it more difficult to cause any rapid movement around the axis; thus masses such as engines far out on the wings result in a resistance to rolling about the longitudinal axis; and a long fuselage with large masses well forward or back will mean a resistance to pitching and yawing.

Can you answer these?

See if you can answer these questions about the various manoeuvres which an aeroplane can perform –

1. What are the six degrees of freedom of an aeroplane?
2. Why is there a definite limit to the smallness of the radius on which an aeroplane can turn?

3. Two aircraft turn through 360° in the same time, i.e. at the same rate of turn, but the radius of turn of one is twice that of the other. Will they have the same angle of bank? If not, which will have the greater?
4. Explain the difference between a gliding and climbing turn from the point of view of holding off bank.
5. Why does an aeroplane spin?

For solutios see Appendix 5.
Numerical examples on manoeuvres will be found in Appendix 3.

Fig 8D Manoeuvres
(By courtesy of the Lockheed Aircraft Corporation, USA)
A rigid-rotor helicopter performing aerobatic manoeuvres.

CHAPTER 9

Stability and control

Meaning of stability and control

The stability of an aeroplane means its ability to return to some particular condition of flight (after having been slightly disturbed from that condition) without any efforts on the part of the pilot. An aeroplane may be stable under some conditions of flight and unstable under other conditions. For instance, an aeroplane which is stable during straight and level flight may be unstable when inverted, and vice versa. If an aeroplane were stable during a nose-dive, it would mean that it would resist efforts on the part of the pilot to extricate it from the nose-dive. The stability is sometimes called **inherent stability**. Note that, nowadays, some military combat aircraft are deliberately made to be inherently unstable, as this increases their manoeuvrability, and can reduce drag. This requires a sophisticated automatic artificial stabilisation system, which has to be totally reliable. Because of the potentially disastrous consequences of system failure, inherent instability is not permitted on civil aircraft, but with adequate safeguard it is possible to relax the level of stability compared to older types, and this has benefits in terms of reduced drag. To the pilot, the artificial stabilisation system makes the aircraft feel and handle as though it were stable.

Stability is often confused with the balance or 'trim' of an aircraft, and the student should be careful to distinguish between the two. An aeroplane which flies with one wing lower than the other may often, when disturbed from this attitude, return to it. Such an aeroplane is out of its proper trim, but it is not unstable.

There is a half-way condition between stability and instability, for, as already stated, an aeroplane which, when disturbed, tends to return to its original position is said to be stable; if, on the other hand, it tends to move farther away from the original position, it is unstable. But it may tend to do neither of these and prefer to remain in its new position. This is called **neutral stability**, and is sometimes a very desirable feature.

Figure 9.1 illustrates some of the ways in which an aeroplane may behave when it is left to itself. Only a pitching motion is shown; exactly the same considerations apply to roll and yaw, although a particular aeroplane may have quite different stability characteristics about its three axes. The top diagram shows complete dead-beat stability which is very rarely achieved in practice. The second is the usual type of stability, that is to say an oscillation which is gradually damped out. The steady oscillation shown next is really a form of neutral stability, while the bottom diagram shows the kind of thing which may easily occur in certain types of aircraft, an oscillation which steadily grows worse. Even this is not so bad as the case when an aeroplane makes no attempt to return but simply departs farther and farther away from its original path. That is complete instability.

The degree of stability may differ according to what are called the **stick-fixed** and **stick**-free conditions; in pitching, for instance, stick-fixed means that the elevators are held in their neutral position relative to the tail plane, whereas stick-free means that the pilot releases the control column and allows the elevators to take up their own positions.

Another factor affecting stability is whether it is considered – and tested – under the condition of **power-off** or **power-on**. On modern aircraft the engine thrust can be comparable with or even greater than the airframe weight and therefore may significantly influence the stability.

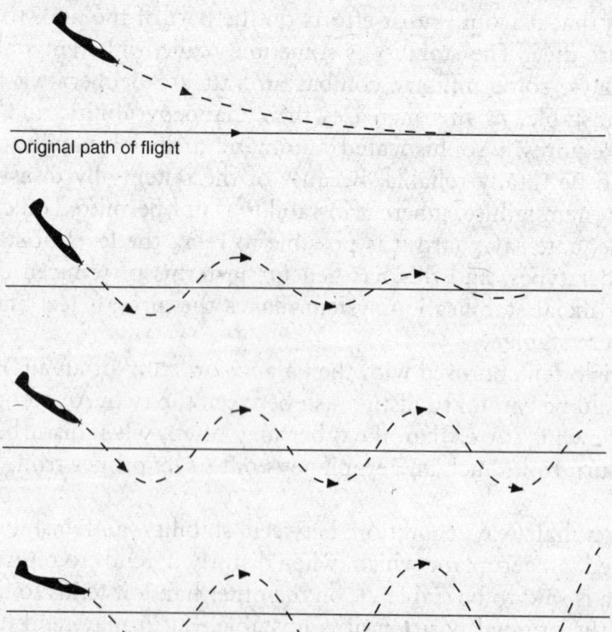

Fig 9.1 Degrees of stability

Control means the power of the pilot to manoeuvre the aeroplane into any desired position. It is not by any means the same thing as stability; in fact, the two characteristics may directly oppose each other.

The stability or control of an aeroplane in so far as it concerns **pitching** about the lateral axis is called **longitudinal stability or control** respectively.

Stability or control which concerns **rolling** about the longitudinal axis is called **lateral stability or control.**

Stability or control which concerns yawing about the normal axis is called **directional stability or control.**

Before we attempt to explore this subject any further we feel it is our duty to warn the reader that the **problems involved in the consideration of stability and control of aeroplanes are considerable.** Any attempt at 'simple' explanation of such problems may, at the best, be incomplete and possibly incorrect. Readers need have no fear of the mathematics, as we shall not even attempt to tackle them, but they must be prepared, if and when they acquire greater knowledge of the subject from more advanced works, to readjust their ideas accordingly.

After this very necessary apology we will try to explain, at any rate, the practical considerations which affect stability and control.

Longitudinal stability

We shall start with **longitudinal stability**, since this can be considered independently of the other two. In order to obtain stability in pitching, we must ensure that if the angle of attack is temporarily increased, forces will act in such a way as to depress the nose and thus decrease the angle of attack once again. To a great extent we have already tackled this problem while dealing with the pitching moment, and the movement of the centre of pressure on aerofoils. We have seen that an ordinary upswept wing with a cambered aerofoil section cannot be balanced or 'trimmed' to give positive lift and at the same time be stable in the sense that a positive increase in incidence produces a nose-down pitching moment about the centre of gravity.

The position as regards the wing itself can be improved to some extent by **sweepback**, by **wash-out** (i.e. by decreasing the angle of incidence) towards the wing tips, by **change in wing section** towards the tips (very common in modern types of aircraft), and by a **reflex curvature** towards the trailing edge of the wing section.

But it is not only the wing that affects the longitudinal stability of the aircraft as a whole, and in general it can be said that this is dependent on four factors –

1. **The position of the centre of gravity**, which must not be too far back; this is probably the most important consideration.

2. **The pitching moment on the main planes**; this, as we have seen, usually tends towards instability, though it can be modified by the means mentioned.
3. **The pitching moment on the fuselage** or body of the aeroplane; this too is apt to tend towards instability.
4. **The tail plane** – its area, the angle at which it is set, its aspect ratio, and its distance from the centre of gravity. This is nearly always a stabilising influence (Fig. 9.2).

Longitudinal dihedral

The tail plane is usually set at an angle less than that of the main planes, the angle between the chord of the tail plane and the chord of the main planes being known as the **longitudinal dihedral** (Fig. 9.3). This longitudinal dihedral is a practical characteristic of most types of aeroplane, but so many considerations enter into the problem that it cannot be said that an aeroplane which does not possess this feature is necessarily unstable longitudinally. In any case, it is the actual angle at which the tail plane strikes the airflow, which matters; therefore we must not forget the downwash from the main planes. This downwash, if the tail plane is in the stream, will cause the actual angle of attack to be less than the angle at which the tail plane is set (Fig. 5.6). For this reason, even if the tail plane is set at the same angle as the main planes, there will **in effect** be a longitudinal dihedral angle, and this may help the aeroplane to be longitudinally stable.

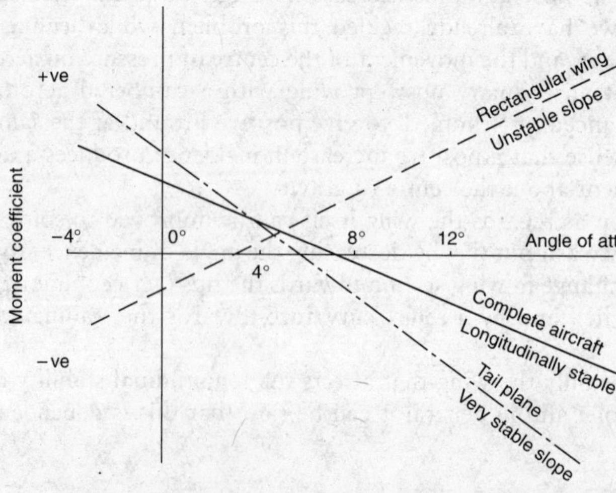

Fig 9.2 Pitching moment coefficient about centre of gravity
Wing, tail plane and complete aircraft.

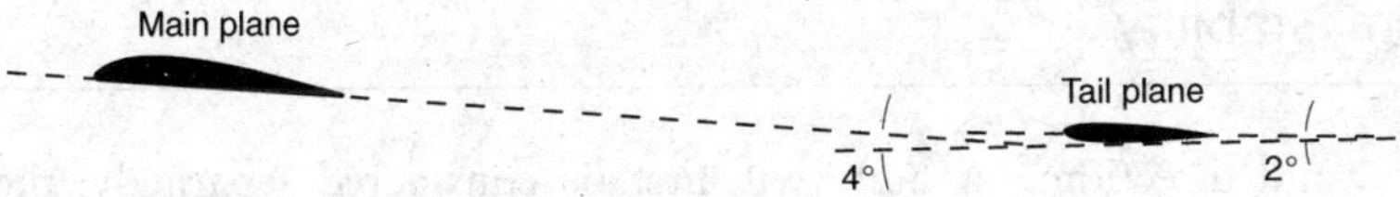

Fig 9.3 Longitudinal dihedral angle

Suppose an aeroplane to be flying so that the angle of attack of the main planes is 4° and the angle of attack of the tail plane is 2°; a sudden gust causes the nose to rise, inclining the longitudinal axis of the aeroplane by 1°. What will happen? The momentum of the aeroplane will cause it temporarily to continue moving practically in its original direction and at its previous speed. Therefore the angle of attack of the main planes will become nearly 5° and of the tail plane nearly 3°. The pitching moment (about the centre of gravity) of the main planes will probably have a nose-up, i.e. unstable tendency, but that of the tail plane, with its long leverage about the centre of gravity, will definitely have a nose-down tendency. If the restoring moment caused by the tail plane is greater than the upsetting moment caused by the main planes, and possibly the fuselage, then the aircraft will be stable.

This puts the whole thing in a nutshell, but unfortunately it is not quite so easy to analyse the practical characteristics which will bring about such a state of affairs; however the forward position of the centre of gravity and the area and leverage of the tail plane will probably have the greatest influence.

It is interesting to note that a tail plane plays much the same part, though more effectively, in providing longitudinal stability, as does reflex curvature on a wing, or sweepback with wash-out of incidence towards the tips.

When the tail plane is in front of the main planes (Fig. 5E) there will probably still be a longitudinal dihedral, which means that this front surface must have greater angle than the main planes. The latter will naturally still be at an efficient angle, such as 4°, so that the front surface may be at, say, 6° or 8°. Thus it is working at a very inefficient angle and will stall some few degrees sooner than the main planes. This fact is claimed by the enthusiasts for this type of design as its main advantage, since the stalling of the front surface will prevent the nose being raised any farther, and therefore the main planes will never reach the stalling angle.

In the tail-less type, in which there is no separate surface either in front or behind, the wings must be heavily swept back, and there is a 'wash-out' or decrease in the angle of incidence as the wing tip is approached, so that these wing tips do, in effect, act in exactly the same way as the ordinary tail plane (Figs 5C and 5D).

Lateral stability

Lateral and directional stability will first be considered separately; then we shall try to see how they affect each other.

To secure lateral stability we must so arrange things that when a slight roll takes place the forces acting on the aeroplane tend to restore it to an even keel.

In all aeroplanes, when flying at a small angle of attack, there is a resistance to roll because the angle of attack, and so the lift, will increase on the down-going wing, and decrease on the up-going wing. But this righting effect will only last while the aeroplane is actually rolling. It must also be emphasised that this only happens while the angle of attack is small; if the angle of attack is near the stalling angle, then the increased angle on the falling wing may cause a decrease in lift, and the decreased angle on the other side an increase; thus the new forces will tend to roll the aeroplane still further, this being the cause of auto-rotation previously mentioned (Fig. 8.11).

But the real test of stability is what happens after the roll has taken place.

Dihedral angle

The most common method of obtaining lateral stability is by the use of a dihedral angle on the main planes (Figs 9.4 and 9A). Dihedral angle is taken as being the angle between each plane and the horizontal, not the total angle between the two planes, which is really the geometrical meaning of dihedral angle. If the planes are inclined upwards towards the wing tips, the dihedral is positive; if downwards, it is negative and called **anhedral** (Fig. 9B, overleaf); the latter arrangement is used in practice for reasons of dynamic stability.

The effect of the dihedral angle in securing lateral stability is sometimes dismissed by saying that if one wing tip drops the horizontal equivalent on that wing is increased and therefore the lift is increased, whereas the horizontal equivalent and the lift of the wing which rises is decreased, therefore obviously the forces will tend to right the aeroplane.

Unfortunately, it is not all quite so obvious as that.

Once the aircraft has stopped rolling, and provided it is still travelling straight ahead, the aerodynamic forces will be influenced only by the airstream passing over the aircraft. This will be identical for both wings and so no restoring moment will result.

What, then, is the real explanation as to why a dihedral angle is an aid to lateral stability? When the wings are both equally inclined the resultant lift on the wings is vertically upwards and will exactly balance the weight. If, however, one wing becomes lower than the other (Fig. 9.5), then the resultant lift on the wings will be slightly inclined in the direction of the lower wing, while the weight will remain vertical. Therefore the two forces will not

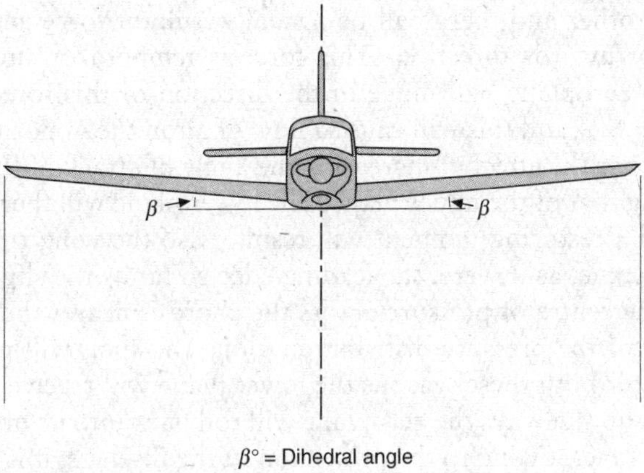

Fig 9.4 Lateral dihedral angle

Fig 9A Dihedral
Pronounced dihedral is evident on this Spitfire: a very late post-war model.

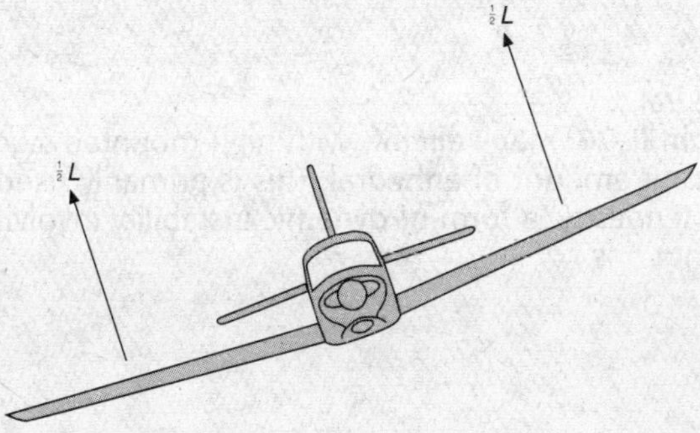

Fig 9.5 Equal lift produced by each wing – no rolling moment due to roll

balance each other and there will be a small resultant force acting in a sideways and downwards direction. This force is temporarily unbalanced and therefore the aeroplane will move in the direction of this force – i.e. it will **sideslip** (Fig 9.6) – and this will cause a flow of air in the opposite direction to the slip. This has the effect of increasing the angle of attack of the lower plane and increasing that of the upper plane. The lower plane will therefore produce more lift and a restoring moment will result. Also the wing tip of the lower plane will become, as it were, the leading edge so far as the slip is concerned; and just as the centre of pressure across the chord is nearer the leading edge, so the centre of the pressure distribution along the span will now be on the lower plane; for both these reasons the lower plane will receive more lift, and after a slight slip sideways the aeroplane will roll back into its proper position. As a matter of fact, owing to the protection of the fuselage, it is probable that the flow of air created by the sideslip will not reach a large portion of the raised wing at all; this depends very much on the position of the wing relative to the fuselage.

Both the leading edge effect on the lower wing, and the shielding of the upper wing by the fuselage, occur on nearly all types of aircraft, and may well mean that an aeroplane has a sufficient degree of lateral stability without any dihedral angle, or too much if some of the following effects also apply. Even if there is no actual dihedral angle on the wings, these other methods of achieving lateral stability may be described as having a '**dihedral effect**'.

Fig 9B Anhedral
Like this Ilyushin IL-76, many aircraft with high-mounted swept wings are given a significant amount of anhedral. This is primarily used to counteract the Dutch roll tendency, a form of dynamic instability involving coupled roll and yaw motions.

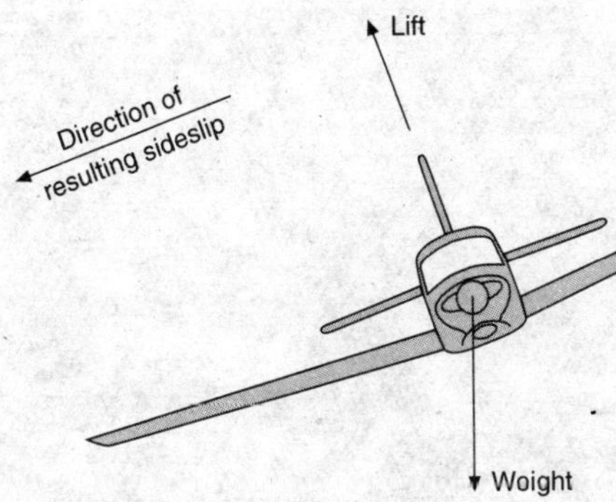

Fig 9.6 The production of a sideslip

High wing and low centre of gravity

If the wings are placed in a high position and the centre of gravity is correspondingly low, the lateral stability can be enhanced. When an aircraft sideslips, the lift on the lower wing becomes greater than that on the higher one. Furthermore, a small sideways drag force is introduced. In consequence, the resultant force on the wing will be in the general direction indicated in Fig. 9.7. You will see that this force does not now pass through the centre of gravity so there will be a small moment which will tend to roll the aircraft back to a level condition. This will occur even on a low-wing aircraft, but is more effective with a high wing because the moment arm is greater. For this reason a high-wing aircraft requires less dihedral than a low-wing type.

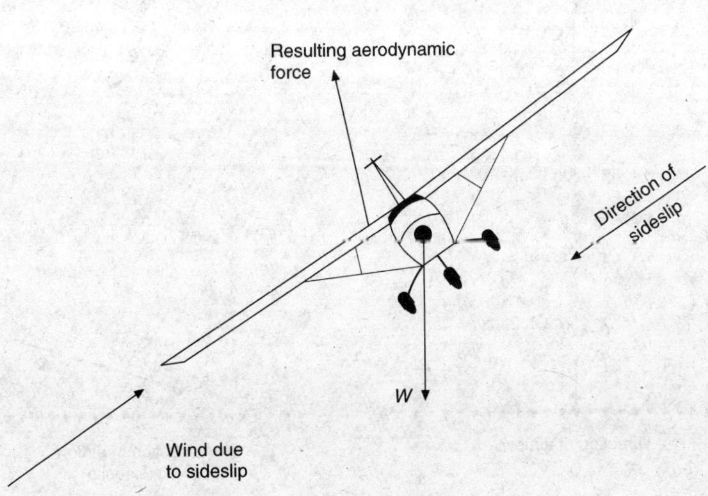

Fig 9.7 High wing and low centre of gravity

Fig 9C Sweepback
An Airbus A330 showing the sweepback of the wings.

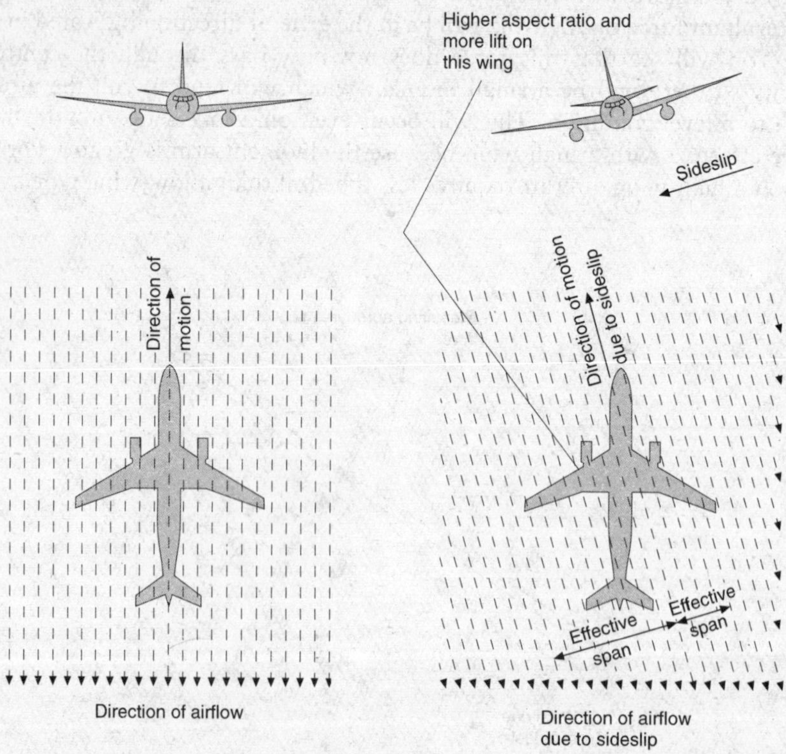

Fig 9.8 Sweepback and lateral stability

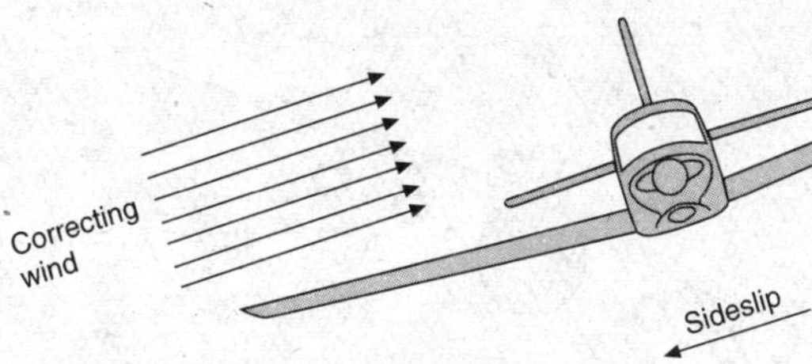

Fig 9.9 Effect of high fin on lateral stability

Sweepback and lateral stability

A considerable angle of sweepback (Fig. 9C) will in itself promote lateral stability, for, supposing the left wing to drop, as in the two previous cases, there will be a sideslip to the left and the left-hand wing will present, in effect, a higher aspect ratio than the right wing to the correcting airflow (Fig. 9.8). It will therefore receive more lift and, as before, recovery will take place after sideslip.

A forward sweep is sometimes used but this is not for reasons of stability (Fig. 9D, overleaf).

Fin area and lateral stability

One factor which may have considerable influence on lateral stability is the position of the various side surfaces, such as the fuselage, fin and rudder, and wheels. All these will present areas at right angles to any sideslip, so there will be pressure upon them which, **if they are high above the centre of gravity**, will **tend to restore** the aeroplane to an even keel; this applies to many modern types which have a high tail plane on top of a high fin (Figs 9.9 and 9E, later) and such types may have anhedral on the main planes to counterbalance this effect and prevent too great a degree of lateral stability; but if the side surfaces are low the pressure on them will tend to roll the aircraft over still more (Figs 9.10 and 9F, later) and so cause lateral instability, although this must be balanced against the effect of high wing compared with the CG position.

The reader will have noticed that, whatever the method of obtaining lateral stability, correction only takes place **after a sideslip** towards the low wing.

It is the sideslip that effects the directional stability.

Directional stability

We shall first try to consider directional stability by itself, if only as a means of convincing ourselves that the two are so interlinked one with the other that

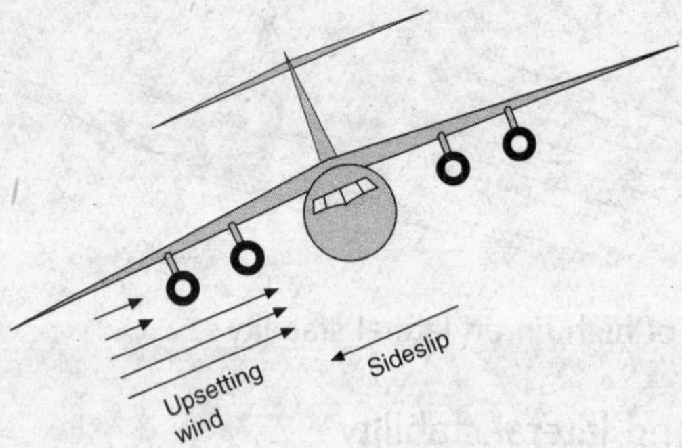

Fig 9.10 Effect of low-slung fuselage and engine pods on lateral stability

they cannot be disposed of separately. In order to establish directional stability we must ensure that, if the aeroplane is temporarily deflected from its course, it will, of its own accord, tend to return to that course again. This is almost entirely a question of the 'side surface' or 'fin area' which has already been mentioned when dealing with lateral stability, but here it is not a question of the relative height of this side surface, but whether it is in front of or behind the centre of gravity (Fig. 9F). When an aeroplane is flying in the normal way the airflow will approach it directly from the front, i.e. parallel to its longitudinal axis. Now imagine it to be deflected from its course as in Fig. 9.11 overleaf; owing to its momentum it will for a short time tend to continue moving in its old direction, therefore the longitudinal axis will be inclined to the airflow, and a pressure will be created on all the side surfaces on one side of the aeroplane.

If the turning effect of the pressures behind the centre of gravity is greater than the turning effect in front of the centre of gravity, the aeroplane will tend to its original course.

If, on the other hand, the turning effect in front is greater than that behind, the aeroplane will turn still farther off its course. Notice that it is the turning effect or the moment that matters, and not the actual pressure; therefore it is not merely a question of how much side surface, but also of the distance from the centre of gravity of each side surface. For instance, a small fin at the end of a long fuselage may be just as effective in producing directional stability as a large fin at the end of a short fuselage. Also, there may sometimes be more side surface in the front than in the rear, but the rear surfaces will be at a greater distance. All the side surfaces of an aeroplane, including that presented by wings with dihedral, affect the directional stability, but to the **fin** is allotted

Fig 9D Sweep forward (opposite)
The Grumman X–29 research aircraft with forward-swept wings. Forward-swept wings have the same advantages in high-speed flight as swept-back wings, but give a better spanwise distribution of lift, leading to lower induced drag.

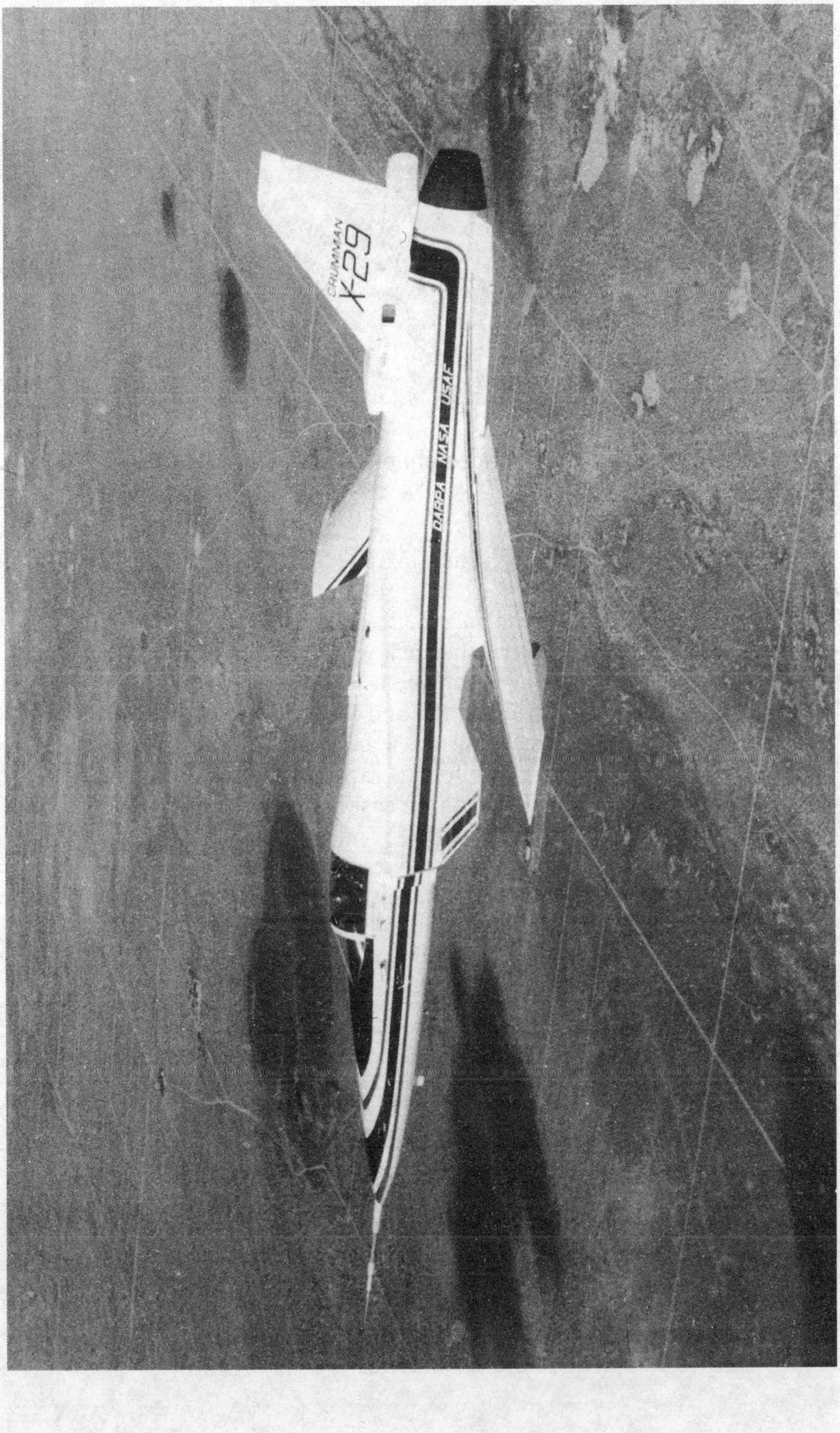
GRUMMAN
X-29
DARPA NASA USAF

Fig 9E High fin
A massive high fin and pronounced anhedral are evident on the McDonnell Douglas C–17.

the particular task of finally adjusting matters and its area is settled accordingly.

There is a very close resemblance between the directional stability of an aeroplane and the action of a weathercock which always turns into the wind; in fact, one often sees a model aeroplane used as a weathercock. The simile, however, should not be carried too far, and the student must remember that there are two essential differences between an aeroplane and a weathercock – first, that an aeroplane is not only free to yaw, but also to move bodily sideways; and secondly, that the 'wind', in the case of an aeroplane, is not the wind we speak of when on the ground, but the wind caused by the original motion of the aeroplane through the air. This point is emphasised because of the idea which sometimes exists that an aeroplane desires to turn head to wind. If such were the case, directional stability would be a very mixed blessing.

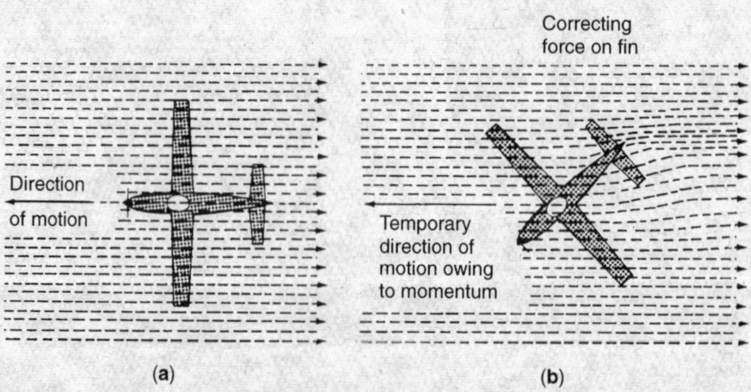

Fig 9.11 Directional stability
(a) Before disturbance; (b) after disturbance

Fig 9F A classic flying boat
(By courtesy of the General Dynamics Corporation, USA)
The large side area of the hull was balanced, from the lateral stability point of view, by the dihedral on the main planes and the high fin at the rear.

Fig 9G Fin and directional stability
A Sukhoi Su-27 blasting down the runway with nosewheel lifting clear of the ground. Note the two large fins with dorsal extension below the tailplane. This aircraft displays exceptional manoeuvrability and control at low speed.

Lateral and directional stability

Now we are, at last, in a position to connect these two forms of stability – the sideslip essential to lateral stability will cause an air pressure on the side surfaces which have been provided for directional stability. The effect of this pressure will be to turn the nose into the relative wind, i.e. in this case, **towards the direction of sideslip**. The aeroplane, therefore, will turn off its original course and in the direction of the lower wing. It is rather curious to note that the greater the directional stability the greater will be the tendency to turn off course in a sideslip. This turn will cause the raised wing, now on the outside of the turn, to travel faster than the inner or lower wing, and therefore to obtain more lift and so bank the aeroplane still further. By this time the nose of the aeroplane has probably dropped and the fat is properly in the fire with all three stabilities involved! The best way of seeing all this happen in real life is to watch a model aeroplane flying on a gusty day; the light loading and slow speed of the model make it possible to watch each step in the proceedings, whereas in the full-sized aeroplane it all happens more quickly, and also the pilot usually interferes by using his controls. If, for instance, the left wing drops and he applies rudder so as to turn the machine to the right, he will probably prevent it from departing appreciably from its course.

We can now explain the technique of turning an aeroplane. Suppose, when we want to turn to the left, instead of applying any rudder we simply bank the aeroplane to the left, as we have already seen it will slip inwards and turn to the left. That is all there is in it. So effective is this method that it is unnecessary to use the rudder at all for turning purposes. So far as the yaw is concerned – and a turn must involve a yaw – the rudder (with the help of the fin) is still responsible, just as (with the help of the fin) it always was. The difference is simply that the rudder and fin are brought into effect by the inward sideslip, instead of by application of rudder which tends to cause an outward skid. The pilot may do nothing about it, but the stability of the aeroplane puts a force on the rudder for him. It should also be emphasised that although it may be most practical, and most sensible, to commence a turn in certain aircraft without application of rudder, such a turn cannot be absolutely perfect; there must be an inward sideslip. The pilot may not notice it, the sideslip indicator may not detect it; but it is there just the same.

Just as a slight roll results in a sideslip and then a yawing motion so if an aircraft moves in a yawed position, as in Fig. 9.11, that is if it moves crabwise (which is really the same thing as slipping or skidding) lateral stability will come into play and cause the aircraft to roll away from the leading wing. Thus a roll causes a yaw, and a yaw causes a roll, and the study of the two cannot be separated.

If the stability characteristics of an aeroplane are such that it is very stable directionally and not very stable laterally, e.g. if it has large fin and rudder and little or no dihedral angle, or other 'dihedral effect', it will have a marked tendency to turn into a sideslip, and to bank at steeper and steeper angles, that it may get into an uncontrollable spiral – this is sometimes called **spiral instability**, but note that it is caused by too much stability (directional).

If, on the other hand, the aeroplane is very stable laterally and not very stable directionally, it will sideslip without any marked tendency to turn into the sideslip. Such an aircraft is easily controllable by the rudder, and if the rudder only is used for a turn the aircraft will bank and make quite a nice turn.

The reader will find it interesting to think out the other characteristics which these two extremes would cause in an aeroplane, but the main point to be emphasised is that too much stability (of any type) is almost as bad as too little stability.

Control of an aeroplane

Where an aeroplane is stable or unstable, it is necessary for the pilot to be able to control it, so that he can manoeuvre it into any desired position.

Longitudinal control is provided by the **elevators**, i.e. flaps hinged behind the tail plane, or movement of the whole tail plane.

Roll control is provided by the ailerons, i.e. flaps hinged at the rear of the aerofoils near each wing tip.

Directional control is provided by the **rudder**, i.e. a vertical flap hinged to the stern post.

The system of control is the same in each case, i.e. if the control surface is moved it will, in effect, alter the angle of attack and the camber of the complete surface, and therefore change the force upon it (see Fig. 9.12, overleaf). On many aircraft the roll can also be controlled by the use of spoilers. These are described in more detail later.

The elevators and ailerons are controlled by movements of a **control column** on which is mounted a handwheel which is usually abbreviated to something rather like a car steering wheel with most of the rim sawn off to leave a pair of small handgrips or 'spectacles'.

Pushing the control column forward lowers the elevators, thus increasing the lift on the tailplane and making the nose of the aircraft drop. Turning the handwheel anti-clockwise lowers the right hand aileron and raises the left, thus rolling the aircraft left-wing down.

In old aircraft and aircraft such as fighter aircraft, where cockpit room is restricted, there is no handwheel and instead the control column moves to left and right as well as backwards and forwards – the left-hand movement is equivalent to turning the handwheel anti-clockwise (left hand down). A small version of this type of 'joystick' control is now used on many aircraft. It resembles the kind of control stick used for model aircraft and computer games. It is placed to one side of the pilot, and consequently referred to as a side-stick.

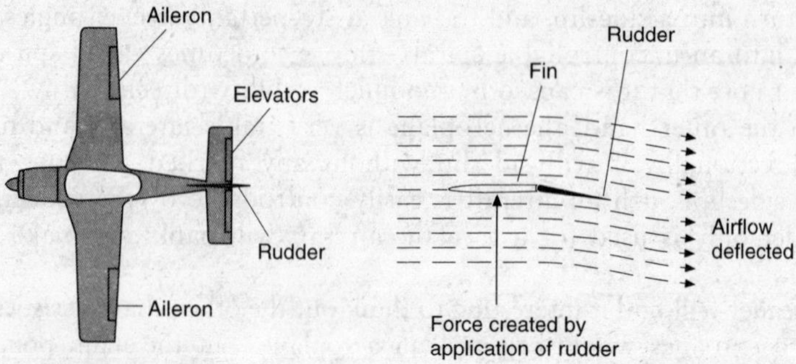

Fig 9.12 Control surfaces

The use of such a small control device is possible because the flight controls are fully power operated, and require no physical force from the pilot.

The rudder is controlled by foot pedals. Pushing the left pedal forward deflects the rudder to the left and therefore turns the nose of the aircraft to the left. This can cause some problems amongst learner pilots as the movement is opposite to that of bicycle handlebars.

In each instance it will be noticed that the control surfaces are placed as far as possible away from the centre of gravity so as to provide sufficient leverage to alter the position of the aeroplane.

On modern aircraft there may also be a secondary set of inboard ailerons which are used in high speed flight where the outboard surfaces could produce excessively large rolling moments, or unacceptable structural loading or wing twist.

Balanced controls

Nowadays most larger aircraft have powered controls, but on smaller aircraft and older types, manual controls are still used. It is perhaps surprising to find that executive jets and even some small regional jet airliners retain manual controls. It should be remembered, however, that aircraft often remain in service for many decades, and quite recent models may only be updated variants of very old designs. The following description applies to manual controls, which are the type that student pilots will initially have to deal with.

Although, in general, the forces which the pilot has to exert in order to move the controls are small, the continuous movement required in bumpy weather becomes tiring during long flights, especially when the control surfaces are large and the speeds fairly high. For this reason controls are often **balanced**, or, more correctly, partially balanced (Fig. 9H).

Several methods have been employed for balancing control surfaces. Figure 9.13 shows what is perhaps the most simple kind of aerodynamic balance. The hinge is set back so that the air striking the surface in front of the hinge causes a force which tends to make the control move over still farther; this partially balances the effect of the air which strikes the rear portion. This is effective but it must not be overdone; over-balancing is dangerous since it may remove all feel of the control from the pilot. It must be remembered that when the control surface is set a small angle, the centre of pressure on the surface is well forward of the centre of the area, and if at any angle the centre of pressure is in front of the hinge it will tend to take the control out of the pilot's hands (or feet). Usually not more than one-fifth of the surface may be in front of the hinge.

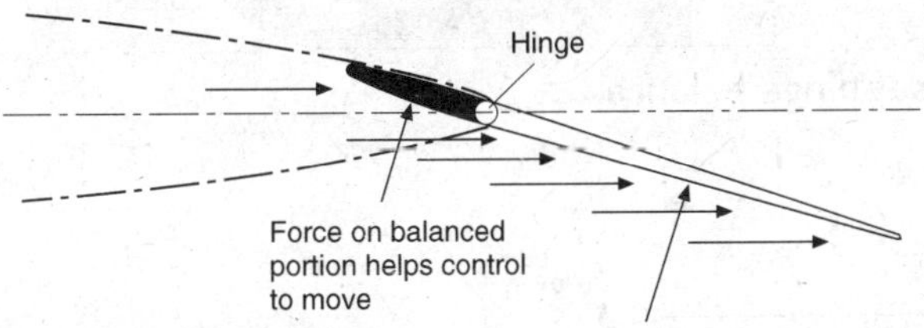

Fig 9.13 Aerodynamic balance

Fig 9H Balanced controls and tabs
(By courtesy of SAAB, Sweden)
Twin-jet training aircraft showing the statically and aerodynamically balanced ailerons with geared servo-tabs; starboard tab adjustable for trimming. Elevators and rudder are also balanced, and there is a trim tab on the rudder, and a servo-tab (adjustable for trimming) on each elevator.

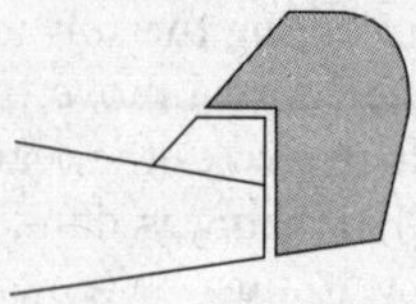

Fig 9.14 Horn balance

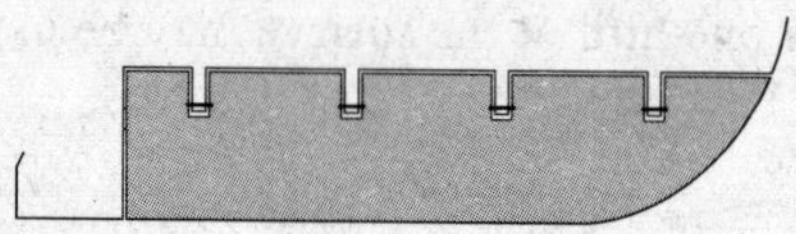

Fig 9.15 Inset hinge balance

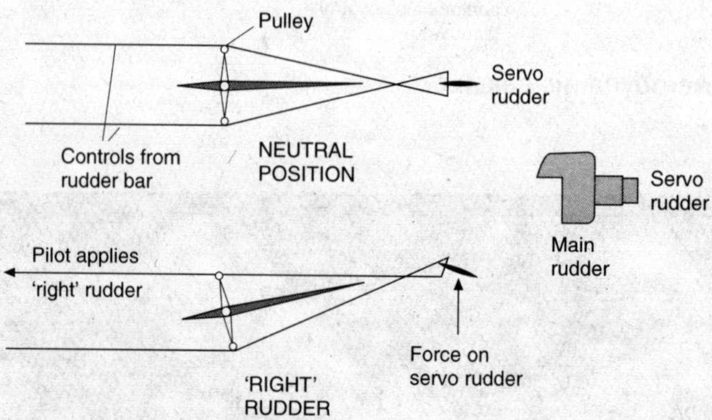

Fig 9.16 Servo system of balance

Figures 9.14 and 9.15 show two practical applications of this type of balance; in each some part of the surface is in front of the hinge, and each has its advantages.

Figure 9.16 shows the servo type of balance which differs in principle since the pilot in this case only moves the small extra surface (in the opposite direction to normal), and, owing to the leverage, the force on the small surface helps to move the main control in the required direction. It is, in effect, a system of gearing.

Perhaps the chief interest in the servo system of balance is that it was the forerunner of the **balancing tabs** and **trimming tabs**. The development of these control tabs was very rapid and formed an interesting little bit of aviation history.

The servo system suffered from many defects, but it did show how powerful is the effect of a small surface used to deflect the air in the opposite direction to that in which it is desired to move the control surface.

The next step was to apply this idea to an aileron when a machine was inclined to fly with one wing lower than the other. A strip of flexible metal was attached to the trailing edge of the control surface and produced the necessary corrective bias.

So far, the deflection of the air was only in one direction and so we obtained a **bias** on the controls rather than a balancing system. The next step gave us both balance and bias; the strip of metal became a **tab**, i.e. an actual flap hinged to the control surface. This tab was connected by a link to a fixed surface (the tail plane, fin or main plane), the length of this link being adjustable on the ground. When the main control surface moved in one direction, the tab moved in the other and thus experienced a force which tended to help the main surface to move – hence the balance. By adjusting the link, the tab could be set to give an initial force in one direction or the other – hence the bias.

Sometimes a spring is inserted between the tab and the main control system. The spring may be used to modify the system in two possible ways –

1. So that the amount of tab movement decreases with speed, thus preventing the action being too violent at high speeds.
2. So that the tab does not operate at all until the main control surface has been moved through a certain angle, or until a certain control force is exerted.

   Tabs of this kind are called **spring tabs.**

The final step (Figs 9.17 and 9.18, overleaf) required a little mechanical ingenuity, but otherwise it was a natural development. The pilot was given the means of adjusting the bias while **in the air**, and thus he was enabled to correct any flying faults, or out-of-balance effects, as and when they occurred.

On small aircraft with manual controls these tabs may be fitted to all of the primary control surfaces. The pilot can adjust their settings from within the cockpit and can thereby arrange the trim so that the aircraft will fly 'hands-off' in almost any flight conditions. On aircraft with power-operated controls, such tabs are unnecessary and the trim wheels are simply used to reset the neutral or hands-off position of either the control column or the actuator system. From the pilot's point of view this feels almost exactly like setting a trim tab.

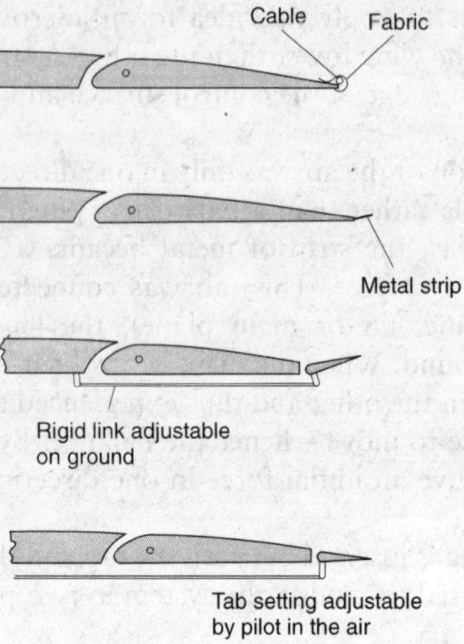

Fig 9.17 Evolution of control tabs

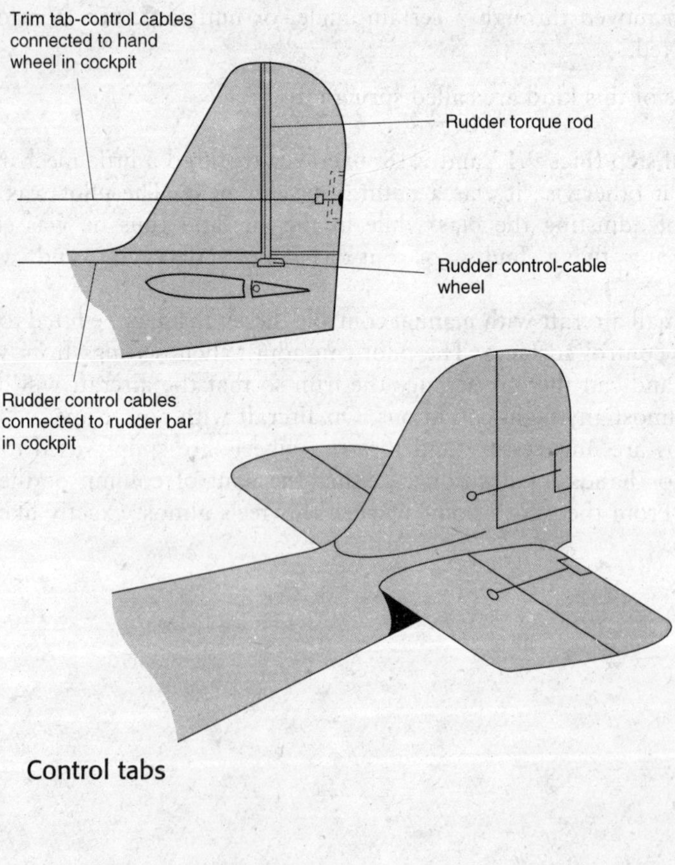

Fig 9.18 Control tabs

Mass balance

Control surfaces are often balanced in quite a different sense. A **mass** is fitted in front of the hinge. This is partly to provide a mechanical balancing of the mass of the control surface behind the hinge but may also be partly to help prevent an effect known as 'flutter' which is liable to occur at high speeds (Fig. 9.19). This flutter is a vibration which is caused by the combined effects of the changes in pressure distribution over the surface as the angle of attack is altered, and the elastic forces set up by the distortion of the structure itself. All structures are distorted when loads are applied. If the structure is elastic, as all good structures must be, it will tend to spring back as soon as the load is removed, or changes its point of application. In short, a distorted structure is like a spring that has been wound up and is ready to spring back. An aeroplane wing or fuselage can be distorted in two ways, by bending and by twisting, and each distortion can result in an independent vibration. Like all vibrations, this flutter is liable to become dangerous if the two effects add up. The flutter may affect the control surfaces such as an aileron, or the main planes, or both. The whole problem is very complicated, but we do know of two features which help to prevent it – a stiff structure and mass balance of the control surfaces. When the old types of aerodynamic balance were used, e.g. the inset hinge or horn balance, the mass could be concealed inside the forward portion of the control surface and thus two birds were killed with one stone; but when the tap type of balance is used alone the mass must be placed on a special arm sticking out in front of the control surface. In general, however, the problems of flutter are best tackled by increasing the rigidity of the structure and control-system components.

Large aircraft and military types now invariably have powered controls and these are much less sensitive to problems of flutter as the actuating system is very rigid.

Perhaps it should be emphasised that the **mass** is **not** simply a weight for the purpose of balancing the control surface statically, e.g. to keep the aileron floating when the control mechanism is not connected; it may have this effect, but it also serves to alter the moments of inertia of the surface, and thus alter the period of vibration and the liability to flutter. It may help to make this clear

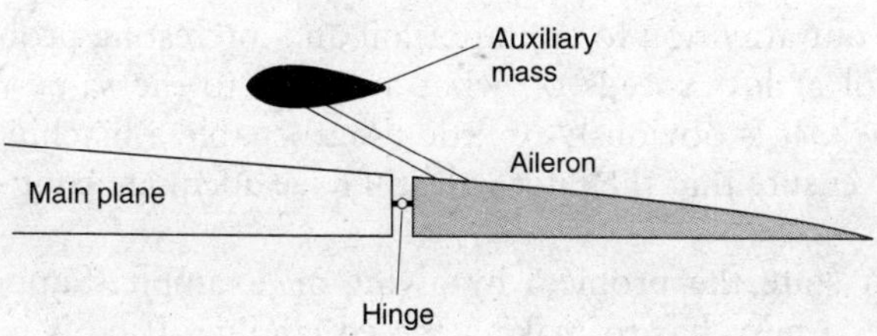

Fig 9.19 Mass balance

Fig 9I Tab control mechanism
(By courtesy of Piaggio, Genoa, Italy)

if we realise that mass balance is just as effective on a rudder, where the weight is not involved, as on an elevator or aileron.

On old military biplane aircraft, the exact distribution of mass on the control surfaces was so important that strict orders had to be introduced concerning the application of paint and dope to these surfaces. It is for this reason that the red, white and blue stripes which used to be painted on the rudders of Royal Air Force machines were removed (they were later restored, but only on the **fixed** fin), and why the circles on the wings were not allowed to overlap the ailerons. Rumour has it that when this order was first promulgated, some units in their eagerness to comply with the order, but ignorant as to its purpose, painted over the circles and stripes with further coats of dope!

Control at low speeds

We now turn our attention to an important and interesting problem – namely, that of control at low speeds or, what amounts to the same thing, at large angles of attack. It is obviously of little use to enable a machine to fly slowly unless we can ensure that the pilot will still have adequate control over it (Fig. 9J).

Let us first state the problem by giving an example. Suppose, owing to engine failure, a pilot has to make a forced landing. If he is inexperienced – and indeed it has been known to happen to pilots of considerable experience

Fig 9J Control at low speed
The massive bulk of the Antonov AN-124 in a tight turn at low speed and low altitude. This photograph was taken from the ground!

– he will often tend to stall his aeroplane in an attempt to reach a distant field or to climb over some obstacle. Now the use of slots or flaps may postpone the stall, may help him to obtain lift at slow speeds, but they will not give him what he most needs – namely, efficient control.

In the first place, owing to the decreased speed of the airflow over all the control surfaces, the forces acting on them will be less and they will feel 'sloppy'. But this is not all. Suppose while he is thus flying near the stalling angle he decides that he must turn to the left, he will move the control column over to the left (which will cause the right aileron to go down and the left one to go up), at the same time applying left rudder. The rudder will make a feeble effort to turn the aeroplane to the left; but what will be the effect of the movement of the ailerons?

The effect of the right aileron going down should be to increase the lift on the right wing, but in practice it may decrease it, since it may increase the angle of attack beyond that angle which gives the greatest lift. But what is quite certain is that the drag will be considerably increased on the right wing, so tending to pull the aeroplane round to the right. This yawing effect, caused by the ailerons, is present at nearly all angles of attack, but it becomes particularly marked near the stalling angle; it is called **aileron drag**.

Meanwhile, what of the left wing? The lift may either have decreased or increased according to the exact angle of attack, but in any case the change in lift will be small. The drag, on the other hand, will almost certainly have decreased as the aileron moved upwards. To sum it all up, the result of the pilot's attempt to turn to the left is that there may or may not be a slight tendency to roll into the left bank required for an ordinary left-hand turn, while at the same time the drag on the wings will produce a strong tendency to turn to the right which may completely overcome the rudder's efforts in the opposite direction (9.20). The conditions are, in fact, very favourable for a spin (both literally and metaphorically); the pilot could hardly have done better had he deliberately attempted to get the aeroplane into a spin.

So much for the problem. What solution can be found? We must endeavour to ensure that when the stalling angle is reached, or even exceeded, the movement of the controls by the pilot will cause the same effect on the aeroplane as in normal flight. The following improvements would all help to attain this end –

1. Increased turning effect from the rudder.
2. Down-going aileron should not increase the drag.
3. Up-going aileron should increase the drag.
4. Down-going aileron should increase the lift at all angles.
5. Up-going aileron should cause a loss of lift at all angles.

A large number of practical devices have been tried out in the attempt to satisfy these conditions; most of them have been partially successful, but none of them has solved the problem completely.

Let us consider a few of these and see to what extent they meet our requirements.

(*a*) The use of very large rudders with sufficient power to overcome the yawing effect of the ailerons in the wrong direction.

The disadvantage is that the size of the rudder required to obtain the desired result is excessive for normal flight. Also this seems to be a method

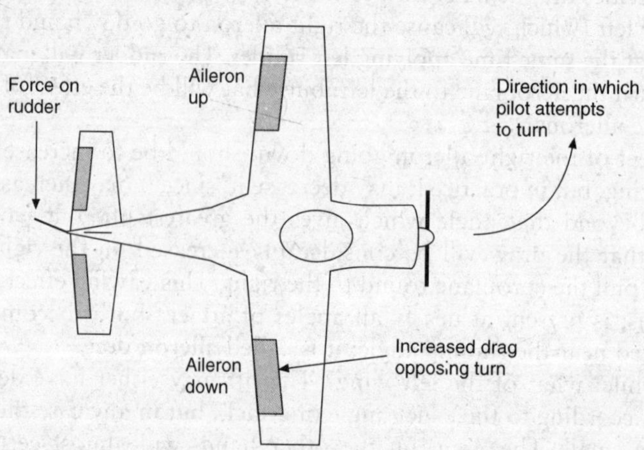

Fig 9.20 Result of an attempt to turn at large angles of attack

of tackling the problem from the wrong standpoint – instead of curing the disease, it allows the disease to remain while endeavouring to make the patient strong enough to withstand it.

(*b*) A **wash-out**, or decrease of the angle of incidence, towards the wing tips.

This will mean that when the centre portions of the wings are at their stalling angle, the outer portions are well below the angle, and therefore the aileron will function in the normal way. The defect of this arrangement is that the wash-out must be considerable to have any appreciable effect on the control, and the result will be a corresponding loss of lift from the outer portions of the wing in normal flight. The same effect can be obtained by rigging up the ailerons so that the trailing edge of the ailerons is above the trailing edge of the wing.

(*c*) **'Frise', or other specially shaped ailerons** (Fig. 9.21). This is a patented device, the idea being so to shape the aileron that when it is moved downwards the complete top surface of the main plane and the aileron will have a smooth, uninterrupted contour causing very little drag, but when it is moved upwards the aileron, which is of the balanced variety, will project below the bottom surface of the main plane and cause excessive drag. This method has the great advantage of being simple, and it undoubtedly serves to decrease the bad yawing effect of the ailerons, and therefore it is often used. Unfortunately, its effects are not drastic enough.

(*d*) **Differential ailerons** (Fig. 9.22). Here, again, is a delightfully simple device suffering only from the same defect that, although it provides a step in the right direction, it does not go far enough to satisfy our needs. Instead of the two ailerons moving equally up and down, a simple mechanical arrangement of the controls causes the aileron which moves upwards to move through a larger angle than the aileron which moves downwards, the idea being to increase the drag and decrease the lift on the wing with the up-going aileron, while at the same time the down-going aileron, owing to its smaller movement, will not cause excessive drag.

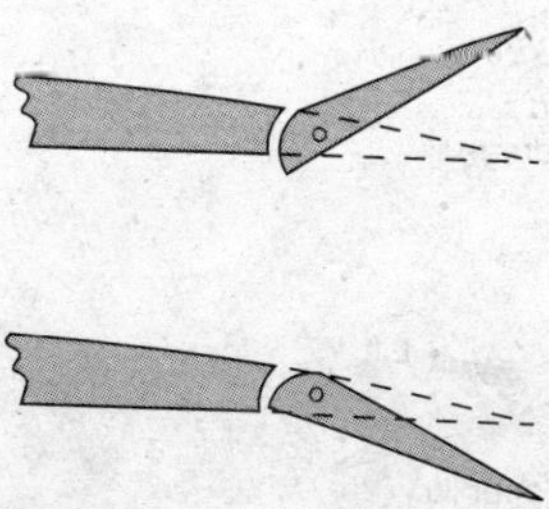

Fig 9.21 Frise ailerons

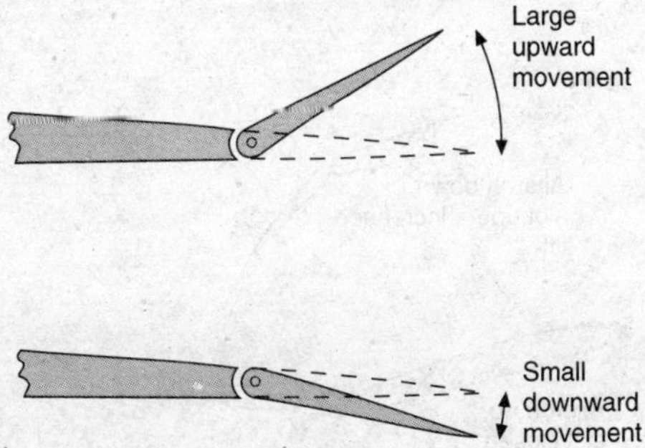

Fig 9.22 Differential ailerons

(*e*) **Slot-cum-aileron control** (Fig. 9.23). The slots, which need only be at the outer portions of the wings in front of the ailerons, may be of the automatic type, or the slot may be interconnected to the aileron in such a way that when the aileron is lowered the slot is opened, while when the aileron is raised, or in its neutral position, the slot is closed. By this means the down-going aileron will certainly serve to increase the lift for several degrees beyond the stalling angle, nor will the drag on this wing become very large since the open slot will lessen the formation of eddies. We shall therefore obtain a greater tendency to roll in the right direction and less tendency to yaw in the wrong direction. This is exactly what is required, and the system proved to be very effective in practice.

(*f*) **Spoiler control** (Fig. 9.24). Spoilers are long narrow plates normally fitted to the upper surface of the wing though they may occasionally be fitted below as well. In the ordinary way they lie flush with the surface, or even inside it, and have no effect on the performance of the aerofoil, but they can be connected to the aileron controls in such a way that when an aileron is moved up beyond a certain angle the spoiler is raised at a large angle to the airflow, or comes up through a slit, causing turbulence, decrease in lift and increase in drag. This means that the wing on which the aileron goes down gets more lift, and very little extra drag, while on the other wing the lift is 'spoilt' and the drag greatly increased. Thus we have a large rolling effect in the right direction combined with a yawing effect, also in the right direction – just what the doctor ordered.

This is what we aimed at, and there is the further advantage that the mechanical operation of the spoiler is easy, since the forces acting upon it are small. This method of control feels strange to the pilot who is unaccustomed to it because the loss of lift caused by the spoiler will result in a decided drop of that wing, which may be alarming when near the ground. But any such strangeness can soon be overcome and the pilot begins to realise the advantages of maintaining good lateral control, up to and beyond the normal stalling angle. The improvement in manoeuvrability is particularly noticeable

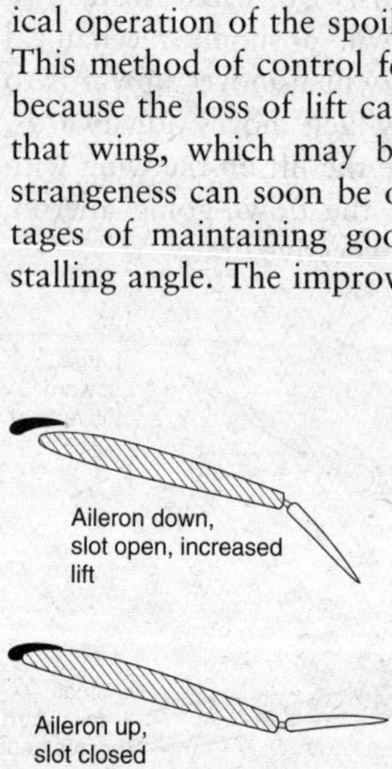

Fig 9.23 Slot-cum-aileron control

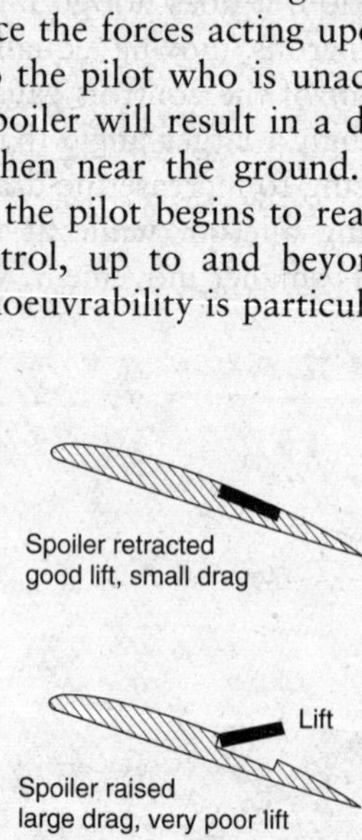

Fig 9.24 Spoiler control

when the aeroplane approaches its ceiling. But, whatever its merits, the spoiler took a long time to become popular as a means of control, though it was, and is, used extensively as an air brake.

It is rather curious that we have been describing the use of spoilers as an aid to lateral control at low speeds; and this indeed was their original purpose, but in many types of modern aircraft it is **at high speed** that the aileron control may result in undesirable characteristics caused by compressibility as discussed in Chapter 11. Modern airliners use a complicated arrangement of spoilers combined with more than one set of ailerons. The control system normally automatically selects the correct combination according to flight speed.

It may be noticed that the elevator control has not been mentioned in dealing with this problem; the elevators usually remain fairly efficient, even at low speeds, since the angle of attack of the tail plane is less than that of the main planes, and therefore there is not the same tendency to stall as with the ailerons. However, on high-speed aircraft, the tail plane has to be able to compensate for the rearward movement of the centre of lift of the wing and it is quite common nowadays for the whole tail plane to be movable in addition to having a hinged elevator. On propeller-driven aircraft, the extra speed of the slipstream normally adds to the effectiveness of both rudder and elevators.

Before leaving the subject of control it should be mentioned that large amounts of sweepback, and even more delta-shaped wings, cause control problems of their own, but since wings of these shapes are nearly always on high-speed aircraft, their consideration will be left to a later chapter.

Powered controls

Nowadays all but small aircraft are usually fitted with power-actuated control surfaces which are very easy to operate even on large airliners. Because such controls offer virtually no natural resistance, they are given some form of artificial 'feel', a resistance which is designed to increase with flight speed (or more precisely with dynamic pressure) so that the control system feels like a direct mechanical linkage. Without such feel it would be quite literally possible to pull the control surfaces off at high speed.

The control column often incorporates a 'stick shaker' which operates when the aircraft approaches a stalled condition. This reproduces the shaking that normally occurs on simple mechanical systems due to the buffeting of the control surfaces caused by turbulence. It is intended to trigger the pilot's conditioned response. On some aircraft, if the pilot fails to respond correctly by pushing the column forward, the controls take over and a 'stick pusher' does the job for him! With the advent of advanced and reliable electronic devices, it has become possible to make control systems of immense complexity that respond smoothly over a very wide range of flight conditions and contain many built-in safety features. In order to prevent loss of control in the event of power failure the systems are usually duplicated, triplicated or even quadrupled.

Dynamic stability

In this chapter we have dealt mainly with what is known as static stability. There are other forms of instability that the designer (and the pilot) has to cope with. These take the form of oscillations or deviations from the desired flight path that vary with time, and are known as dynamic instabilities. They can be a nuisance, nauseating or downright dangerous if left unchecked. An example is the spiral instability referred to earlier in this chapter.

A full description of dynamic stability is beyond the scope of this book, but the reader should be aware that the problems of stability and control are far more complex than the simple outline given here.

Can you answer these?

Questions on stability and control –

1. What would be the characteristics of an aircraft with extreme directional stability and little lateral stability?
2. What would be the characteristics of an aircraft with extreme lateral stability and little directional stability?
3. What is the object of balancing controls?
4. Distinguish between 'mass' balance and 'aerodynamic' balance.
5. Explain the difficulty of obtaining satisfactory lateral control at large angles of attack.
6. Describe some of the methods which have been tried with a view to overcoming this difficulty.
7. Explain why spoilers are sometimes used at low speeds, and sometimes at high speeds, as an aid to lateral control.

For solutions see Appendix 5.

CHAPTER 10

A trial flight

In the preceding chapters I have tried to explain the principles on which the flight of an aeroplane depends. However, flying an aeroplane is a different thing altogether and in this chapter we will explore what it is actually like to fly a light, low-speed aircraft of the sort that most pilots use when they start their training. During the flight we will use the knowledge we have gained previously to explain what is happening.

The aircraft we will fly is shown in Fig. 10A. It is a simple aircraft with fixed undercarriage and a fixed pitch propeller driven by a piston engine. Let us have a quick look in the cockpit to see the instruments and controls (Fig. 10B, overleaf). First we notice the main flying instruments – an air-speed indicator (ASI, calibrated in knots), an altimeter (calibrated in ft), a rate of climb

Fig 10A Cessna 152, the aircraft for our flight

indicator (ft/min), and a turn and slip indicator (Fig. 8.6). Then we have an artificial horizon and a direction indicator (DI). Finally, tucked away at the top of the windshield, we have a small magnetic compass.

Next we come to the engine and fuel instruments; the tachometer, which measures the engine speed, is situated on the right hand side of the cockpit, then comes the ammeter (which shows if our electrical generator is working properly) and next to this is a suction gauge. The suction gauge is needed because the artificial horizon and the DI are gyroscopic instruments, which derive their power from engine-supplied suction. If the suction drops below a certain level, these instruments will become unreliable as the gyroscopes slow down and we need warning if this is likely to happen! Under the cluster of main flight instruments we find the fuel gauges (one for each wing tank) and indicators for engine temperature and oil pressure.

The flying controls are quite simple. In front of us is the main yoke. We push this forward to lower the elevators and nose of the aircraft, or pull to raise the nose. We turn the 'wheel' to operate the ailerons and bank the aircraft to left or right. At our feet are the pedals, which are used to operate the rudder, and above the pedals the toe-operated differential brakes are situated. On a quadrant, in front of us, there is the flap lever and position indicator. On this aircraft the flaps are electrically operated and so they will not work until the electrical supply is turned on.

Fig 10B The cockpit

The engine controls are quite simple. A push-pull throttle control is provided on the dashboard – push forward to increase engine revs. Apart from this there is a mixture control to compensate for the reduced air density as we climb and a carburettor heat control which we use, particularly at reduced engine speeds, to prevent the carburettor icing up – possibly with disastrous results! As we will only be flying at or below 3000 ft, we will leave the mixture at rich until we eventually want to stop the engine.

To complete the story we need to take a quick look at the navigational equipment. On this aircraft we have a communications radio, a navigational radio (which we shall not use, as we are flying in good visibility and not far from the airfield) and a transponder to allow Air Traffic Control to track our position. Under the fuel gauges and engine instruments is the key-operated switch for the magnetos and engine starter (rather like the ignition switch on a car), together with the light switches and pitot heater switch. Like the carburettor heater, this prevents the formation of ice but this time on the pitot-static tube, which is vital if we are going to know our correct airspeed. There is also a master switch, which isolates all the electrical systems apart from the magnetos. These provide the spark at the spark plugs and keep the engine running. If there is an electrical fault and we have to switch off the master, at least the engine will not stop!

So far we have spent a long time and the aircraft has not even moved yet. Well, that's flying! If something goes wrong in a car you can always pull off the road to sort it out. With an aeroplane we have to make sure everything is OK before we start off.

Pre-flight checks

Again, unlike a car, we cannot just throw our cases into the back and hope for the best. The basic aircraft is carefully weighed so that the weight and position of the centre of gravity are both known. We have to worry about the moveable weights and do weight and balance calculations for the loads we are carrying on this flight. Every aircraft has a flight manual and it is this that will tell us if our loading is within safe limits for the aircraft. We are not going far – we will be in uncontrolled airspace and visibility and forecast are good, so we will not need to file a flight plan. It is still a good idea to plan what we intend to do, though. Runway in current use is 22L (the 22 means the heading is about 220^0 on the DI), so we will take off southwest towards the city, turn left towards the east before we get to the built-up area and try some turns, stalls and a few other manoeuvres. We will then rejoin the circuit and land – a total flight time of perhaps 50 minutes.

Now for our detailed pre-flight checks. We first inspect the paperwork to make sure our particular aircraft has up-to-date service records, insurance, etc. Taking our checklist, we then go to the aircraft. First we switch off the radios and then switch on the master switch and the switches for all the systems we are going to test – the flashing anti-collision beacon, the navigation lights, the landing lights and the pitot heater. We walk round the aircraft, checking that all these systems are working, then we return to the cockpit and test the flaps before switching everything off again.

Now we walk slowly round the aircraft, checking such things as the condition of the tyres and brakes, the condition of the surfaces, the control hinges, etc. We also make sure that, when we waggle the controls, the main yoke moves in the correct direction – misrigging has been known! As we come to the leading edge of the starboard wing we notice a small hole. This is the stall-warning device. As an aerofoil approaches the stall the suction at the leading edge gets stronger. The stall warning device sounds a buzzer in the cockpit when the aerofoil is dangerously close to the stall. We test that this is working by sucking gently at the hole. Now we are at the nose we check the engine oil level and the condition of the propeller. We must also check the fuel. First we take samples to make sure it has not been contaminated by water and then we clamber on to a step on the strut of each wing in turn to dip the tank and make sure we have enough fuel for the flight – the electrical gauges may fail and there are no filling stations at 3000 ft. Clearly the last pilot filled it up before he parked it, so we do not have to worry about a trip to the pumps.

Nearly there now! We get into the cockpit, plug our headsets in ready for use and check that the instruments and controls are all in good condition and the first aid and fire extinguisher are to hand and in date. Then we adjust the seats and strap ourselves in. Time to start the engine. Fuel on, three strokes of the primer as the engine is cold, mixture rich, carburettor heat cold, throttle to idle, master switch and anti-collision beacon on, check that no circuit breakers have popped out, open the window and shout, 'Clear prop' as a warning to the unwary and then turn the key and slowly push the throttle forward. As the engine starts we check that the starter motor has disengaged, that the oil pressure is rising satisfactorily, that the ammeter is reading satisfactorily and that the vacuum gauge is reading in the green area. Now we set the engine speed to 1200 rpm, put on our headsets, make sure the intercom is switched on and that we can hear each other.

A few checks still to do. I told you flying was a thorough business. Magnetos are duplicated for safety, so we make sure that the engine will run happily on either one before returning the switch to both on. Now we set the DI to agree with the magnetic compass and the altimeter to read the approximate height of the airfield (200 ft). Then if we do forget to set it later on it will not be too far out. In this flight we will be measuring altitude with respect to mean sea level (known as QNH). The engine temperature now looks OK and it is time to contact the tower. We are told that runway 22 is in use, that the surface wind is 8 knots from 270° and that the QNH is 1009.0 millibars (remember them?). We repeat back the information, check the altimeter setting and now we are ready to taxi to the runway.

Taxying

Our aircraft is parked into wind. We pull back on the yoke to lighten the load on the nose wheel and push the throttle forward until we start to move. More checks now. Do the brakes work? We throttle back and try them, repeating for both left hand and right hand control positions. Now we park into wind for the power checks. First we check that oil pressure and temperature are OK. Next set 1700 rpm and make sure these remain within limits and another check on the magnetos before making sure the engine does not quit when we set it to idle.

Now we are almost ready. We make a final check that all switches are set correctly, the flaps are up, mixture rich and carburettor heat off and another quick check to see the controls move correctly as we pull back for the elevators and turn the wheel for the ailerons. Now, at last, we can taxi to the holding point on the runway. As we taxi we position the ailerons to keep the into-wind wing down and the elevators set to reduce the load on the nose wheel as much as possible. At the holding point we stop to tell the tower that we are about to line up on runway 22L, look for any aircraft on the approach, release the brakes and line up on the runway centre line.

Take off

Engine to 2000 rpm and a quick final check on engine and instruments, release brakes and throttle fully forward. A quick glance to make sure the ASI registers the increasing speed and, as the aircraft gathers speed, we use the pedals to counter any swing and keep the nose pointing down the runway. The controls begin to become more effective as the speed increases, so we ease the ailerons and the backpressure on the elevators. As the speed reaches 55 knots, we pull back gently and now we are climbing away at full throttle and adjust the elevators to give a steady climbing speed of 60 knots, close to the speed that gives us the best angle of climb, to clear any obstacles.

Once things have settled down, to remove the force on the yoke, we trim, using the trim wheel that operates a small tab on the right hand elevator, and wait until the altimeter shows that we are 500 ft above the airfield. Then we push forward slightly and settle to 70 knots (the speed for best rate of climb) and retrim, keeping the horizon at a constant position in the windscreen. It might seem odd to increase the rate of climb by lowering the nose and increasing the speed but a quick glance at the vertical speed indicator shows that we are, in fact, climbing faster than before. It will be a good idea to look again at Chapter 7, when we get down, to work out the reason why.

Some simple exercises

Now we are properly airborne, it is time to leave the airfield and try a few manoeuvres. We start by banking about 10° to the left to make a gentle turn on to a course of 90°. We are now approaching 3000 ft and need to level off into straight and level flight. Nose down, let the speed build and, as our cruising speed of 90 knots approaches, we gently throttle back to the cruise engine setting of 2300 rpm. Finally, when things are nicely settled, we use the trim to remove the control force and find the aircraft flies quite happily 'hands off'. Every 15 minutes or so we do the 'FREDA' checks to make sure that all is well:

Fuel – have we got enough?

Radios – on correct frequencies?

Engine – temperature and pressures OK? Pull out the carburettor heat to the on position and check for a small fall in engine speed and a return to the previous speed when the carburettor heat is turned off. This check ensures that we haven't inadvertently started icing up the engine.

Direction finder – does it read the same as the magnetic compass? (The DF is a gyroscopic instrument and tends to drift slightly with time).

Altimeter – is it set correctly and are we at the correct height?

Checks complete. Now to try a few manoeuvres. First we pull back slightly. After an initial climb, the aircraft settles to a lower speed of 86 knots. We can reset the throttle to remove the subsequent tendency to slowly gain height but as we were flying fairly near the minimum power setting, not much movement is necessary. To speed up we do the opposite; yoke forward and increase throttle slightly to prevent sink. Unlike a car, we do not use the throttle as the speed control, although it may need some slight adjustment. The main thing is to control the angle of attack with the elevators – pull back to slow up and push forward to increase speed.

Now let's try a descent to 2500 ft. Throttle back to 1500 rpm. When we do this we have to remember that carburettor icing can be a real danger, even on a fairly hot day. We don't want the engine quitting on us; we have only got one! So we pull on the carburettor heat control, which directs the engine air over the hot exhaust manifold, before it enters the carburettor. We need some slight control input to cope with the change in engine torque and the change in pitching moment, due to its offset thrust line. When we have settled into a steady descent (a descent rate of about 500 ft/min on the rate of climb indicator) we can again trim the aircraft, so that it will fly 'hands off' in the steady descent. In spite of these other small effects, the main effect of the power change is that the aircraft settles into a steady descent. As 2500 ft approaches, all we need to do is to pull back slightly to level out and return the throttle to the cruise setting (don't forget to turn off the carburettor heat) and finally retrim in level flight.

Now climb back to 3000 ft. Full power first – we don't want to lose speed. Pull back slightly until the speed falls to 70 knots (best rate of climb). Trim when things have settled. As we reach 3000 ft, push forward into level flight. Then, as we reach the cruising speed of 90 knots retrim when things have settled. Quite simple in a well-behaved aircraft like this!

Next a stall. Throttle back to idle (carburettor heat to on). Gradually pull back so that no height is lost. Notice as the speed reduces that the controls start to have a 'soggy' feel. As the speed drops off and we continue to raise the nose the increased angle of attack causes the suction peak on the top of the wing to move towards the leading edge. Eventually we here a buzz as it triggers the stall warning device which detects the pressure near the wing leading edge. When we hear this, we are only a few knots above stalling speed. Normally we would push the yoke forward to reduce the angle of attack (yes, even if we are near the ground!). Now, however, we press on raising the nose, as we wish to experience a stall. We are now down to 45 knots, and, in spite of our efforts to raise the nose, it drops and we have stalled. We entered the stall flying straight and level, so nothing very exciting happens. To recover; push the yoke forward, turn carburettor heat off and select full throttle. As the speed builds up to the cruising speed we pull back to level out and reduce the throttle to the cruise setting.

We tried a gentle turn when we were leaving the airfield circuit in a climb. Now we will try to do some level turns. As with most flying, the secret is to keep your eyes on the horizon. We check for other aircraft from right to left (the direction in which we are going to turn), bank gently to 10° (indicated by the scale outside the artificial horizon) and try to keep the line of the horizon constant in the windshield. Note that, as we get to the required bank angle, we have to remove the input to the ailerons and even reverse them slightly to hold the correct angle. This is because the wing on the outside of the turn has a slightly higher airspeed, so the centre of lift moves slightly towards this wing, promoting a steeper bank than was aimed for. However, the correction is made instinctively and careful observation is needed to see what has happened.

A glance at the altimeter shows a slight descent, so a little backpressure to bring the nose up. (Note that the rate of climb indicator has a response that is too slow to be of use to us here). A glance at the ball in the turn and slip indicator shows a small degree of sideslip. Because the wing symmetry has been disturbed by the increase in speed of the outer wing, there is a slight tendency for the aircraft to yaw away from the turn due to the asymmetric vortex drag. This is countered by a bit of left rudder. It sounds simple but things can get a bit confusing with an over-correction sending the sideslip the other way. Thankfully a simple rule resolves the problem; if the ball is on the right, add right rudder and *vice versa*.

Now let's be a bit more adventurous and try a steep (60^0 bank angle) turn. This seems a lot more hairy as we are pulling a load factor of 2 (remember how the load factor goes up in a turn). The principles are the same but a lot more difficult to control under these circumstances. Additionally we will need to add some power or we will not be able to sustain speed and height. As we know, the power required goes up in a turn. In the shallower turn we were able to leave the throttle as it was and trade off a bit of speed (as we were above the speed for minimum power). Now we need a bit more from the engine to keep us going round at constant height. A check for other aircraft and round we go. The increased load on our bodies makes it quite difficult and, even when we have established a nice level turn, it is not easy to roll out on exactly the right heading and throttle back to re-establish straight and level flight.

Circuit, approach and landing

Now it's time to make our way back. We are to the north east of the airfield, after all our manoeuvrings, and at present flying west. Away to our left we can see the trunk road that passes just north of the airfield. We will steer to intercept it and cruise back to the airfield, keeping just to the right of the road (the rule is keep to the right when following such a feature). We will also gently descend to 2000 ft.

Before we get too close to the airfield, let's plan carefully what we are going to do, while we have time to think and there is no other traffic to avoid. When

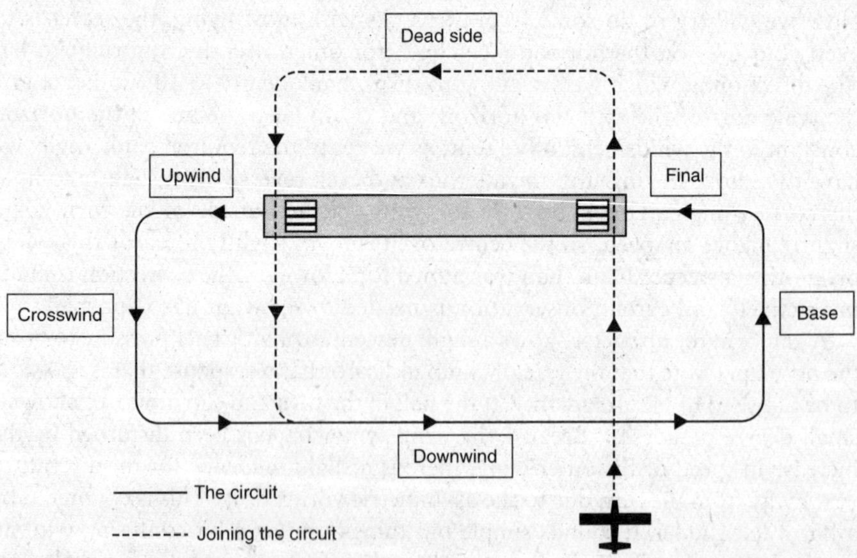

Fig 10.1 Joining the circuit for a landing

we get close to the airfield we will be very busy! Runway 22R is in use, so we will need to overfly the circuit (Fig. 10.1), before joining it for a landing. The circuit height is 1000 ft above the runway, so we will overfly the northern end of the runway, crossing above the runway numbers 500 ft higher than the circuit height. Remember, our altimeter is set to QNH (measuring altitude above sea level) so we will need to add the elevation of the airfield to give a reading of 1700 ft. This means we will cross the end of the runway well out of the way of any aircraft landing or taking off and, after this, we will turn parallel to the runway, on the opposite side to the circuit (the 'dead side') and descend to circuit height. Then, we can re-cross the runway, at the other end and, if the circuit is clear, join it. Sounds complicated but this procedure will keep us clear of other aircraft in a relatively crowded airspace. With any luck Fig. 10.1 will make things a bit clearer. Now we have sorted things out in our minds and the airfield is just a few miles away; we give them a call to say that we are intending to rejoin the circuit.

Now to put into practice all that we have just planned. First we throttle back slightly, get our altitude down to 1700 ft and aim to cross the runway at right angles at its northern end. We cross at 1700 ft and soon it is time to turn left so that we are flying parallel to the runway on the 'dead side' (Fig. 10.1). As we reach the circuit height of 1200 ft, we level off by applying slightly more throttle and then turn to cross the other end of the runway. Another left turn finally brings us into the circuit and we call 'downwind' to let everybody know where we are and run through the landing checklist to make sure everything is prepared.

Now things start to get really busy! We are well past the end of the runway so we make another left turn on to 'base leg' and immediately throttle back to 1500 rpm, put carburettor heat on (we don't want the engine to quit on us now!) and, at the same time, pull back on the yoke so we lose speed but keep altitude. As the airspeed falls, we see that we are just entering the 'white zone' on the airspeed indicator, which means it is now safe to lower the flaps. The flaps are extended by two stages on the base leg. As they extend there is a marked change in the trim and it becomes necessary to push forward in order to keep the nose down. When the flaps are extended, the speed is still reducing and, as it approaches 70 knots, we adjust the forward pressure on the yoke to keep it at this value. Now we are in a steady descent at 70 knots and we can trim the aircraft so that it will fly in this condition by itself.

While this is happening we need to keep a close check on our height and where the airfield is. Eventually we turn towards the runway for our final approach and landing. As we do so we lower the final stage of flap. This time there is very little trim change but because of the extra drag, the speed falls to 65 knots. We call 'final' to inform other traffic and the airfield and concentrate on aiming at the numbers at the beginning of the runway but keeping strictly to 65 knots. If our speed drops we must push forward and, if the rate of descent seems too high, we apply a bit more throttle. When we increase the throttle, note how we have to adjust to keep the nose pointing in the right direction. Because of the offset thrust line, an increase in throttle setting tends to pitch the nose up and also roll the aircraft because of the extra torque.

As we approach, we find a tendency to drift off to the left. Remember that crosswind when we took off? We must compensate by rolling slightly right and keeping the nose pointing down the runway by putting in a bit of opposite rudder to induce a slight sideslip. Now we are nicely on the approach but we must be alert to correct for anything which might disturb the aircraft, such as the wind over the group of trees, which causes us to sink too rapidly as we pass overhead.

At last we arrive at the runway threshold, a few feet above the ground. Now we can stop worrying about the speed. We set the throttle to idle and pull back into the 'flare' until we are flying just above the runway. We now need to change where we are looking so that we can judge height and pull back as the speed falls off to keep flying parallel to the runway. Slightly too much pull back and we find ourselves a bit too high. So we freeze on the yoke (if we push forward we shall probably overcorrect!) until we sink to the proper height again. Eventually the stall warning goes off, the yoke is right back and we sink gracefully on to the main wheels. As the nose wheel touches, we gently apply brakes and, when the speed is low enough, pull off the runway to the left to get out of the way of any following aircraft. Next we come to a halt and clean the aircraft up – flaps up, carburettor heat to cold, lights off apart from the anti-collision beacon, radios off apart from the communications (we will want to know what is happening until we have actually parked the aircraft).

When this is done we open the throttle and taxi back to our parking space. Finally, we let the engine run at 1200 rpm for half a minute, so it's not too cold after all that idling, check that both magnetos are still working, put the throttle to idle and pull out the mixture to lean. This finally stops the engine and we turn off the master switch and remove the keys and all our belongings from the aircraft to finally complete our flight.

Now we have got down in one piece, you might like to try the numerical questions in Appendix 3. They are all connected with our flight but you will need to look at the other chapters before you will be able to tackle them.

CHAPTER 11

Flight at transonic speeds

Introduction

The earlier chapters of this book have dealt with flight at speeds well below that at which sound travels in air – in short, at subsonic speeds. It is true that frequent references have been made to the problems of high-speed flight, but detailed consideration of these problems has been deferred until now.

From what has already been said, it is clear that the speed of sound has an important influence on the flow, and this is what we examine first.

The speed of sound

When a body moves through the air at speeds well below that at which sound travels in air, there is, as it were, a message sent ahead of the body to say that it is coming. When this message is received, the air streams begin to divide to make way for the body, and there is very little, if any, change in the density of the air as it flows past the body. The way in which air is thus 'warned' of the presence of the body which is approaching, or of changes in the shape or attitude of that body, can be clearly illustrated in a smoke tunnel in which the air flows over an aerofoil fitted with a flap (Fig. 11.1, overleaf). If the smoke streams are allowed to settle when the flap is in the closed position, and the flap is then lowered, all the streams of smoke many chords length in front of the aerofoil immediately change course, and streams which previously flowed below the aerofoil, now flow above it. Once again, if we could 'see the air' in front of an approaching aeroplane, this fact would be immediately obvious, and the air would begin to be disturbed perhaps 100 metres or more in front of the aeroplane of the approach of which it must have had some warning.

What we have called a 'message' or 'warning' is really due to a wave-motion in the air set up by the areas of increased and decreased pressures

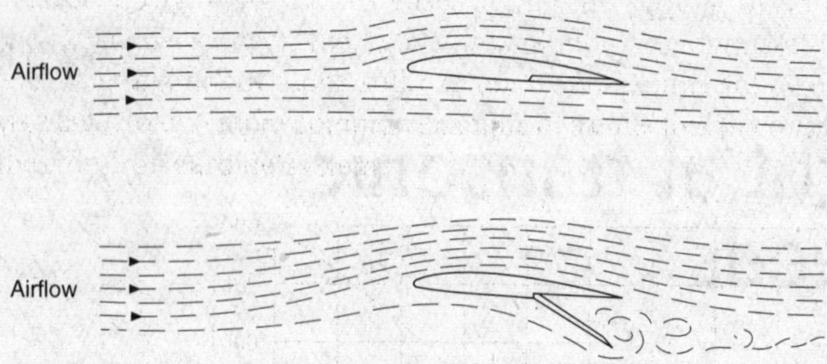

Fig 11.1 Effect of lowering flap on airflow in front of aerofoil

around the body. These pressures are communicated in all directions to the surrounding air by means of 'waves'. These waves are similar to sound waves, and they travel at the speed of sound, which is about 340 m/s (or 1224 km/h) in air at sea-level conditions. There is no mystery in this relationship between pressure waves and sound waves because sound is a pressure wave set up by some local compression of the air, and the speed of sound is simply the speed of propagation of rarefactions and compressions of small amplitude in the air.

So if a body travels through the air at the speed of sound, there will be no time for the message as given by the wave to get ahead and the air will come up against the body with a 'shock'.

Compressibility and incompressibility

Early chapters emphasised that air is compressible; it was also emphasised, though perhaps not emphatically enough, that though it is compressible it does, in fact, behave at ordinary speeds almost as though it were incompressible. Of course such an assumption is not true, air is really compressible or, what is sometimes more important, expandable at all speeds and the density does change, i.e. increases or decreases, as the wings and bodies of aeroplanes move through air at quite ordinary speeds, but the point is that the error in making the assumption is so small as to be negligible, while the simplification that the assumption gives to the whole subject is by no means negligible.

As we approach the speed of sound the error in making this assumption of incompressibility can no longer be justified, the air is definitely compressed, or expanded. We are now dealing with a compressible and expandable fluid.

It should be clear from this that the change is gradual, not sudden; it is all a question of deciding when the error becomes appreciable, and a rough idea of the error involved may be obtained from Table 11.1, which represent the error in assuming the ordinary laws of aerodynamics when estimating the drag of a body moving through air at the speeds mentioned.

It will be sufficiently clear from this that we must begin to change our ideas at speeds considerably lower than 340 m/s.

Table 11.1 Error introduced from assuming air is incompressible

Speed			*Error in assuming incompressibility*
m/s	*knots*	*km/h*	
45	87	161	About $\frac{1}{2}$%
90	175	322	Less than 2%
134	260	483	4%
179	347	644	7%
224	436	805	11%
268	522	966	16%

Approaching the speed of sound

Consider a point A sending out pressure waves. If the point is not moving, successive waves 1, 2, and 3 will form circles round the point just like the waves which radiate from a point on the surface of water when a stone is dropped into the water at that point (Fig. 11.2*a*). Now suppose that the point A is moving in the direction shown, but at a speed less than the speed of propagation of the waves, and that it sends out the wave 1 when it is A1, 2 at A2, 3 at A3; the picture will now take the form of Fig. 11.2*b*. Now increase the speed of A until it is the same as the speed of the propagation of the waves (Fig. 11.2*c*), and we soon see why the air ahead gets no warning that A is approaching – we get a pretty good idea too how the waves pile up and what causes the 'shock'.

Some time ago this 'shock', which occurred before the speed of sound was actually reached, and which in some instances had rather alarming effects on the flight of the aeroplane, was considered as something to be avoided at all costs; and the rapid increase in drag seemed a formidable enough 'barrier' at that time, if only because there seemed little hope of providing engine power to match it.

Thus it was that the transonic range of speeds, and especially the approach to the speed of sound, came to acquire rather a bad reputation in the minds of both designers and pilots, and the difficulty in solving the problems was all the more frustrating because we knew, from experience of the flight of bullets and shells, that flight beyond the speed of sound, flight at supersonic speeds, was not only possible but apparently free from some of the troubles of transonic flight.

To-day, of course, we know a great deal more about flight at transonic speeds, the 'shock' and the 'barrier' are no longer obstacles to be avoided but rather to be got over, or through, and the supersonic region is within the reach of aeroplanes as well as bullets and shells; but with all our increase of

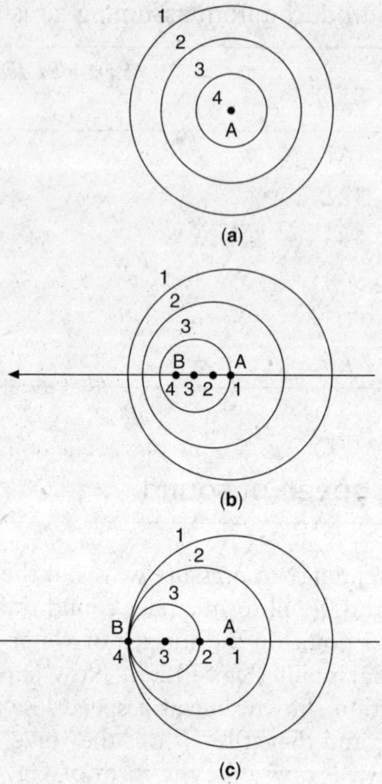

Fig 11.2 Propagation of pressure waves
(*a*) Point stationary – sending out pressure waves 1, 2, 3, 4.
(*b*) Point moving below speed of sound.
(*c*) Point moving at the speed of sound.

knowledge, it is interesting to note that flight at transonic speeds still presents special problems of its own. That is why it deserves a whole chapter to itself.

Shock waves

Let us see if we can find out a little more of what actually happens during the change from incompressible flow to compressible flow, and so discover the cause of the mounting error in making the assumption of incompressibility. Let us also investigate the 'shock', together with its cause and effects.

As the speed of airflow over say a streamline body increases, the first indication that a change in the nature of the flow is taking place would seem to be a breakaway of the airflow from the surface of the body, usually some way back, setting up a turbulent wake (Fig. 11.3). This may occur at speeds less than half that of sound and has already been dealt with when considering the boundary layer. It will, of course, cause an increase of drag over and above that which is expected at the particular speed as reckoned on the speed-squared law.

As the speed increases still further, the point of breakaway, or separation point tends to creep forward, resulting in thicker turbulent wake starting forward of the trailing edge.

This happens because, when we reach about three-quarters of the speed of sound, a new phenomenon appears in the form of an incipient shock wave (Figs 11.4 and 11A, overleaf). This can be represented by a line approximately at right angles to the surface of the body and signifying a sudden rise in pressure and density of the air, thus holding up the airflow and causing a decrease of speed of flow. There is a tendency for the breakaway and turbulent wake to start from the point where the shock wave meets the surface which is usually at or near the point of maximum camber, i.e. where the speed of airflow is greatest.

Observation of shock waves

The understanding of shock waves is so important to the understanding of the problems of high-speed flight that it is worth going to a lot of trouble to learn as much as we can about them.

As we remarked so often in dealing with flight at subsonic speeds, it would all be so much easier to understand if we could see the air. Well, fortunately we almost can see shock waves; they are not merely imaginary lines, lines drawn on diagrams just to illustrate something which isn't actually there, they

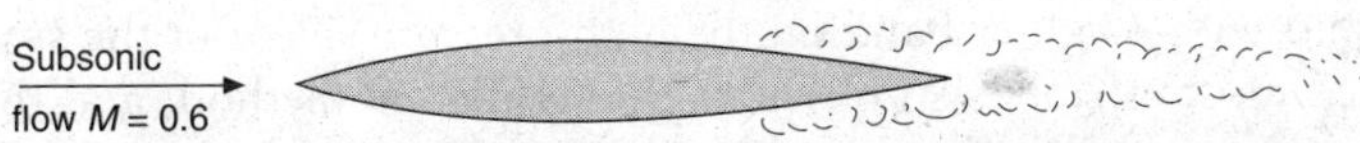

Fig 11.3 Breakaway of airflow

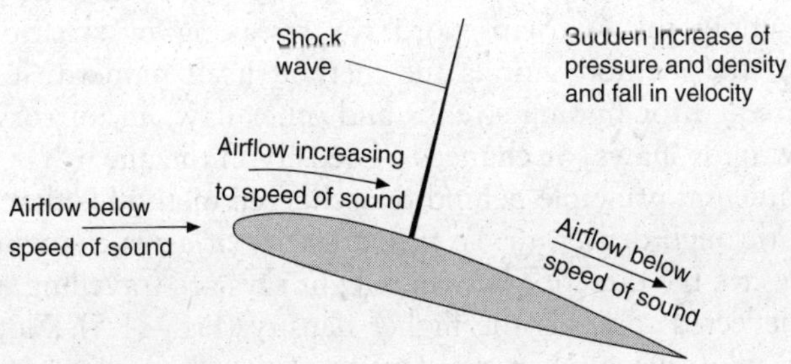

Fig 11.4 Incipient shock wave

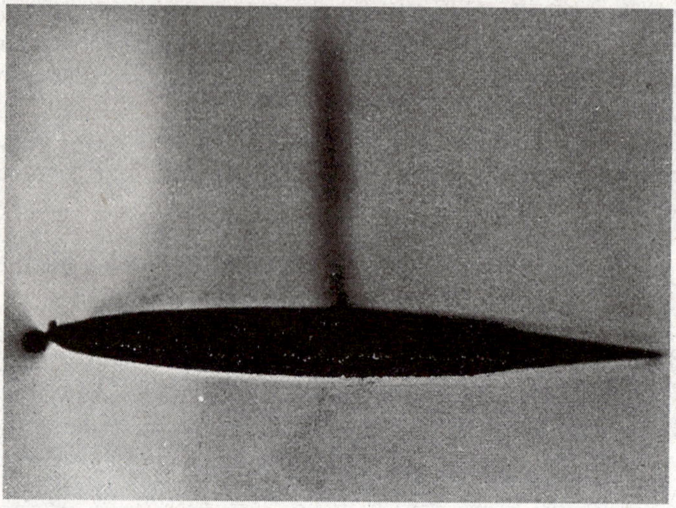

Fig 11A Incipient shock wave
(By courtesy of the Shell Petroleum Co Ltd)
An incipient shock wave (taken by schlieren photography) has formed on the upper surface; the light areas near the leading edge are expansion regions, separated by the stagnation area which appears as a dark blob at the nose.

are physical phenomena which can be photographed in the laboratory by suitable optical means, and in certain conditions they may even be visible to the naked eye.

The photographic methods depend on the fact that rays of light are bent if there are changes of density in their path. In the 'direct shadow' method the shock wave appears on a ground-glass screen, or on a photographic plate, as a dark band with a lighter band on the high-pressure side. But the system that has proved most effective is known as the **'schlieren'** method, and the results obtained by it are so illuminating, and have contributed so much to our understanding of the subject, that the method itself deserves a brief description. The word 'schlieren', by the way, is not the name of some German or Austrian scientist, but simply the German word for streaking or striation, which is descriptive of the method; nor is the method itself modern, it was used a hundred years ago for finding streaks and other flaws in mirrors and lenses, just as it now finds 'flaws' or changes of density in the air.

The fundamental principle behind the schlieren method is that light travels more slowly through denser air; so if the density of air is changing across, or at right angles to, the direction in which light rays are travelling, the rays will be bent or deflected towards the higher density (Fig. 11.5). Notice that the bending only takes place when the density of the air is changing (across the path of the rays) – there is no bending when the rays pass through air density which is constant across their path, nor if the density changes along their path; in those cases the rays are merely slowed up by high density.

Figure 11.6 (overleaf) shows a typical arrangement of mirrors and lenses as used in the schlieren method. From a light source A the rays pass through a lens B, to a concave mirror C, which reflects parallel rays through the glass walls of a wind tunnel to another concave mirror D, which in turn reflects the rays on to a knife edge at E, where an image of the light source is formed. Rays that pass through changing density near the model in the tunnel are bent, those passing through falling density being deflected one way, and those passing through rising density the other way. At E either more or less light will be let through depending on whether the ray has been deflected onto or away from the knife edge by the density changes. Thus the image of the working section at F will show light or dark areas (Fig. 11A). In a film on High-speed Flight produced by the Shell Petroleum Company, to whom I am indebted for these schlieren pictures, a colour filter was used at E and in this ingenious way increasing density was shown on the screen in one colour, decreasing density in another, and unchanging density in yet another; thus giving a real live picture of changes of density as the air flowed over different shapes at different speeds.

It is interesting to note that whereas the schlieren method reveals the change of density in a certain distance, the direct shadow method shows changing rates of density change, and other methods, such as the Mach Zender and laser interferometers, the actual density in different places. So each method has its advantages, but each needs rather different interpretation if we are to realise exactly what it is showing us.

Figure 11A is a photograph of a shock wave obtained by the schlieren method, and the fact that it appears as a narrow strip at right angles to the top surface of the aerofoil shows that across the strip there is a sudden increase in

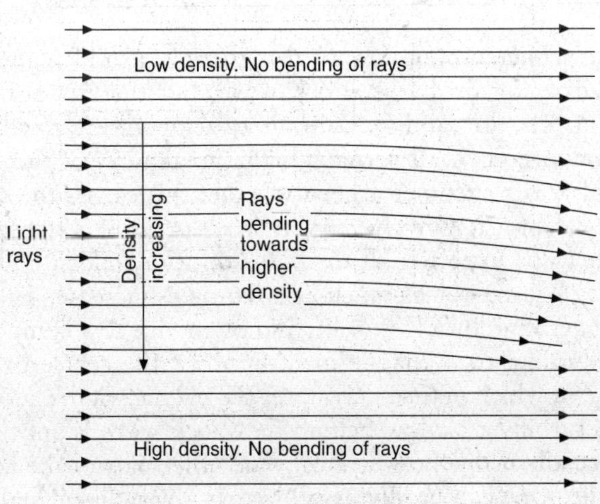

Fig 11.5 Bending of light rays through changing density

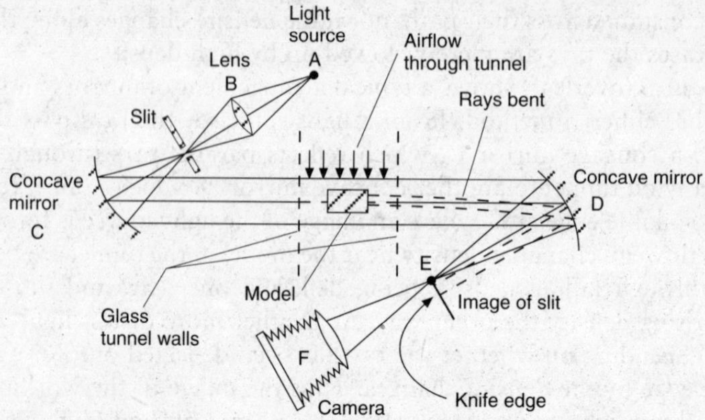

Fig 11.6 Arrangement for schlieren photography
View from above.

density. Immediately behind and immediately in front of the strip the density is reasonably constant, though greater of course behind than in front. Colour photography reveals another small area of increasing density immediately in front of the nose of the aerofoil as we might expect, and a large area above the nose and a smaller area below in which the density is decreasing.

Schlieren pictures of airflow at speeds well below that of sound show no appreciable changes of density at all, so now we see the fundamental change that takes place as we approach the speed of sound; the air begins to reveal its true nature as a very compressible – and expandable – fluid.

Effects of shock waves – the shock stall

It is clear from schlieren photographs that there is a sudden and considerable increase in density of air at the shock wave, but there is also, as has been stated, a rise in pressure (and incidentally of temperature), and a decrease in speed. Most important of all perhaps is the breakaway of the flow from the surface, though it is sometimes argued whether this causes the shock wave or the shock wave causes it. Whichever way it is the result is the same.

As is rather to be expected all this adds up to a sudden and considerable increase in drag – it may be as much as a ten times increase. This is accompanied, if it is an aerofoil, by a loss of lift and often, due to a completely changed pressure distribution, to a change in position of the centre of pressure and pitching moment, which in turn may upset the balance of the aeroplane. At the same time the turbulent airflow behind the shock wave is apt to cause severe buffeting, especially if this flow strikes some other part of the aeroplane such as the tail plane. One can hardly avoid saying – very like a stall. Yes, so like the stall that it is called just that – a **shock stall**.

But the similarity must not lead us to forget the essential difference – no it isn't the speed, we have already made it clear, or tried very hard to make it clear, that the ordinary stall can occur at any speed; the essential difference is that the 'ordinary stall' occurs at a large angle of attack and, to avoid confusion, we shall in future call it the **high-incidence stall** to distinguish it from the **shock stall** which is more likely to occur at small angles of attack.

From what has already been said the reader will probably have realised that the formation of shock waves is not a phenomenon that occurs on the wings alone; it may apply to any part of the aeroplane. Even the shock stall, which may first become noticeable owing to the sudden increase of drag and onset of buffeting – it is sometimes called the **buffet boundary** – may be caused by the formation of shock waves on such parts as the body or engine intakes, rather than on the wings (Fig. 11B, overleaf).

Shock drag

The sudden extra drag which is such a marked feature of the shock stall has two main components. First the energy dissipated in the shock wave itself is reflected in additional drag (wave drag) on the aerofoil. Secondly, as we have seen, the shock wave may be accompanied by separation, or at any rate a thickening of and increase in turbulence level in the boundary layer. Either of these will modify both the pressure on the surface and the skin friction behind the shock wave.

So this shock drag may be considered as being made up of two parts, i.e. the wave-making resistance, or **wave drag**, and the drag caused by the thick turbulent boundary layer or region of separation which we will call **boundary layer drag.**

As has already been explained the shock wave and the thickened turbulent boundary layer or separation are like the chicken and the egg – we don't know which comes first; what we do know is that when one comes so does the other. That is not to say that they are by any means the same thing, or that they have the same effects, or that a device which reduces one will necessarily reduce the other.

Mach Number

The time has come to introduce a term that is now on everybody's lips in connection with high-speed flight – **Mach Number**. This term is a compliment to the Austrian Professor Ernst Mach (1838–1916), who was professor of the history and theory of science in the University of Vienna, and who was observing and studying shock waves as long ago as 1876. Incidentally, his

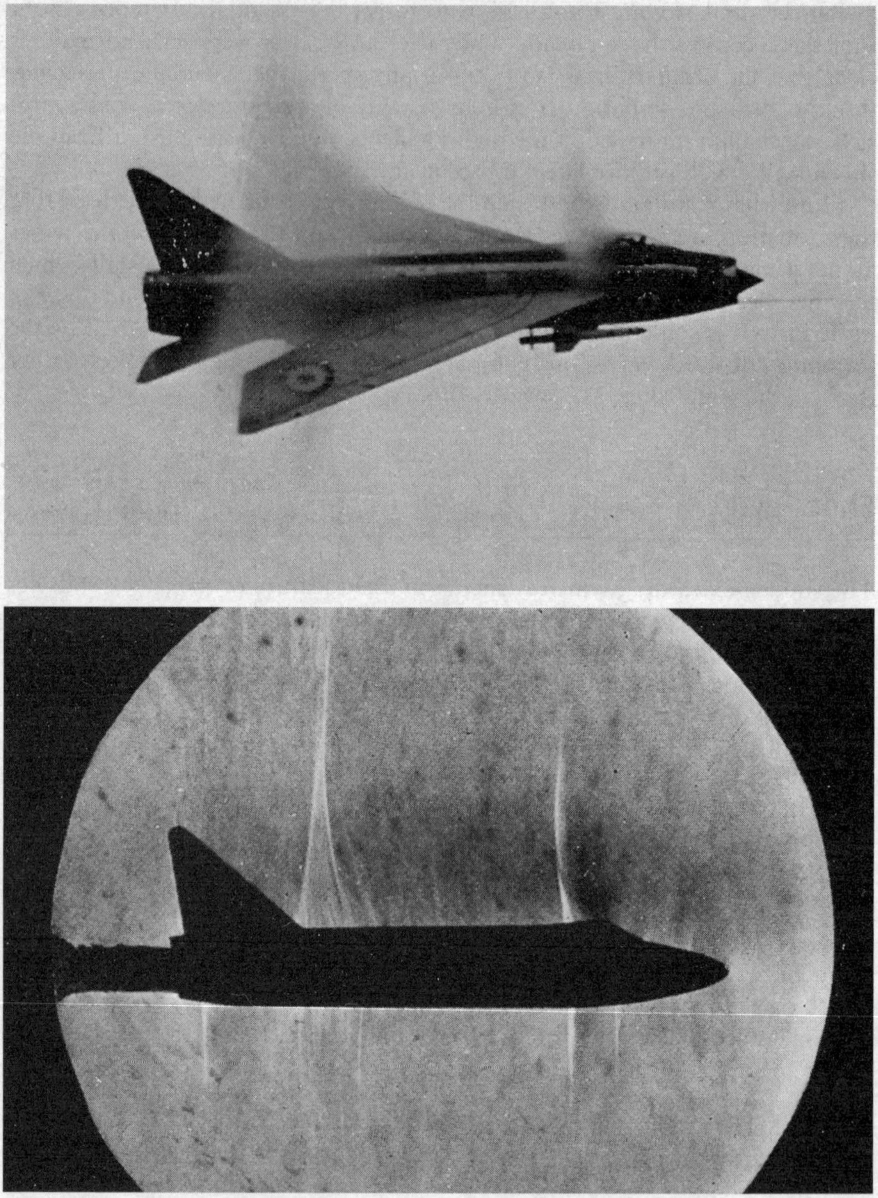

Fig 11B Shock waves
(By courtesy of the former British Aircraft Corporation, Preston)
Top: Lightning at M 0.98; low pressure regions above canopy and wing cause condensation, the rear limit of the condensation marks the shock wave.
Bottom: Schlieren photograph of model of Lightning at M 0.98; note the extraordinarily close resemblance to actual flight.

Science of Mechanics, which was published in English in 1893, throws much light on the work of Newton and others, and is well worth reading.

Fortunately the definition of Mach Number is simple. The **Mach Number** (M) refers to the speed at which an aircraft is travelling in relation to the speed of sound. Thus a Mach Number of 0.5 means that the aircraft is travelling at half the speed of sound. Both the speed of the aircraft and the speed of sound are true speeds.

Variation of speed of sound

There is one small complication that must be introduced into the definition of Mach Number even at this stage. The speed of sound varies according to the temperature of the air, and therefore we must add to the definition the fact that the speed of sound must be that corresponding to the temperature of the air in which the aircraft is actually travelling. People are often surprised to hear that the speed of sound in air depends on temperature alone.

The actual relationship is that the speed of sound is proportional to the square root of the absolute temperature.

We have seen that temperature falls with height in the atmosphere, and in the stratosphere where the temperature is about −60°C (213 K) the speed of sound will have fallen from about 340 m/s at sea-level to about 295 m/s.

Perhaps it should be emphasised again that this drop in the speed of sound is not really a function of the height at all; at a temperature of −60°C such as may occur at sea-level in, say, the North of Canada in winter, the speed of sound would also be about 290 m/s, while in tropical climates it might be well over 340 m/s even at considerable heights.

This variation of the speed of sound with temperature accounted for the rather surprising feature of speed record attempts of some years back in that the pilots waited for hot weather, or went to places where they expected hot weather, in order to make the attempts. Surprising because it had always been considered, and was in fact true, that high temperatures act against the performance of both aircraft and engine. The point, of course, is that the record breakers wanted to go as fast as possible while keeping as far away as possible from the speed of sound – so they wanted the speed of sound to be as high as possible. Nowadays in breaking speed records the aim of the pilot is just the opposite, i.e. to get through the speed of sound as quickly as possible – but we are anticipating.

Critical Mach Number

It has already been made clear that the onset of compressibility is a gradual effect, and that things begin to happen at speeds considerably lower than the speed of sound, that is at Mach Numbers of less than 1. One reason for this is that, as explained in earlier chapters, there is an increase in the speed of airflow over certain parts of the aeroplane as, for instance, over the point of greatest camber of an aerofoil. This means that although the aeroplane itself may be travelling at well below the speed of sound, the airflow relative to some parts of the aeroplane may attain that value. In short, there may be a local increase in velocity up to beyond that of sound and a shock wave may form at this point. This in turn, may result in an increase of drag, decrease of lift, movement of centre of pressure, and buffeting. In an aeroplane in flight the results may be such as to cause the aircraft to become uncontrollable, in much the same way as it becomes uncontrollable at the high incidence stall at the other end of the speed range.

All this will occur at a certain Mach Number (less than 1), which will be different for different types of aircraft, and which is called the **critical Mach Number** (M_{cr}) of the type.

The reader who has followed the argument so far will not be surprised to learn that the general characteristic of a type of aircraft that has a high critical Mach Number is slimness, because over such an aircraft the local increases of velocity will not be very great. This was well illustrated by the Spitfire, a 'slim' aircraft that was originally designed without much thought as to its performance near the speed of sound, yet which has proved to have a critical Mach Number of nearly 0.9, one of the highest ever achieved.

We had some difficulty in deciding whether the ordinary stalling speed should be defined as the speed at which the lift coefficient is a maximum, or at which the airflow burbles over the wing, or at which the pilot loses control over the aircraft. They are all related, but they do not necessarily all occur at the same speed. So now with the critical Mach Number – is it the Mach Number at which the local airflow at some point reaches the velocity of sound? or at which a shock wave is formed? or at which the air burbles? or when severe buffeting begins (this is sometimes called the 'buffet boundary' of the aircraft)? or at which the drag coefficient begins to rise? – or, again, when the pilot loses control? I do not know – nor, apparently, does anyone else! Authorities differ on the matter, each looking at it according to their own point of view, or sometimes according to whether they want to claim a high critical Mach Number for a pet type of aircraft. However, it doesn't matter very much; they are really all part of the same phenomenon.

Is it possible for an aircraft to fly at a Mach Number higher than its critical Mach Number? Is it possible for an aircraft to have a critical Mach Number higher than 1? These two questions may at first sound silly, but they are not. The answers to both depend entirely on which of the many definitions of critical Mach Number we adopt. If the critical Mach Number is when the pilot loses control, then he can hardly fly beyond it; but if it is when a shock wave is formed, or when the drag coefficient begins to rise, why not? The pilot may not even know that it has happened, any more than he knows whether he is at the maximum lift coefficient in an ordinary stall. Graphs of lift and drag coefficients are all very well, but one cannot see them on the instrument panel when flying. Supposing the pilot can maintain control through all the shock waves, increases of drag coefficient and so on, then the critical Mach Number is higher than 1 or, to be more correct, the aircraft has not got a critical Mach Number in any of the senses that we have so far defined it, except for the one relating to the first appearance of supersonic flow locally or the first appearance of shock waves.

Drag rise in the transonic region

The behaviour of the drag coefficient, for a thin aerofoil shape at constant angle of attack, can best be illustrated by a diagram (Fig. 11.7). This shows that up to a Mach Number of about 0.7 the drag coefficient remains constant – which means that our elementary principles are true – then it begins to rise. According to one definition the Mach Number at which it begins to rise, in this example 0.7, is the critical Mach Number. At *M* of 0.8 and 0.85 C_D is rising rapidly. Note that the curve then becomes dotted and the full line is resumed again at an *M* of about 1.2. The reason for this is interesting. For a long time, although it was possible to operate high-speed wind tunnels up to an *M* of about 0.85, and again at *M* of 1.2 or more, in the region of the speed of sound a shock wave developed right across the wind tunnel itself, the tunnel became 'choked' and the speed could not be maintained. Thus there were no reliable wind tunnel results in this region, and the dotted part of the curve was really an intelligent guess. This difficulty has now been overcome, and experiments have also been made by other means, by dropping bodies, or propelling them

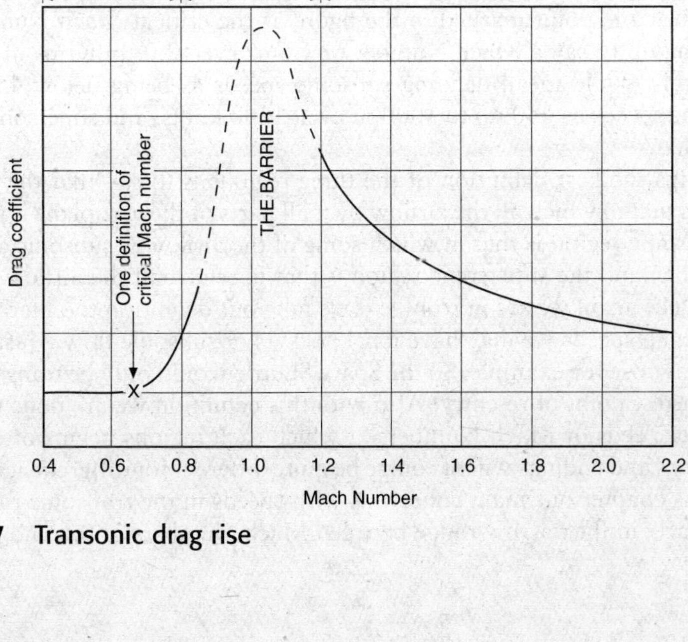

Fig 11.7 Transonic drag rise

with rockets and also, of course, by full-scale flight tests, and the guess can now be confirmed. Previous experiments on shells had of course suggested that there was nothing much wrong with it. The curve is still left dotted, partly to remind us of this bit of aeronautical history, but more now to emphasise the strange behaviour of the drag coefficient in the transonic region.

After a Mach Number of about 1.2, C_D drops and eventually, at M of 2 or more becomes nearly constant again though at a higher value than the original, variously quoted as 2 or 3 times.

The diagram shows that there is a definite hurdle to be got over. But it also shows that conditions on the other side are again reasonable and that supersonic flight, as we now of course know only too well, is a practical proposition.

The reader is advised to work through Question 253 (Appendix 3) in which he will plot the actual drag, as distinct from the coefficient, and it will then be clear that the drag also actually falls after $M = 1$, but that the reduction is not quite so evident in terms of drag as in terms of drag coefficient.

Subsonic – transonic – supersonic

We have already talked about flight at subsonic, transonic, and supersonic speeds, and it should now be clear that the problems of flight are quite different in these three regions, but the dividing lines between the regions are of necessity somewhat vague. Figure 11.7 shows the subsonic region as being below a Mach Number of 0.8, the transonic region from M 0.8 to M 1.2, and the supersonic region above M 1.2. There are arguments in favour of considering the transonic region as starting earlier, say at a Mach Number of about 0.7 or near the point marked in the figure as the critical Mach Number, and extending up to say a Mach Number of 1.6 or even 2.0. In terms of sea-level speeds this would mean defining subsonic speeds as being below 450 knots, transonic speeds as 450 up to 1000 or even 1200 knots, and supersonic speeds above that.

Perhaps the best definition of the three regions is to say that the subsonic region is that in which all the airflow over all parts of the aeroplane is subsonic, the transonic region is that in which some of the airflow is subsonic and some supersonic, and the supersonic region is that in which all the airflow is supersonic. Once again we are in trouble if we take our definition too literally. Even at very high speeds we may have local pockets of subsonic flow – just in front of a blunt nose for example. So the Space Shuttle would only be transonic even at the fastest point of re-entry! Also with this definition we are none the wiser as to the speeds or Mach Numbers at which each regions begins or ends; the beginnings and endings will of course be quite different for different aeroplanes.

In this chapter our main concern is with speeds in the transonic range, and particularly in the narrow range between Mach Numbers of 0.8 and 1.2. This range, as is probably already evident, presents us with some of the most baffling but fascinating problems of flight; it is the range in which most of the change takes place, the change from apparent incompressibility to actual compressibility, the gradual substitution of supersonic flow for subsonic flow; it is the range about which we are even now most ignorant.

Flight at transonic speeds – drag and power required

There was a time when the prospects of supersonic flight seemed poor owing to the lack of engines with the necessary power to overcome the rapid rise in drag which begins at the critical Mach Number. Even without compressibility effects the drag would rise with the square of the velocity, and the power – which is Drag × Velocity – with the cube of the velocity. The effect of compressibility is to increase these values still further in accordance with Table 11.2, which shows the approximate figures for a Spitfire.

Table 11.2 Power required with compressibility

Velocity (*knots*)	250	350	450	550	650
Velocity (m/s)	129	180	232	283	335
Velocity (km/h)	464	649	834	1019	1205
Power (kW) *without compressibility*	750	1900	2430	6000	9700
Actual power required (kW)	750	1900	2510	7500	22 000

Actually the problem was much more serious than this, if we assumed that the aircraft would be driven by propellers. As was explained in Chapter 4, the efficiency of a good propeller is about 80 per cent at its best, but its 'best' is at speeds of 129 to 180 m/s, after which the efficiency falls off very rapidly (although a new generation of propellers has pushed this up) – this happens for various reasons, but chiefly because the propeller tips are the first part of the aircraft to suffer from compressibility. With the high forward speed of the aeroplane combined with the rotary speed of the circumference of the propeller disc, Mach Number troubles begin to occur at aircraft speeds of about 180 m/s, and the result is so disastrous that the power which would have to be supplied to the propeller by the engine in order to attain the speeds given in Table 11.2 would look something more like Table 11.3 a conventional propeller.

Table 11.3 Power required accounting for propellor efficiency

Velocity (m/s)	130	180	230	280	335
Power (kW) *to be given to propeller*	820	2390	6000	15 000	100 000?

When one looks at these figures one realises why it was that people who knew what they were talking about forecast not so many years ago that it would be a long, long time before we could exceed 300 m/s.

Yet they were wrong. And for one simple reason – the advent of the gas turbine and the first flight of a jet-driven aircraft in 1941. This made all the difference, partly because of the elimination of the propeller and its compressibility problems (it is true that there are similar problems with the turbine blades in the jet engine), but mainly because, the efficiency of the jet increases rapidly over just those speeds, 154 to 257 m/s, when the efficiency of the propeller is falling rapidly. The net result is that whereas the reciprocating-engine-propeller combination requires nearly twenty times as much power to fly at 500 compared with 250 knots, the jet engine only requires about five times the thrust, and it is thrust that matters in a jet engine. Further, the weight of the jet engine is only a small fraction of that of the reciprocating-engine-propeller combination, and at this speed even the fuel consumption is less.

Maybe the prophets ought to have foreseen the jet engine – but they didn't, at least not within anything like the time during which it actually appeared. Of course, there were good reasons too why they didn't foresee it, for no metal could then possibly stand up to the temperatures of the gas turbine blades.

The jet engine, then, was the first step in solving the problem of high-speed flight. And while on the subject of engines, the rocket system of propulsion takes us even a step further and no one now, who knows the subject dares predict the limits of speed that may be reached with rockets – outside the atmosphere there really isn't any limit.

As hinted above, it is interesting to note that lately the wheel looks like turning a full circle. Improvement in propeller design means that the tip problems can be largely overcome and a gas turbine/propeller combination promises better efficiency in the future than a turbojet at transonic speeds.

Flight at transonic speeds – the pilot's point of view

We have so far discussed the problems of approaching the speed of sound very much from the point of view of the designer – but what about the pilot? Well, to find out what is going on, or what is likely to happen, the first thing needed is an instrument to measure at what Mach Number an aircraft is flying. Various types of machmeter are already in existence, and no doubt they will be improved in accuracy and reliability. For a machmeter to give a reliable indication of the Mach Number it must measure, in effect, the true speed of the aircraft and the true speed of sound for the actual temperature of the air. The first is usually done via the indicated speed, which can be corrected for air density by a compensating device within the instrument, but which still includes position error. A temperature compensating device can be used to give the true speed of sound, but in modern instruments this has been eliminated, and a combination of aneroid barometer and air speed indicator gives all the correction required – except that of position error. The term 'Indicated Mach Number' is sometimes used for the reading of the machmeter, but it is an unsatisfactory term since it differs from indicated air speed in that the main correction, that of density, has already been made in the instrument itself.

In the absence of a machmeter the pilot will find that the air speed indicator is apt to give very misleading ideas – even more so than usual. This assumes, of course, that the pilot is not one of those who have already discovered that what the air speed indicator reads is not an air speed at all! Without such knowledge, a pilot may reason that if the speed of sound is around 340 m/s it is impossible to run into trouble with say 103 m/s on the clock; a shock stall at this speed might come as a real shock.

Yet such a shock is possible because –

In the first place, the speed of sound is a real speed, a true speed.

Secondly, it decreases with fall of temperature, down to 295 m/s, or less, in the stratosphere.

Thirdly, the speed as indicated is not the true speed, and the error is more than 100 per cent in the stratosphere, so that at a real speed of 340 m/s at say 40 000 ft, the indicator will read less than 154 m/s.

Fourthly, trouble begins not at the speed of sound but at the critical Mach Number, and if this is 0.7 (a low value but by no means unknown), a shock stall may occur at 0.7×154, i.e. 108 m/s on the clock.

Fifthly – and a point not so far mentioned – a shock stall occurs at an even lower critical Mach Number during manoeuvres, so that in a turn the 108 might be reduced to less than 103.

And there we are!

The figures are all possible, they might even be worse.

There is one compensation; the pitot head is one of the first parts to experience the effects of compressibility, which may cause the air speed indicator to over-read at very high speeds – but this is usually allowed for in calibration.

Table 11.4 Air speed indicator readings and truespeed

Height		*True speed of sound*		*True speed at which shock stall will occur assuming M_{cr} 0.7*	*Reading of ASI at this speed*	
feet	*metres*	*knots*	*m/s*		*knots*	*m/s*
0	0	661	340	463	463	238
10 000	3048	640	329	448	385	198
20 000	6096	614	316	430	315	162
30 000	9144	589	303	412	253	130
40 000	12 192	573	295	401	200	103
50 000	15 240	573	295	401	156	80

Table 11.4 may be a help to a pilot in realising what is going on; the figures are calculated on the assumption of the International Standard Atmosphere and will vary to some extent according to how much actual conditions differ from this.

The figures in the last column are the readings of the air speed indicator at which a shock stall may occur (it may even occur at lower indicated speeds because the figures given do not allow for manoeuvres), and they are apt to be rather alarming. There is, however, another way of looking at it – and one that is much more heartening.

Suppose one dives from 50 000 ft at a constant true speed of, say, 450 knots; that is at a rapidly increasing speed on the clock of –

176 knots at 50 000 ft,

225 knots at 40 000 ft,

275 knots at 30 000 ft,

328 knots at 20 000 ft,

387 knots at 10 000 ft,

and 450 knots at sea-level,

strange as it may seem, the actual Mach Numbers would be decreasing as follows –

450/570 or 0.77 at 50 000 ft,

450/570 or 0.77 at 40 000 ft,

450/590 or 0.76 at 30 000 ft,

450/616 or 0.73 at 20 000 ft,

450/640 or 0.70 at 10 000 ft,

450/666 or 0.68 at sea-level.

This means that if the critical Mach Number were 0.7 an aircraft that was shocked stalled at 50 000 ft would become unstalled at a height of about 10 000 ft.

Even if the true speed were to increase during the dive, as would probably happen in practice, there might still be a drop in Mach Number.

This consoling feature of the problem is based on the assumption of rise in temperature with loss of height – if the temperature does not rise, that is to say, if there is an inversion, well the reader – and the pilot – can calculate what will happen!

Behaviour of aeroplane at shock stall

All this rather assumes that there is something to be feared about a shock stall, and that pilots try to avoid it. After all, there was a time when we looked upon the high incidence stall in the same way – something to be avoided at all costs. Now, however, it is practised by all pilots in the very early stages of learning. Much the same is happening to the shock stall – it is all a question of knowledge, and many aircraft currently cruise safely well into the transonic region.

By far the most important effect is a considerable change of longitudinal trim – usually, but not always, towards nose-down, and sometimes first one way then the other. Unfortunately the change of trim is made even worse by the very large forces required to move the controls, and the ineffectiveness of the trimmers. There is also likely to be buffeting, vibration of the ailerons, and pitching and yawing oscillations which may become uncontrollable, and which are variously described as snaking (yawing from side to side), porpoising (pitching up and down), and the **Dutch roll** (a combination of roll and yaw).

These effects can, though, be alleviated by the use of power controls and automatic stability augmentation systems. The best way of avoiding the difficulties is to keep an eye on the machmeter – if there is one – and, if the worst comes to the worst, to get into regions of higher temperature. The best ways of getting out of trouble are to stop going so fast – or to go faster! In a climb it is easy to stop going so fast – just throttle back. That is why the safest and best research work can often be done in climbing flight. In level flight it may not be quite so easy to lose speed, especially if the controls cannot be moved; and in a dive, which is where these troubles are most likely to occur, it will be even more difficult. It is essential therefore that all aircraft which are capable of these speeds, and which have undesirable characteristics in the transonic range, should have some kind of dive brake, or spoiler, which can safely be used at high Mach Numbers. We have had to go very fast on aeroplanes before the need for a brake was recognised! As to going faster; well, that will take us into the region of supersonic flight which will be dealt with in the next chapter.

Height and speed range

It was explained in Chapter 7 that for a piston-engined aircraft owing to limitations of power the speed range of an aircraft narrows with height, until, at the absolute ceiling, there is only one possible speed of flight. This speed, however, was not the stalling speed, but rather the speed of best endurance, and the absolute ceiling was simply a question of engine power; for a given aircraft, the greater the power supplied, the higher would be the ceiling.

Now, however, with almost unlimited thrust available in the form of jets or rockets, or eventually perhaps atomic energy, there is an altogether new aspect of the limitation of height at which an aircraft can fly without stalling – one way or the other. For the true speed of the high incidence stall will increase with height, while the true speed of the shock stall will fall from sea-level to the base of the stratosphere, and then remain constant. The result, assuming a sea-level stalling speed (high incidence) of 46 m/s and a critical Mach Number of 0.8, is shown in Fig. 11.8, the shaded portions being the regions in which flight is not possible without stalling. It will be evident that if this aeroplane is to avoid both kinds of stall, it cannot fly above 23 000 m, whatever the power available, while if it flies at 23 000 m, it can only fly at the stalling speed. If it flies any slower it will stall (high incidence), and if it flies any faster it will stall (shock). Surely a modern interpretation of being between the devil and the deep sea! Figure 11.9 shows exactly the same thing from the point of view of indicated speeds. Thus, quite apart from engine power, there is a limitation to the height of subsonic flight, and a narrowing of the speed range as the limiting height is approached.

This curious coming together of the two stalls will occur at considerably lower heights during manoeuvres, which will cause the high incidence stalling speed to increase, and the shock stalling speed to decrease. The figures in Table 11.5 are quite reasonable for 40 000 ft (say 12 000 m).

Table 11.5 Effect of altitude on stalling speeds

High incidence stall (true air speeds)					*Shock stall (true air speeds)*			
	Normal stall, sea-level	Normal stall at 40 000 ft	Stall at 4g, at 40 000 ft	Further increased at high M to, say,	At 4g, this may be reduced to, say	$M = 0.7$ 40 000 ft	$M = 1$ 40 000 ft	$M = 1$ Sea-level
	90	180	360	380	380	400	570	661
knots	→	→	→	→	←	←	←	← knots
Increase of high incidence stalling speed					Decrease of shock stalling speed			

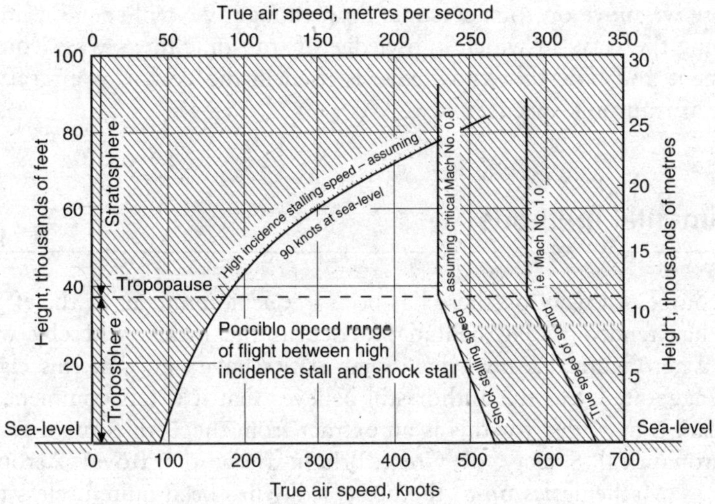

Fig 11.8 Subsonic speed range of flight – true speeds

One must always be careful in interpreting results of this kind. The above discussion must not be taken to imply that all aircraft are limited by shock stall or buffet at the maximum speed. This will be true for aircraft designed for economical cruise at transonic speed (as are most commercial airliners) but clearly many aircraft are designed to pass through the transonic speed range and cruise at supersonic speed – but more of that in the next chapter.

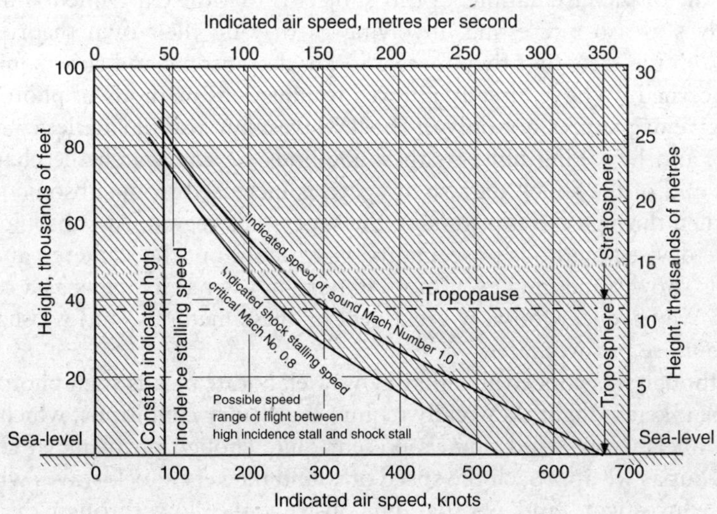

Fig 11.9 Subsonic speed range of flight – indicated speeds

Before we move on to supersonic flow, though, we will spend a little time examining the ways in which knowledge of transonic flows was acquired by experiment and how this knowledge is used in the design of aircraft which operate at transonic speeds.

Experimental methods

In the course of this book there has been frequent comment to the effect that theory has tended to lag behind practice as the means whereby we have acquired aeronautical knowledge. Some critics have said that this claim has been exaggerated, but the author still believes that it is fair comment. It certainly isn't a new idea for this is an extract from the 14th Annual Report of the Aeronautical Society of Great Britain (now the Royal Aeronautical Society) – 'Mathematics up to the present day has been quite useless to us in regard to flying' – the date of that report, 1879!

But whatever may have been the relative importance of theory and experiment in the acquisition of knowledge about subsonic flight, even the most theoretically-minded critic will surely agree that we have had to rely almost entirely on experiment in solving the problems of transonic flight, for here we have a mixture of subsonic and supersonic flow, compressibility and incompressibility, two completely different theories all mixed up. Curiously enough, in real supersonic flight, theory comes into its own; supersonic theory is simpler and older than that of subsonics or transonics – Newton's theories used to calculate air resistance of a body at 515 m/s give an answer nearer the truth than if they are used to calculate its resistance at 51.5 m/s.

But one of the fascinations of this subject is that the experimental methods themselves are so interesting, involving as they do their own theories quite apart from the facts that they reveal. So far as transonic and supersonic flight are concerned we have already referred to ingenious methods of photography which give us pictures not only of shock waves, but also of smaller changes of density; and by taking films by these methods we can watch the changes of density and of the shock pattern as speed is increased from subsonic to transonic, then through the transonic to the supersonic region. At subsonic speeds we have devised methods of seeing the flow of air, but the schlieren and other methods show something even more important – what happens as a result of airflow. We have learnt a very great deal by these methods, and we shall look at some more pictures later.

But though it is conceivable that by very elaborate means such photographs could be taken in flight they really require laboratory conditions, which means wind tunnels. Now wind tunnels present quite enough problems at subsonic speeds; but as we approach the speed of sound the very shock waves which we want to investigate and photograph, obstruct the flow through the tunnel (even when it hasn't got a model in it), and raise a barrier so effective that the tunnel is virtually choked and the high-speed flow cannot get through. Even if the tunnel design can be modified so as to allow the flow, the shock waves on even the smallest of models will cause very severe interference between the model and the walls of the tunnel, and so make the results of the tests valueless. This choking of wind tunnels, which was particularly difficult to overcome between Mach Numbers of 0.85 and 1.1, is the explanation of two rather curious facts in aeronautical history, that a truly supersonic wind tunnel became a practical proposition before a transonic one, and that flight at supersonic speeds took place before such speeds were reached in wind tunnels.

The problem of the transonic wind tunnel has now been largely overcome (though rather late in the day) by using slotted or perforated walls, and there are many types of supersonic tunnel in use today. Some of these are very similar to subsonic types in general outline; the extra speed has simply been obtained by more power together with suitable profiling of the duct – perhaps simply is not quite the right word, because the increase in power required, and consequent cost, is tremendous. It might be thought that fans of the propeller type would not be suitable for such tunnels; but in fact they can be used because they are situated at a portion of the tunnel where the speed is comparatively low, the cross-section of the tunnel, and therefore the propeller, being correspondingly large. The great size of the propeller, often larger than any used on aircraft, presents problems of its own, but none the less this conventional type of tunnel, which may be straight through, return flow or even open jet, is probably the most satisfactory where a large tunnel is required.

For smaller types, and higher speeds, it is more usual to employ some kind of reservoir of compressed air and, by opening a valve, to allow this to blow through the tunnel to the atmospheric pressure at the exhaust (Fig. 11.10). By arranging for the exhaust to be into a vacuum tank, even higher speeds can be obtained. One great disadvantage of this method is that continuous running for long periods is impossible; in fact constant speed is only achieved for a very

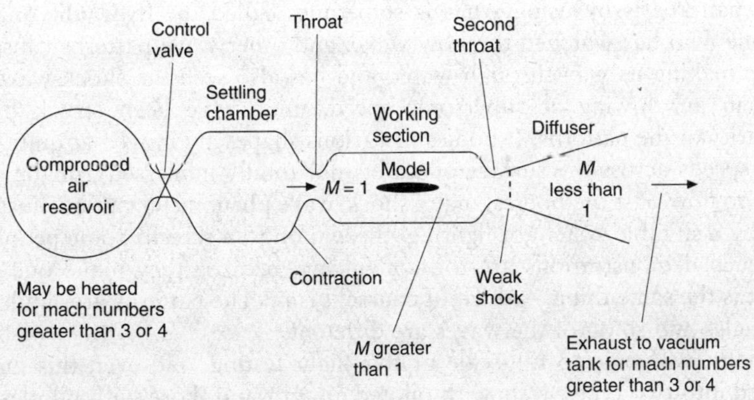

Fig 11.10 High-speed wind tunnel: blow-through type

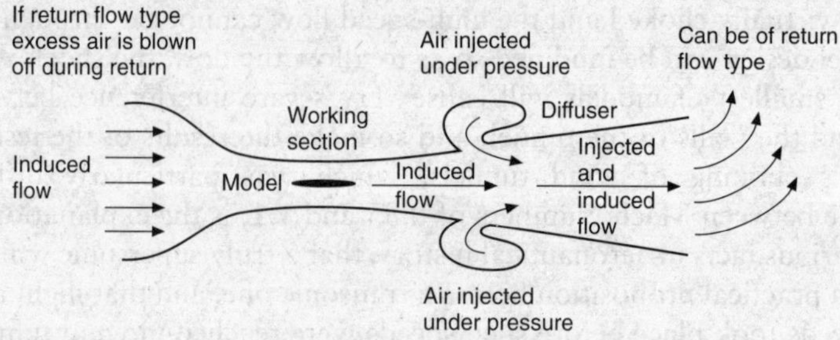

Fig 11.11 High-speed wind tunnel: induced flow type

short time. It sounds rather primitive, and in some ways it is, but Mach Numbers of 4 or more may be reached by this method, though the cross-sectional area of the working section is usually small.

More efficient, at some Mach Numbers, than this **straight blow-down** type of tunnel is the **induction or induced flow type**, in which air is blown in or injected just down stream of the working section (Fig. 11.11), thus 'inducing' a flow of air from the atmosphere through the mouth of the tunnel. Notice that in this type the compressed air does not flow over the model at all, only the induced air does. The injected flow can be the jet of a turbo-jet engine; in fact the jet engine itself can be in the tunnel. An induction type tunnel may be of the return flow variety, in which case provision must be made for the excess air to be blown off from the return passage.

It is one thing to obtain a high Mach Number in a wind tunnel – it is quite another thing also to obtain a high Reynolds Number (see Appendix 2) and so eliminate scale effect. This brings in the question of high-density tunnels, low-density tunnels, cryogenic (low temperature tunnels) and even the use of gases other than air; this interesting problem will be touched on in Appendix 2, but it is really beyond the scope of this book.

A simple and fascinating way of observing patterns very similar to shock-wave patterns is by using what is sometimes called the **hydraulic analogy**. Anyone who has watched the bow wave, and other wave patterns caused by a ship making its way through water, and has also seen the shock-wave patterns in air flowing at supersonic speed, must have been struck by the similarity of the patterns. If bodies of various shapes are moved at quite moderate speeds across the surface of water (not totally immersed), or the water made to flow past the bodies, many shock-wave phenomena can be illustrated and, by a suitable system of lighting, thrown on to a screen. Some people say that such demonstrations are too convincing, because they make one think that it is the same thing – which of course it isn't. The patterns are similar but the angles and so on of the waves are different.

Finally, we come to full-scale or free-flight testing, and even this may be divided into two types, tests with piloted aircraft and those with pilotless aircraft or missiles. Not only are these the eventual tests, but in the case of transonic flight, in particular, they have also been the pioneer tests and most of them have been made in piloted aircraft. In the early investigations, piloted aircraft certainly had their limitations because we could only approach the speed of sound in a dive; this had rather obvious and rather serious disadvantages. It took a long time (and a lot of height lost) to reach the critical speeds and then, if the symptoms were alarming – as they sometimes were – we were unpleasantly near the ground by the time a recovery could be made; moreover the making of such a recovery was not made any easier if the symptoms were severe buffeting, or worse – a further dropping of the nose and steepening of the dive – or worse still – heaviness of the controls so severe that they could not be moved. In such circumstances the point made in an earlier paragraph that the Mach Number might be decreasing as we hurtled towards the ground was hardly sufficient consolation.

As the thrust of jet engines increased – and it was this thrust which made even the approach of the speed of sound possible – the shock stall could be reached in level flight in some types of aircraft, and later still in climbing flight. This was an altogether different proposition from the pilot's point of view and, as testing at transonic speeds lost its terrors, our knowledge increased correspondingly more rapidly.

Test on bullets and shells moving through the air at supersonic speeds have been made on ballistic ranges since before the days of practical flight, and photographs of shock waves were taken more than 60 years ago, but it was only when aircraft themselves began to approach the speed of sound that the significance of such tests in respect of aircraft design began to be realised; and it was the development of the ramjet and rocket as means of driving missiles, and the parallel development of electronic instruments which could not only guide the missiles but take readings and keep records during the flight – in some cases even transmitting them back to earth – it was these that contributed most of all to our knowledge, and to the solution of the problems of transonic and supersonic flight.

So we can sum up the experimental methods that have been used to investigate these problems as coming under the following headings –

(*a*) Photography of shock waves.

(*b*) High-speed wind tunnels.

(*c*) The hydraulic analogy.

(*d*) Free flight in piloted aircraft.

(*e*) Rockets and missiles.

Now let us see what all this has taught us.

Shock–wave patterns

In an earlier paragraph we described how a shock wave is formed at a speed of about three-quarters of the speed of sound, i.e. at about $M = 0.75$. On a symmetrical wing at zero angle of attack the incipient shock wave appears on both top and bottom surfaces simultaneously, approximately at right angles to the surfaces, and, as one would expect, at about the point of maximum camber (Fig. 11.12*b*). On a wing at a small angle of attack, even if the aerofoil section is symmetrical, the incipient shock wave appears first on the top surface only (Fig. 11A) – again as one would expect, because it is on the top surface that the speed of the airflow first approaches the speed of sound.

Figure 11.12 shows how the shock-wave pattern changes (on a symmetrically shaped sharp-nosed aerofoil at zero angle of attack) as the speed of airflow is increased from subsonic, through the transonic range to fully supersonic flow.

Between the formation of the incipient wave (at a Mach Number of about 0.75 or 0.8) and the time when the wing as a whole is moving through the air at a speed of sound ($M = 1.0$), the shock wave tends to move backwards, but in doing so becomes stronger and extends farther out from the surface, while there is even more violent turbulence behind it (Fig. 11.12*c*). At a speed just above that of sound another wave appears, in the form of a bow wave, some distance ahead of the leading edge; and the original wave, which is now at the trailing edge, tends to become curved, and shaped rather like a fish tail (Fig. 11.12*d*). As the speed is further increased the bow wave attaches itself to the leading edge, and the angles formed between both waves and the surfaces become more acute (Fig. 11.12*e*). Still further increases of speed have little effect on the general shock-wave pattern – but here we are trespassing on supersonic flight, which is the subject of the next chapter.

At each wave there is a sudden increase of pressure, and density, and temperature, a decrease in velocity, and a slight change in direction of the airflow. The thickness of a shock wave, through which these changes take place, is only of the order of 2 to 3 thousandths of a millimetre – they look thicker on photographs because it is not possible to get a perfectly plane shock wave in the experiment. The changes at the shock wave are **irreversible**, which is another way of saying that the losses, which lead to wave drag, cannot be recovered. It is interesting to note, however, that the incipient waves only extend a short distance from the surface, and leaks are possible round the ends of the waves; as speed increases the waves extend and there is less and less possibility of such leaks. It is interesting to note, too, that the decrease in velocity, which occurs behind the shock wave, means that when an aircraft is moving through the air, and a shock wave is formed, the air behind the shock wave begins to move in the direction in which the aircraft is travelling.

In addition to showing the shock-wave patterns, Fig. 11.12 also indicates the areas in which the flow is subsonic or supersonic. In (a) at $M = 0.6$ it is all

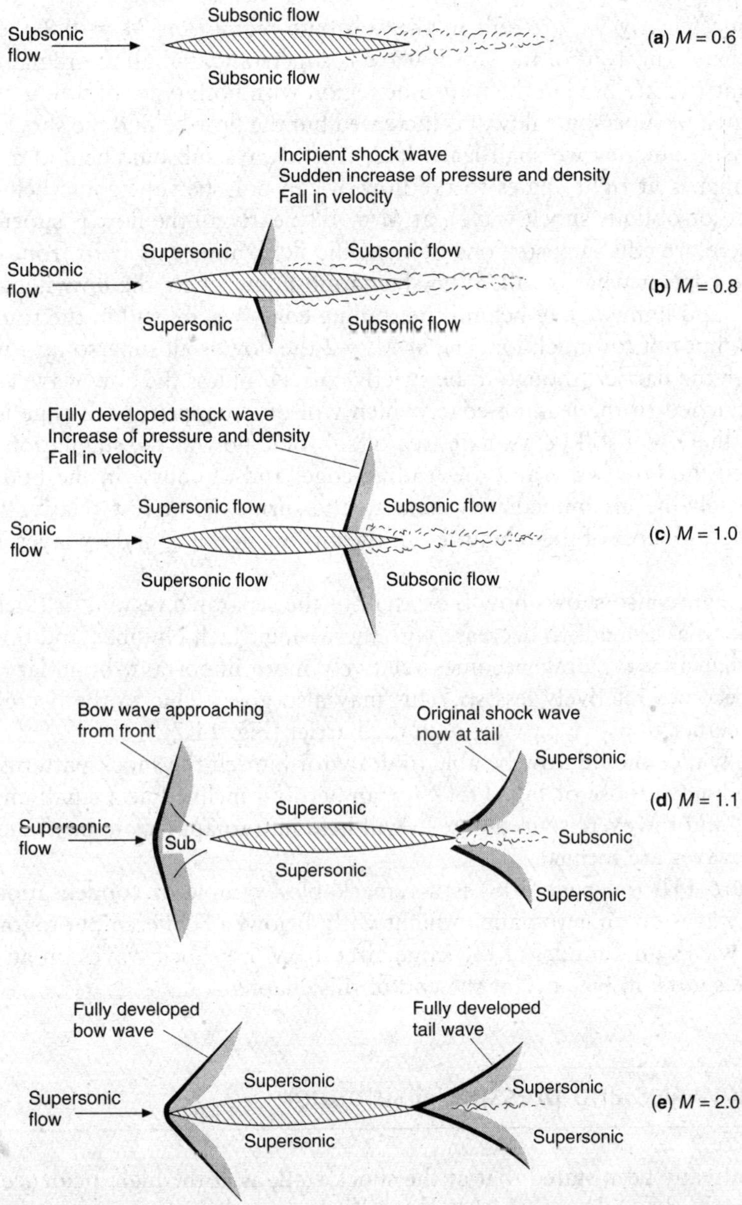

Fig 11.12 Development of shock waves at increasing mach numbers
(*a*) Subsonic speeds. No shock wave. Breakaway at transition point.
(*b*) At critical Mach Number. First shock wave develops.
(*c*) At speed of sound. Shock wave stronger and moving back.
(*d*) Transonic speeds. Bow wave appears from front. Original wave at tail.
(*e*) Fully supersonic flow. Fully developed waves at bow and tail.

subsonic (clearly we are still in the subsonic region); at $M = 0.8$ the flow immediately in front of the shock wave is supersonic, but all the remainder is subsonic (we are now in the transonic region with both types of flow); at $M = 1$ the area of supersonic flow has increased but the flow behind the shock wave is still subsonic (as we shall learn later it is always subsonic behind a shock wave that is at right angles to the flow, it can only be supersonic behind an inclined or oblique shock wave); at $M = 1.1$ nearly all the flow is supersonic, but there are still small regions of subsonic flow, immediately in front of the leading edge at what is called the **stagnation point** where the flow is brought to rest, and immediately behind the trailing edge (we are still in the transonic region, but not for much longer); at $M = 2$ the flow is all supersonic – we are through the barrier (though to be strictly correct, unless the bow wave is actually attached to the leading edge, which will only happen if the edge is very sharp, there will still be a small area of subsonic flow at the stagnation point between the bow wave and the leading edge; and of course in the boundary layer itself the air immediately next to the surface is at rest relative to the surface, and most of the remainder of the airflow in the boundary layer is subsonic).

The figure also shows how the extent of the separated region, or thickened boundary layer tends to decrease with increasing Mach Number, and this suggests that as wave drag becomes relatively more important, boundary layer drag becomes relatively less so. This may also give a clue to the decrease in drag coefficient as we pass through the barrier (Fig. 11.7).

The reader should now be able to draw for himself the shock patterns, corresponding to those of Fig. 11.12 for an aerofoil inclined at a small angle of attack, and the exercise in doing so will help him to appreciate how and why shock waves are formed.

Figure 11B (on page 306) is a remarkable example of condensation and shock waves on an aeroplane in flight with, below, a Schlieren photograph of shock waves on a model of the same aircraft. More shock waves on an aerofoil are shown in Fig. 11F at the end of this chapter.

Shock waves and pressure distribution

It has already been stated that at the shock stall, as at the high incidence stall, there are sudden changes both in lift and drag; and it is only to be expected that these are due in the main to changes in the pressure distribution over the wings or other surfaces. So pressure plotting has been as important a feature of research into the problems of high-speed flight as into those of subsonic flight, and the connection between the shock patterns and pressure distribution is naturally of great interest and importance.

It must be remembered that when we were originally considering the shapes of aerofoil sections in Chapter 3, the so-called laminar flow aerofoil (Fig 3.22) proved of great value at speeds of 140 m/s upwards. The characteristics of this section were comparative thinness and gently graduated camber, with the point of maximum camber farther back than on slow-speed types. As a result of this shape – in fact it was the purpose of it – the airflow speeded up very gradually and the distribution of decreased pressure over the upper surface was much more even than for the slower types on which there was a marked peak of suction quite near the leading edge (Fig. 3.8). We naturally approached the transonic region with aerofoil sections of this type, and so in considering the pressure distribution diagrams we must expect a fairly even distribution of decreased pressure on the top surface before the shock waves appear (this will be evident in Figs 11.13 and 11.14, overleaf).

Figure 11.13 shows in a very realistic way how the decrease in pressure, or suction, on the upper surface of a wing is affected by the formation of a shock wave when the wing as a whole is moving near the critical Mach Number. It shows the local Mach Number of flow across the surface of the wing; pressure of course depends on the speed, and hence Mach Number; the higher the local speed the less the pressure (back to Bernoulli again). The pressure scale is not given since compressibility effects complicate the issue considerably, but qualitatively Figure 11.13 gives the right idea.

Figure 11.14 is rather more complicated, but it is worth trying to understand because it demonstrates so clearly the practical effect of the shock waves on the pressure distribution – this time on both top and bottom surfaces, and for a symmetrical wing section at a small angle of attack – and the resultant effect on the lift and drag coefficient of the wing over the transonic range.

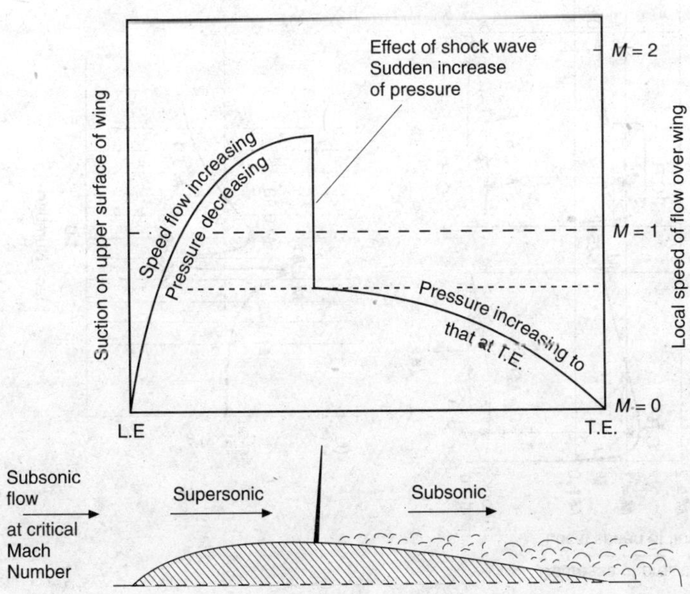

Fig 11.13 Shock wave and pressure distribution

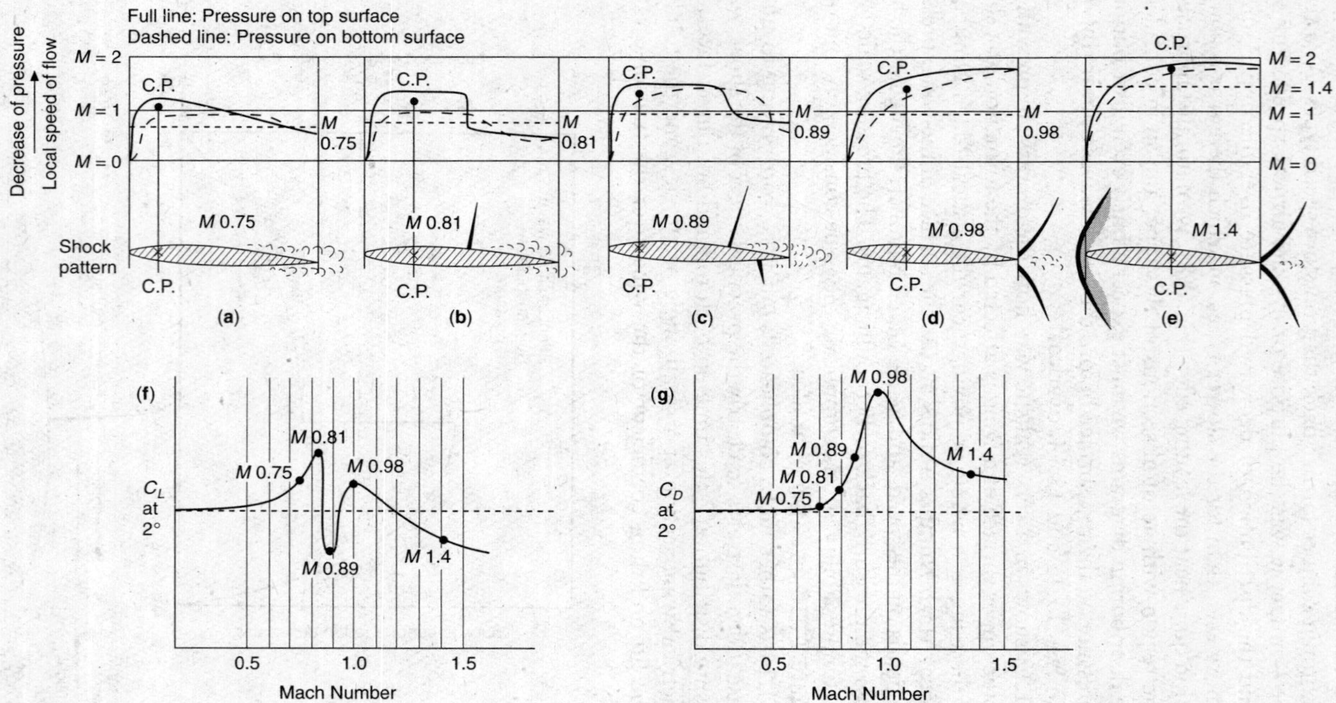

Fig 11.14 Shock-wave patterns and pressure distribution
Symmetrical bi-convex section at 2° angle of attack.

Incidentally, too, the figure gives a partial answer to the exercise suggested at the end of the previous paragraph, but perhaps the reader tackled that exercise before reading on!

At (*a*), (*b*), (*c*), (*d*), and (*e*) in the figure, which illustrate what happens at Mach Numbers of 0.75, 0.81, 0.89, 0.98, and 1.4 respectively, we see first the shock patterns at these speeds. All refer to a symmetrical bi-convex aerofoil section at an angle of attack of 2°. Above the shock pattern is shown the corresponding Mach Number distribution, across the chord; the full line representing the upper surface and the dotted line the lower one; decreased pressure is indicated upwards, as this corresponds to increasing Mach Number. It is the difference between the full line and the dotted line which shows how effective in providing lift is that part of the aerofoil section; if the dotted line is above the full line the lift is negative. The total lift is represented by the area between the lines, and the centre of pressure by the centre of area. Increasing speed, which is proportional to the decreasing pressure, is also shown upwards.

On the vertical scale on each diagram is a Mach Number of 1, i.e. the speed of sound, and the free stream Mach Number, i.e. the Mach Number at which the aeroplane as a whole is moving through the air. Thus we can see at a glance over what parts of the surfaces, upper or lower, the local airflow is subsonic or supersonic, and over what parts its speed is above or below that at which the aerofoil is travelling.

Parts (*f*) and (*g*) of the figure show the lift coefficient and drag coefficient corresponding to each diagram. Of these the lift curve is most interesting, partly because we have already seen the drag curve in Fig. 11.7, but more because the drag is not revealed by the pressure distribution to the same extent as the lift is, and in fact the pressure distribution does not show the important part of the drag that is tangential to the surface at all.

Figure 11.14 should be studied at leisure; it tells a long story – too long for me to point out every detail. But let us just look at some of the more important points.

Part (*a*) of course, is the subsonic picture, except that separation has already become apparent near the trailing edge and there is practically no net lift over the rear third of the aerofoil section; the centre of pressure is well forward and, as (*f*) shows, the lift coefficient is quite good and is rising steadily; the drag coefficient, on the other hand, is only just beginning to rise.

In (*b*) the incipient shock wave has appeared on the top surface; notice the sudden increase of pressure (shown by the falling line) and decrease of speed at the shock wave. The centre of pressure has moved back a little, but the area is large, i.e. the lift is good (see (*f*)), and the drag (*g*) is rising rapidly.

The pressure distribution in (*c*) shows very clearly why there is a sudden drop in lift coefficient (see (*f*)) before the aerofoil as a whole reaches the speed of sound; on the rear portion of the wing the lift is negative because the suction on the top surface has been spoilt by the shock wave, while there is still quite good suction and high-speed flow on the lower surface. On the front portion there is nearly as much suction on the lower surface as on the upper. The centre of pressure has now moved well forward again; the drag is increasing rapidly (*g*).

Part (*d*) is particularly interesting because it shows the important results of the shock waves moving to the trailing edge, so no longer spoiling the suction or causing separation. The speed of flow over the surfaces is nearly all supersonic, the centre of pressure has gone back to about half chord, and owing to the good suction over nearly all the top surface, with rather less on the bottom, the lift coefficient has actually increased (see (*f*)). The drag coefficient is just about at its maximum.

At (*e*) we are through the transonic region. The bow wave has appeared. For the first time the speed of flow over about half of both surfaces is less than the Mach Number of 1.4 of the aerofoil as a whole. The lift coefficient has fallen again, because the pressures on both surfaces are nearly the same; and this time – for the first time since the critical Mach Number – the drag coefficient has fallen considerably.

Sonic bangs

We are now all too familiar with the noises made by aircraft 'breaking the sound barrier', not to mention those unfortunate people who have suffered damage to property as a result. These so-called **sonic bangs**, or **booms**, are of course, caused by shock waves, generated by an aircraft, and striking the ears of an observer on the ground, or his glasshouses or whatever it may be; but there has been considerable argument as to the exact circumstances which result in the shock waves being heard, why there are often two or more distinct bangs, whether the second one came first, and so on.

Strangely enough many people don't seem to realise that we were familiar with sonic bangs, and their effects on us and our property, long before aircraft flew at all. A crack of the whip is probably the oldest man-made example; it may not have been responsible for breaking glasshouses, but in the hands of a circus performer it can be a pretty shattering noise. A roll or a clap of thunder is an example from nature of a series of shock waves; and one must have noticed how the bangs produced by aircraft often resemble a short roll of thunder. Explosions, too, produce shock waves, and, although a bombing raid is hardly an appropriate time to analyse such things, there were many unfortunate enough to experience during the war the disastrous effects on their ears and property of the shock wave of an exploding bomb, as distinct from the damage caused by the bomb itself. Some, too, may even have noticed the rather weird way in which the different bangs arrived at different times depending on where the observer was relative to the launching and explosion of the bomb. But the nearest thing to the sonic bangs produced by aircraft are the crack of a rifle bullet, or of a shell going overhead, or of the V2 rocket of wartime memory.

If an aircraft were to fly at supersonic speed at a height of a few feet over one's head the shock waves from wings, body, tail, etc., would strike one's ears in rapid succession, so rapid that one couldn't distinguish between them, and (if one remained conscious at all) the impression, so far as noise is concerned, would probably be of a short roll of thunder. The higher the aircraft flew, the less violent would be the noise produced by the shock until it would hardly be noticed at all from the ground; the noise of the engine, and of the aircraft itself, are of course continuous noises which are quite distinct from that of the shock.

An aircraft diving towards the earth at supersonic speed, and at an angle of say 45°, then suddenly slowing up and changing direction, will 'shed' its shock waves, which will travel on towards the earth and strike any observers who may happen to be in their path. It is certainly quite clear from schlieren photographs that a bow wave approaches from the front as the speed of sound is approached, and, conversely, goes ahead of the aircraft when it decelerates below the speed of sound.

So far as effects at ground level are concerned, we know that these become less intense with the height of the aircraft; more intense with Mach Number, though not anything like in proportion; that they are affected by the dimensions of the aircraft, increasing with its weight and volume, and being of longer duration according to its length; that they are more intense during accelerated flight (when the shocks tend to coalesce) than in steady flight; and that they decrease very rapidly with lateral displacement from the line of flight of the aircraft, in fact they only extend over a certain lateral distance. All this is rather what one might expect, but the problem is complicated because shocks of different intensities may be generated by the body, wings, tail and other parts of the aircraft, hence sometimes the roll as of thunder rather than one or two sharp bangs. One factor that one might not expect is the extent to which the bangs vary according to the conditions in the atmosphere between the aircraft and the ground, the winds, temperature, turbulence and so on. The actual pressures created at ground level are not so great as is sometimes thought; the overpressure in the Concorde boom, for instance, was only of the order of 96 newtons per square metre.

Can anything be done about it? Not much; if we insist on flying at these speeds. We can legislate against supersonic flight other than over the sea or thinly populated areas, but even so aircraft have to reach these areas. Moreover the tendency must be for the weight and size of such aircraft to increase rather than the reverse. Some alleviation can be obtained by control of climbing speeds; and at certain heights the speed of sound may be exceeded without creating a boom at all, the shock wave being dissipated before it reaches the ground. So really there is nothing for it but to fly as much as possible over the sea, and as high as possible, perhaps even really high – as we shall mention in the last paragraphs of this book.

Finally it should be mentioned that the publicity that has been given to sonic booms has tended to make us forget all the other noises created by aircraft, those from the engines, propellers or jets, and from the motion of the aircraft itself; these noises are probably more objectionable than sonic booms to those who live on landing or take-off paths, they are by no means confined to transonic and supersonic aircraft, and there are better prospects of reducing some of these noises as, for example, by using quieter engines.

Raising the critical Mach Number – slimness

When increase in engine thrust – due to the rapid development of jet engines – first made transonic flight possible, research was concentrated on the problem of raising the critical Mach Number, of postponing the shock stall, of getting as near to the barrier as possible without getting into it – in short, of keeping out of trouble rather than facing it.

There are two main ways of raising the critical Mach Number. The first is **slimness**. The need for slimness will be abundantly clear from all that has been said about shock waves and their effects – and the slimness applies to all parts, the aerofoil section, the body, the engine nacelles, the fin, tail plane and control surfaces, and perhaps most of all to small excrescences (if there must be such things) on the aircraft. The aerofoil section must be of the low-drag laminar-flow type already referred to, and must have a very low ratio of thickness to chord. The Spitfire of the Second World War has already been mentioned as an example of slimness, and of a high critical Mach Number – all the more remarkable in that it was not designed for transonic speeds.

The full line in Fig. 11.15 shows very clearly the effect of thickness/chord ratio on the critical Mach Number for a straight wing (the dotted line will be referred to in the next paragraph); at a t/c ratio of 10 per cent this wing has a critical Mach Number of only just over 0.8, at a t/c ratio of 8 per cent it is raised to 0.85, and at 4 per cent it is over 0.9. Not long ago the t/c ratios of wings for fighter aircraft were from 9 to 12 per cent, but they have now been reduced to 7 or 8 per cent, and may yet be still further reduced to a figure as low as 3 per cent, though there are of course very great design and manufacturing difficulties in producing such thin wings. In thus speaking of thin wings it is important to keep in mind that what really matters is not the actual thickness, but **the ratio of thickness to chord.**

Raising the critical Mach Number – sweepback

The second main way of raising the critical Mach Number (and this applies only to the wings, tail, fin and control surfaces) is **sweepback** – not just the few degrees of sweepback that was sometimes used, rather apologetically and for various and sometimes rather doubtful reasons, on subsonic aircraft, but 40°, 50°, 70° or more.

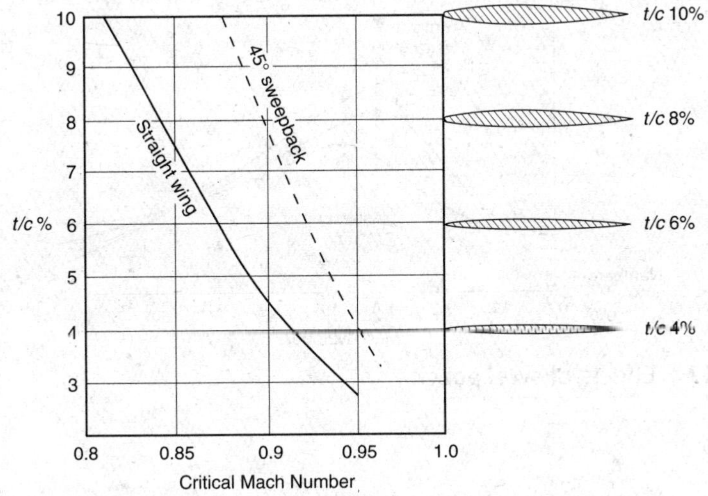

Fig 11.15 Critical Mach Number and t/c ratio

Sweepback of this magnitude not only delays the shock stall, but reduces its severity when it does occur. The theory behind this is that it is only the component of the velocity across the chord of the wing ($V \cos \alpha$) which is responsible for the pressure distribution and so for causing the shock wave (Fig. 11.16); the component $V \sin \alpha$ along the span of the wing causes only frictional drag. This theory is borne out by the fact that when it does appear the shock wave lies parallel to the span of the wing, and only that part of the velocity perpendicular to the shock wave, i.e. across the chord, is reduced by the shock wave to subsonic speeds. As the figure clearly shows, the greater the sweepback the smaller will be the component of the velocity which is affected,

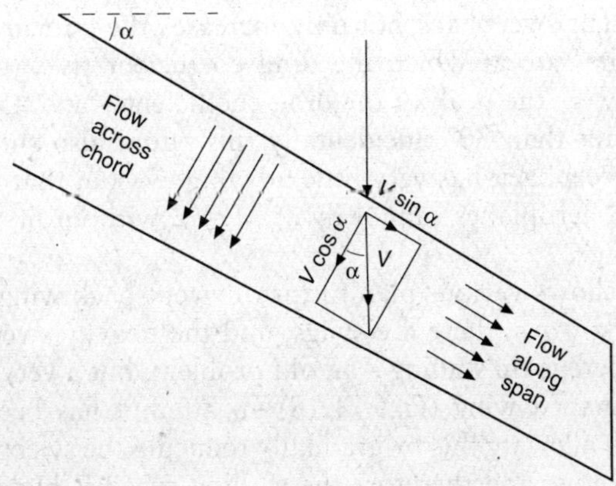

Fig 11.16 Sweepback – components of velocity

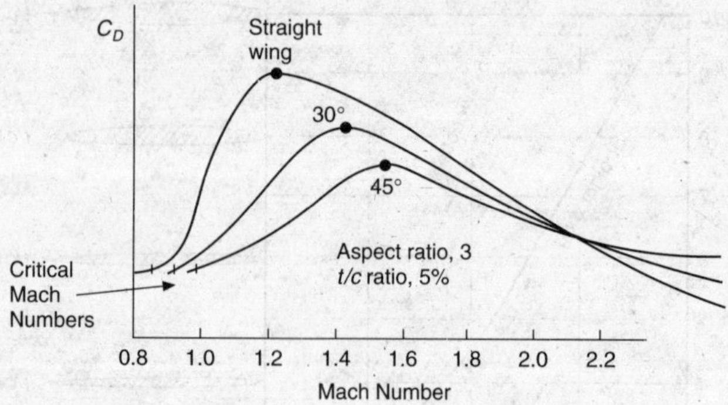

Fig 11.17 Effects of sweepback

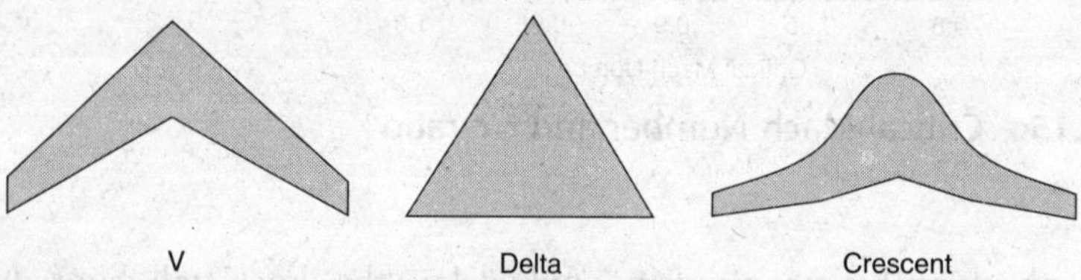

Fig 11.18 Swept-back wings for transonic speeds

and so the higher will be the critical Mach Number, and the less will be the drag at all transonic speeds of a wing of the same *t*/*c* ratio and at the same angle of attack.

Experiment confirms the theoretical advantages of sweepback, though the improvement is not quite so great as the theory suggests. The dotted line in Fig. 11.15 shows how a wing swept back at 45° has a higher critical Mach Number than a straight wing at all values of *t*/*c* ratio, the advantage being greater for the wings with the higher values of *t*/*c*. Figure 11.17 tells us even more; it shows that sweepback not only increases the critical Mach Number, but it reduces the rate at which the drag coefficient rises (the slope of the curve), and it lowers the peak of the drag coefficient – and 45° of sweepback does all this better than 30°. Incidentally this figure also shows that, above about $M = 2$, sweepback has very little advantage – but that is another story and, in any case, aeroplanes cannot fly at $M = 2$ without first going through the transonic range.

Figure 11.18 shows various plan forms of swept-back wings.

Of course, as always, there are snags, and the heavily swept-back wing is no exception. There is **tip stalling** – an old problem, but a very important one; in the crescent-shaped wing (Fig. 11.18) an attempt has been made – with some success – to alleviate this by gradually reducing the sweepback from root to tip. $C_L max$ is low, and therefore the stalling speed is high, and $C_L max$ is obtained at too large an angle to be suitable for landing – another old problem, and one that can generally be overcome by special slots, flaps or suction devices. There are also control problems of various kinds, and the designer doesn't like the extra bending and twisting stresses that are inherent in the heavily swept-back wing design. But whatever the problems sweepback seems to have come to stay – at least for aircraft which are designed to fly for any length of time at transonic or low supersonic speeds.

Control problems

Reference has already been made to the unpleasant things that may happen to aircraft as they go through the speed of sound – violent changes of trim, up or down, oscillations, buffeting, and so on. In such circumstances the first essential from the pilot's point of view is that he should have complete control; and so be master of the aircraft and its movements.

Unfortunately this is by no means easy to provide – and for several reasons.

Consider, for instance, an ordinary tail plane and elevator. At subsonic speeds an elevator depends for its effectiveness on the complete change of flow which occurs over both tail plane and elevator when the elevator is raised or depressed; the actual forces which control the aircraft are in fact much greater on the tail plane than they are on the elevator itself. Now as soon as a shock wave is formed on the tail plane – and the most likely place for its formation is at the hinge between tail plane and elevator – a movement of the elevator cannot affect the flow in front of the shock wave, so we have to rely entirely on the forces on the elevator itself to effect control. But the elevator will be in the turbulent flow behind the shock wave, and so may itself be very ineffective in producing the control forces required.

At higher speeds, when the shock wave moves back over the elevator, the opposite trouble may occur, and the forces on the elevator may be so great that it becomes almost immovable.

The answer to these troubles has been found in the all-moving slab type of tail plane (Fig. 5H) – sometimes given the rather curious name of a flying tail – and in making this **power-operated.** In this way we have what at the same time is an adjustable tail plane and an elevator.

The same principle can be applied to the ailerons – by having **all-moving wing tips** – and, more rarely, to the rudder, by having a combined fin and rudder.

But this is not the only high-speed control problem. Another is the possibility of **reversal of the controls**, owing to the distortion of the structure by the great forces on the control surfaces. This is most likely to happen on the ailerons. Suppose, for instance, that the starboard aileron is lowered in order to raise the starboard wing – if the force on the aileron is very great it will tend to twist the wing in such a way as to reduce its angle of attack, and so reduce the lift instead of increasing the lift on that wing; and the net result is that the wing will fall instead of rise. The aileron is acting on the wing just as a control tab is intended to act on a control surface; but that is little consolation to the pilot, and it doesn't take much imagination to realise his horror when a movement of the control column has exactly the opposite effect to that intended! Nor is it an easy problem for the designer to solve since the extra weight involved in providing sufficient stiffness, especially on thin and heavily swept-back wings, may be prohibitive. So as well as requiring power-generated controls, special high-speed ailerons may be needed – these are situated well inboard where the structure is stiffer instead of at the tips. Another solution is to use 'spoilers' instead of ailerons. These are small flaps which come out of the top surface of the wing and disrupt the local flow, thus reducing lift on that wing and giving rolling control.

The introduction of power-operated controls has in itself caused a new problem in that the pilot no longer 'feels' the pressure resisting the movement of the controls; this feel was always a safety factor in that it made the pilot conscious of the forces he was applying, and in fact there was some advantage in that there was a limit to what he could do to the aeroplane owing to the sheer limitation of his strength. So important is this matter of feel that when power-operated controls are used it has been necessary to incorporate artificial or synthetic 'feel'; and this is made even more real by grading it so that it varies not only with the movement of the control surface but with the density of the air and the air speed, in other words with the dynamic or stagnation pressure, $\frac{1}{2}\rho V^2$ or q – it is sometimes called 'q-feel'. Quite apart from the safety aspect, this synthetic feel gives the pilot a sense of control over the aeroplane which restores something of the art of flying.

But this is not the only problem resulting from the use of powered controls. If the power fails, and if there is no means of reverting to manual operation, the control may lock solid and the pilot be denied the use of rudder, elevators or ailerons. The answer to this is to introduce a safety factor by having more than one control surface, each having a separate power control; thus the Concorde had two rudders, one above the other, and six elevons (combined elevators and ailerons, which is the usual arrangement on delta or highly-swept wings). The Russian counterpart even had eight elevons.

Again on delta and highly-swept wings there is sometimes an interesting use of **spoilers**, this time as an aid to longitudinal control. Large aircraft flying at high speed, with their considerable inertia, are slow to respond to the elevators, just as they are to the ailerons. If one set of spoilers is fitted inboard (i.e. close to the fuselage) on each wing, and so forward of the aerodynamic centre; and another set is fitted well outboard and so, owing to the sweepback, behind the aerodynamic centre; and these are then linked to the elevator control in such a way that the fore and aft sets can be operated differentially, they will cause a movement of the centre of pressure which will aid the elevator control, just as when they are operated differentially on the port and starboard wings they assist the ailerons in providing lateral control, as described above.

Area rule

We should by now realise that if the drag is to be kept to a minimum at transonic speeds, bodies must be slim and smooth, and have 'clean lines'. What is the significance of clean lines? Well, it is often said to be in the eye of the beholder, what looks right is right – yes, but it depends on who looks at it; and a little calculation, a little rule, formula, or whatever it may be will often aid our eyes in designing the best shapes for definite purposes. The **area rule** (Fig. 11C) is simply one of these rules, and put in its simplest form it means that **the area of cross-section should increase gradually to a maximum, then decrease gradually**; in this sense a streamline shape obeys the area rule, though for transonic speeds, and indeed for high subsonic speeds, the maximum cross-sectional area should be about half-way, rather than one-third of the way back, this giving a more gradual increase of cross-sectional area with an equally gradual decrease. The body in Fig. 11.19 (overleaf) obeys the area rule – but it hasn't got any wings. If we add a projection to a body, such as the wings to a fuselage, we shall get a sudden jump in the cross-sectional area – and that means that the area rule is not being obeyed. What then can we do? – the answer is that we must decrease the cross-sectional area of the fuselage as we add the cross-sectional area of the wings in such a way that the total cross-sectional area of the aeroplane increases gradually. Similarly behind the point of maximum cross-sectional area it is the total cross-sectional area that must be gradually decreased.

Fig 11C Transonic area rule
The transonic British Aerospace Buccaneer which finally saw action in the Gulf War shortly before retirement. The bulge in the rear fuselage is for purposes of area rule.

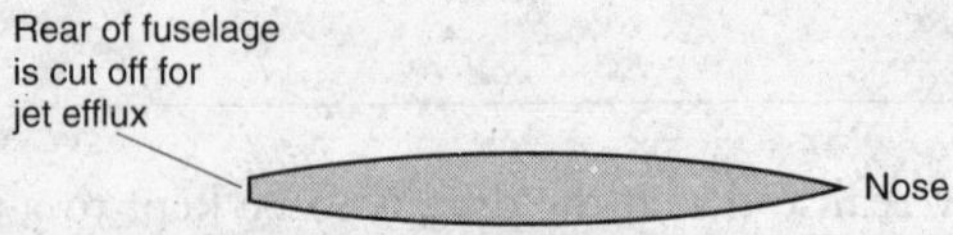

Fig 11.19 Area rule – plan view of fuselage without wings or tail

It will be realised that the application of this rule gives a waist to the fuselage where wings or other parts such as the tail plane are attached (Figs 11.19 and 11.20). It will be realised, too, that sweepback – in addition to its other advantages – is to some extent an area rule in itself so far as the wings are concerned, the cross-sectional area being added gradually, and so the waisting of the fuselage will be less marked with swept-back wings than with straight wings.

Vortex generators

Many devices are used by the designer to control the separation or breakaway of the airflow from the surface of the wing – all these devices, in one way or another, over one part of the wing or another, have this in common, that they are intended to prevent or delay this breakaway. How? Well, that depends to some extent on the device, and we will consider **vortex generators** first (Fig. 11D).

The fundamental reason for the breakaway is that the boundary layer becomes sluggish over the rear part of the wing section, flowing as it is against the pressure gradient. The formation of a shock wave makes matters worse; the speed in the boundary layer is still subsonic which means that pressure can be transmitted up stream, causing the boundary layer to thicken and, if the pressure rise is too steep, to break away from the surface. Now vortex generators are small plates or wedges, projecting an inch or so from the top surface of the wing, i.e. three or four times the thickness of the boundary layer. Their purpose is to put new life into a sluggish boundary layer; this they do by shedding small lively vortices which act as scavengers, making the boundary layer turbulent and causing it to mix with and acquire extra energy from the surrounding faster air, thus helping it to go farther along the surface before being

Fig 11.20 Area rule – effect of wings and tail on plan view of fuselage

slowed up and separating from the surface. In this way the small drag which they create is far more than compensated by the considerable boundary layer drag which they save, and in fact they may also weaken the shock waves and so reduce shock drag also; and the vorticity which they generate can actually serve to prevent buffeting of the aircraft as a whole – a clever idea indeed, and so simple. The net effect is very much the same as blowing or sucking the boundary layer, but the device is so much lighter in weight and simpler. The greater the value of the thickness/chord ratio the more necessary does some such device become.

There are various types of vortex generator; Fig. 11.21 illustrates the bent-tin type, which may be co rotating or contra-rotating. The plates are inclined at about 15° to the airflow, and on a wing are usually situated on the upper surface fairly near the leading edge.

Other devices to prevent or delay separation

There are several other devices which have been used to prevent or delay separation of the boundary layer, and so allay the rapid increase in drag at the sonic barrier, or the buffeting, or violent changes in trim which are liable to occur as a result of shock waves or separation, or some or all of these troubles.

Fig 11D Vortex generators
The Buccaneer again, showing vortex generators on the outer wing. This 'fix' was used to improve wing flow attachment on many early swept-wing aircraft.

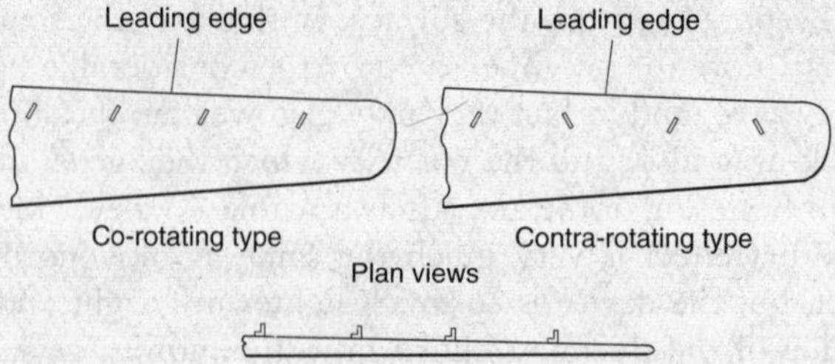

Fig 11.21 Vortex generators: bent-tin type

A thickened trailing edge is sometimes employed; this causes vortices which have much the same effect as those created by vortex generators, though naturally the effect is not felt so far forward on the wing surface.

On heavily swept-back wings **fences** (Figs 11.22) are often fitted; these are vanes of similar height to vortex generators, but running fore and aft across the top surface of the wing, and designed to check any spanwise flow of air along the wing, for this in turn is likely to cause a breakaway of the flow near the wing tips and so lead to tip stalling, particularly on swept wings.

Another problem with highly swept wings is the tendency for the flow to separate in the tip region first. This causes all sorts of problems, for example large changes in pitching moment. This effect may be reduced by introducing a notch or saw-tooth in the leading edge (Figs 11.23 and 11E). The notch also generates a strong vortex which controls the boundary layer in the tip region.

Leading-edge droop and **leading-edge flaps** are becoming quite common features of high-speed aircraft, but these are to prevent separation of the flow at the low-speed end of the range, i.e. at large angles of attack, and so help to solve one of the main problems of aircraft designed for transonic and supersonic speeds, that of making them fly safely slowly. A permanent droop is called **leading-edge droop** or **droop-snoot**; when it is adjustable it is called a leading-edge flap. Either may be combined with trailing-edge flaps and other devices, and Fig. 11.24 (overleaf) illustrates a combination of leading-edge droop, double slotted trailing-edge flaps, and air brakes – all helping to the same end.

But to conclude the problems of flight at transonic speeds on an optimistic note, it can generally be said that once one has a good transonic shape, it

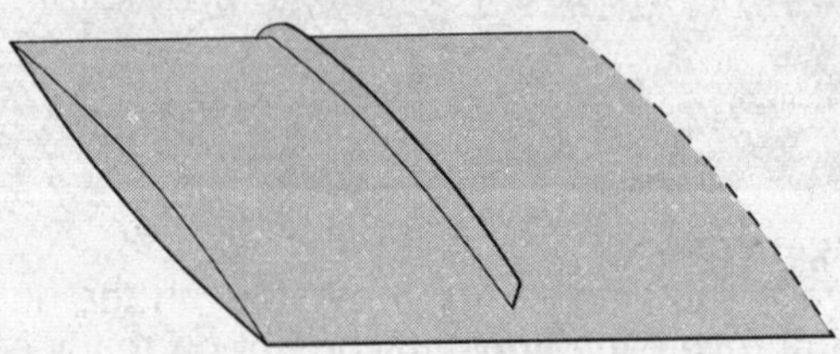

Fig 11.22 A boundary layer fence

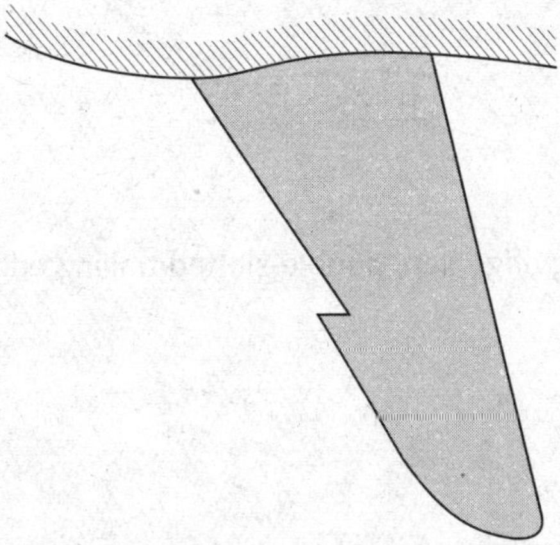

Fig 11.23 Leading edge saw-tooth or dog-tooth

Fig 11E Saw-tooth or dog-tooth
(By courtesy of McDonnell Douglas Corporation, USA)
The Phantom (RAF version), showing clearly the dog-tooth on the leading edge; the outer wings have 12° dihedral; there are blown leading and trailing edge flaps; the slab tail has 15° anhedral and a fixed slot; the rudder is inter-connected with the ailerons at low speeds.

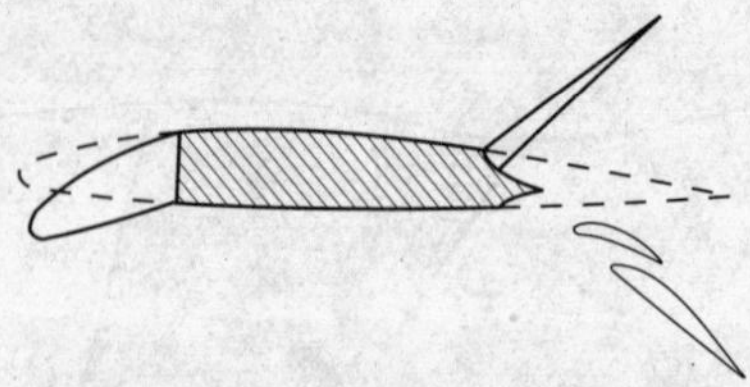

Fig 11.24 Leading edge flap, double-slotted trailing edge flap and air brake

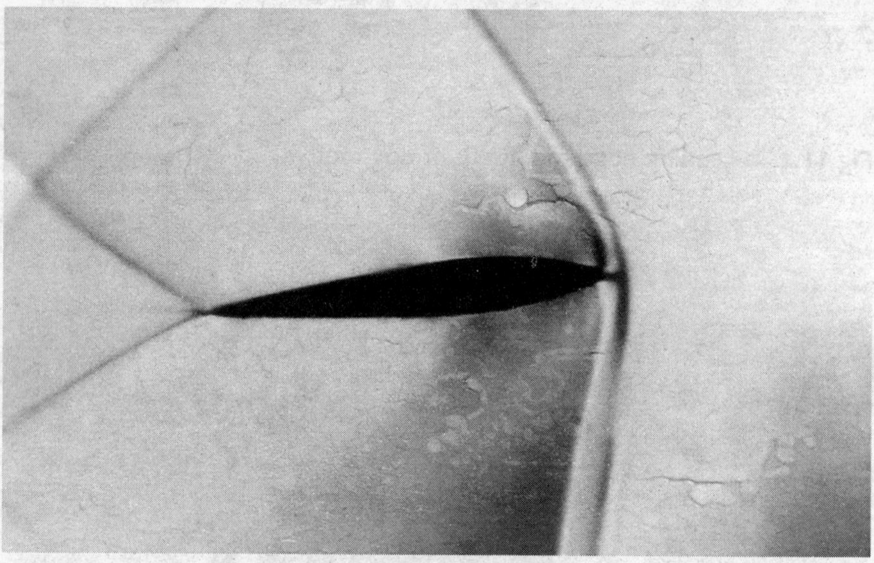

Fig 11F Shock waves
Shock waves from an aerofoil at incidence to the flow. Note the stronger leading edge shock on the underside. The wave in the top left-hand corner is a reflection from the wind-tunnel wall.

remains good, and the flow around it changes little between subsonic and transonic speeds.

Can you answer these?

1. Is the speed at which sound travels in water higher or lower than that at which it travels in air?
2. Does the speed of sound change with height – if so, why?
3. At what part of a wing does a shock wave first form?

4. What is the buffet boundary of an aircraft?
5. What is a Mach Number, a critical Mach Number, and a machmeter?
6. How does the appearance of a shock wave on a wing affect the pressure distribution over the wing?

For solutions see Appendix 5.

CHAPTER 12

Flight at supersonic speeds

Introduction

As explained in the previous chapter subsonic, transonic and supersonic flight merge one into the other, and it is not easy to define where one ends and the other begins. But the change from subsonic to transonic does at least involve some outward and visible signs – in the laboratory there is the formation of shock waves, and in flight there is the shock stall with its varying effects according to the type of aircraft – whereas the change from transonic to supersonic is not accompanied by any such signs whether in the laboratory or the air, so the dividing line between the two is even more vague. In general we can only say that supersonic flight begins when the flow over all parts of the aeroplane becomes supersonic. But at what Mach Number does that happen? Does it in fact ever happen? Are there not always likely to be one or two stagnation points? And what about the boundary layer where the flow near the surface is certainly subsonic? Perhaps after all it is better to say at about *M*1.2, or 1.5, or maybe 2? Figure 12A (overleaf) illustrates an aircraft designed for supersonic speeds.

Supersonic shock pattern

We have already had a look at the supersonic shock pattern in Fig. 11.12e, and except that the angles of shock waves become rather more acute as the speed increases there is very little change in this pattern over the supersonic range of speeds. To see why the angles of the shock waves change, we must understand the meaning and significance of the **Mach Angle**.

Mach angle

Figure 11.2 illustrated the piling up of the air in front of a body moving at the speed of sound, and explained how the incipient shock wave is formed. This incipient shock wave is at right angles to the direction of the airflow, and this means as near as matters at right angles to the surface of a body such as a wing.

Now suppose a point is moving at a velocity V (which is greater than the speed of sound) in the direction A to D (Fig. 12.1). A pressure wave sent out when the point is at A will travel outwards in all directions at the speed of sound; but the point will move faster than this, and by the time it has reached D, the wave from A and other pressure waves sent out when it was at B and C will have formed circles as shown in the figure, and it will be possible to draw a common tangent DE to these circles – this tangent represents the limit to which all these pressure waves will have got when the point has reached D.

Now AE, the radius of the first circle, represents the distance that sound has travelled while the point has travelled from A to D, or, expressing it in velocities, AE represents the velocity of sound – usually denoted by a – and AD represents the velocity of the point V.

$$\text{So the Mach Number } M = \frac{\text{Speed of point}}{\text{Speed of sound}} = \frac{V}{a}$$

(as illustrated in the figure this is about 2.5).

The angle ADE, or a, is called the **Mach Angle** and by simple trigonometry it will be clear that

$$\sin a = \frac{a}{V} = \frac{1}{M}$$

in other words, the greater the Mach Number the more acute the angle α. At a Mach Number of 1, α of course is 90°.

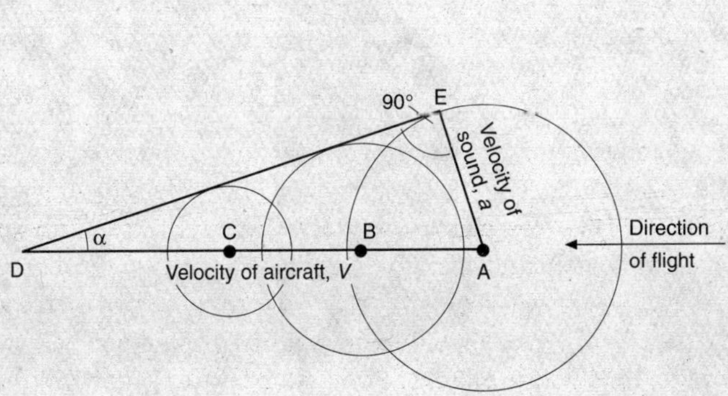

Fig 12.1 Mach angle

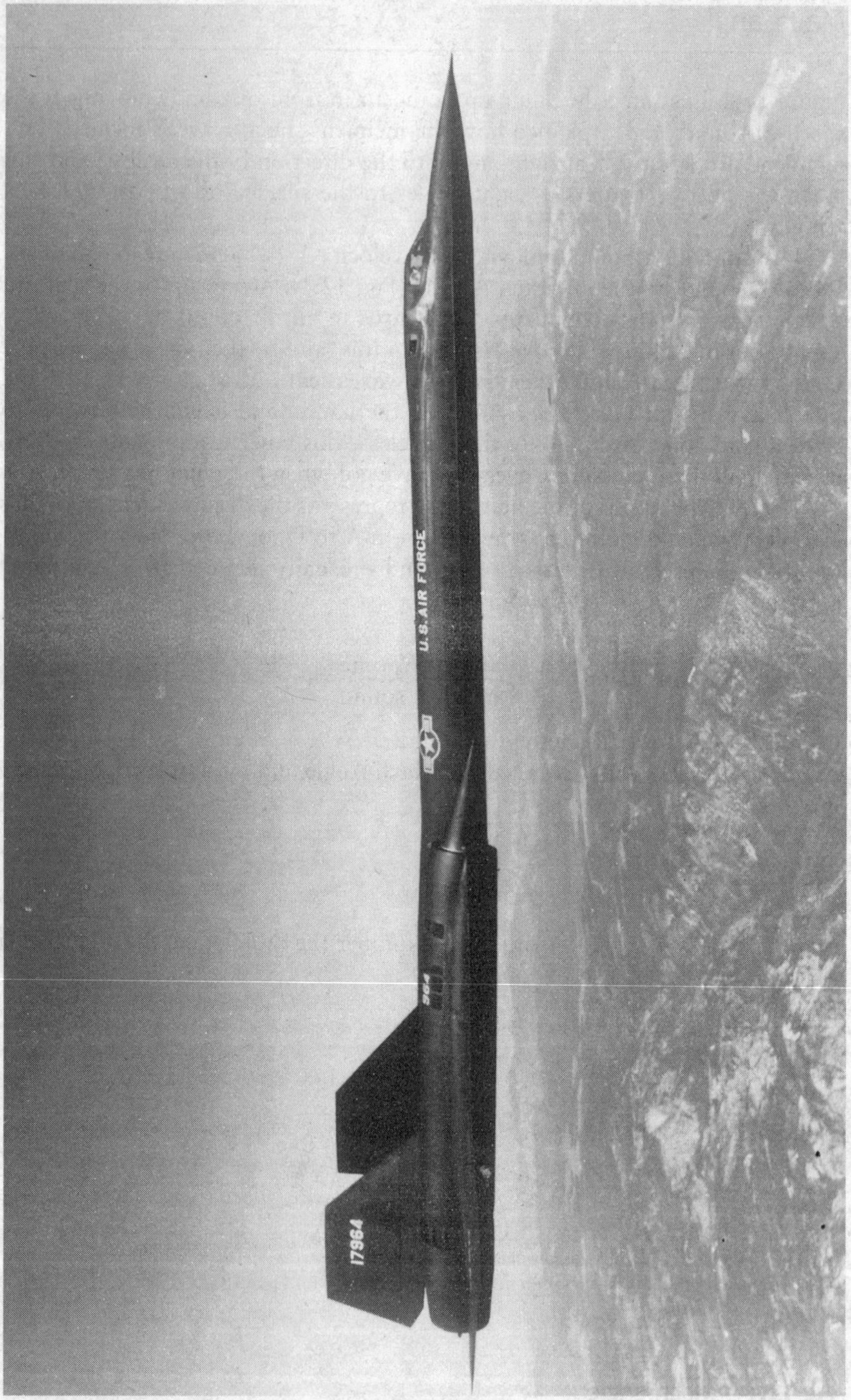
U.S. AIR FORCE
964
17964

If the moving point is a solid 3-dimensional body, such as a bullet, a complete cone – called the **Mach Cone** – will be formed, the angle at the apex being 2α. If the moving point represents a straight line such as the leading edge of a wing, a wedge will be formed, again with an angle 2α at the leading edge.

The tangent line DE is called a **Mach Line**, and it clearly represents the angle at which small wavelets are formed; the velocity of the airflow can even be calculated by measuring the angle on photographs of the wavelets.

Again the hydraulic analogy may be useful, since similar effects are seen when a ship passes through water or a thin stick is placed in a fast-moving stream of water. Only the region within the wedge formed by the bow waves is affected by the stick; the water outside this region flows on as if nothing was there. And the faster the flow, the sharper is the angle of the wedge.

It might be thought that the Mach Line represented the inclination of the shock waves – but this is not so. Disturbances of small amplitude travel at the speed of sound, but shock waves, which are waves of larger amplitude, actually travel slightly faster than sound, and therefore they form at a rather larger angle to the surface. This fact is difficult to explain without going into the mathematics of fluid flow, which is quite beyond the scope of this book, but the following explanation of how shock waves are formed may help us to understand how their slope is determined.

Imagine a supersonic flow of air over a flat surface. This surface can never be perfectly smooth, and may be considered as consisting of a very large number of particles or slight bumps. At each of these bumps a Mach Line will be formed; its angle to the surface depending on the speed of the flow in accordance with the formula $\sin \alpha = 1/M$.

If the speed of flow remains constant, the Mach Lines will all be parallel as in Fig. 12.2*a* (overleaf).

If the speed of flow is accelerating, the Mach Lines will diverge as the angle becomes more acute with the increasing speed (Fig. 12.2*b*).

But if the speed of flow is decelerating the Mach Lines will converge, add up as it were, and form a more intense disturbance or wave, one of greater amplitude (Fig. 12.2*c*).

This is one way of explaining how a shock wave is formed at all, but it also gives some indication of how its slope is determined. Unfortunately it is not very convincing from this point of view, and it could even be argued that the slope of the shock wave is less steep than some of the Mach Lines. Also, is a shock wave formed because the air is slowing up, or is it the shock wave that slows up the flow?

Fig 12A Flight at supersonic speed (opposite)
(By courtesy of the Lockheed-California Company, USA)
The SR-71 Blackbird was capable of flight at Mach numbers in excess of 3. Delta wings with high leading-edge sweep, lifting chines on the forward fuselage, and two turbo-ramjet engines: bypass turbojets that effectively function as ramjets in high-speed flight.

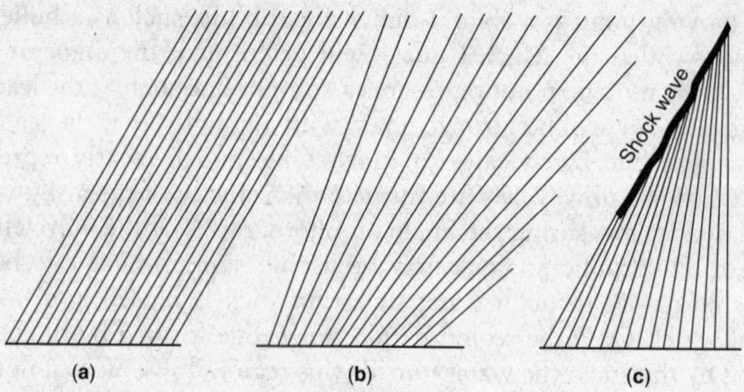

Fig 12.2 Mach lines and shock wave
(*a*) Speed of flow constant.
(*b*) Accelerating flow.
(*c*) Decelerating flow.

So perhaps we must fall back on the argument that shock waves travel faster than sound, and even more on the fact that the shock wave is at a steeper angle than the Mach Lines, for very fortunately that it is a fact can easily be seen on photographs; e.g. in Fig. 11B it is clear that the bow and tail main shock waves are at a coarser angle than the small wavelets which result from small pressure disturbances due to the surface roughness.

Other examples of the practical effect of large disturbances travelling faster than the speed of sound are the hearing of sonic bangs before the noise of the aircraft, and the way in which the shock wave of an explosion is followed by the other noises.

But has all this got any practical significance in aircraft design? Yes, as a matter of fact, it has; but to understand what the practical significance is we must study the nature of supersonic flow.

Supersonic flow

There are fundamental differences between supersonic flow and subsonic flow, and perhaps these differences are best illustrated by the different ways in which the two kinds of flow turn corners, or – what comes to much the same thing – pass through contracting or expanding ducts.

Although we may not have put it in that way we have already studied this in the case of subsonic flow; and perhaps we may sum up the results by saying –

1. That subsonic flow anticipates the corner or whatever it may be, and so the pattern of the flow changes before the corner is reached.
2. That the change of flow takes place **gradually** on **curved paths.**

3. That at what we might call a concave corner, or in a contracting duct, the flow speeds up and the pressure falls.
4. That at what we might call a convex corner, or in an expanding duct, the flow slows down and the pressure rises.

There are of course complications (as, for instance, if we overdo the suddenness of the change and the flow breaks away from the surface), but in the main we have established these four principles, and have seen numerous examples of how they are applied in practice.

Now let us look at supersonic flow.

We have already made it sufficiently clear that the first principle is different – and we have explained why it is different. Supersonic flow does not, and cannot, anticipate a corner or anything else that lies ahead, because there is no means by which it can know that it is there.

What, then, of the other three principles?

Compressive flow

Let us consider first what happens when supersonic flow meets what we have called a concave corner, or putting it more practically, a sharp, small-angled wedge. One way of describing this kind of corner is to say that if the flow were to go straight on it would intersect the body (Fig. 12.3).

Figure 12.4 (overleaf) shows what happens. The flow will in fact go straight on until it hits something – but what it hits will not be the wedge itself, but the shock wave which is formed by the slowing up of the flow as a result of the point of the wedge being inserted in the flow, and the consequent converging of the Mach Lines (perhaps that wasn't such a bad explanation after all).

In this type of flow there will be an inclined or oblique shock wave. Now it has already been explained that a shock wave at right angles to the flow causes a sudden reduction in the speed of flow, but a shock wave oblique to the flow causes both a reduction in the magnitude of the velocity, and a change in its direction. The change of direction is a result of the fact that it is only the component of the velocity at right angles to the shock wave which is reduced; the

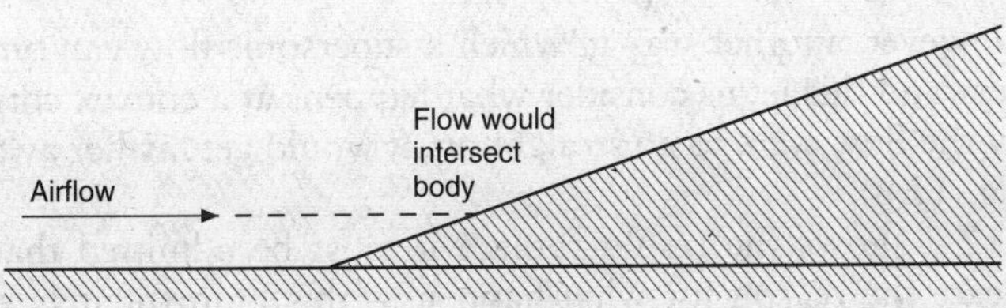

Fig 12.3 Compressive flow – concave corner

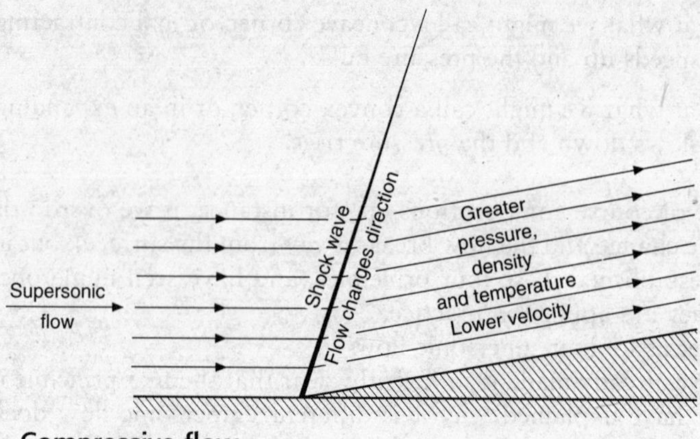

Fig 12.4 Compressive flow

other component (along the shock wave) remains unchanged in passing through the shock wave. This is illustrated in Fig. 12.5, and it is clear that the new direction of flow will be parallel to the new surface.

So the flow has turned the corner; the change of direction was sudden and occurred entirely at the shock wave. The flow after the corner is at a reduced velocity (though it may still be supersonic), the lines of flow are closer together, the pressure is higher, the density is higher (the air is compressed, possibly quite appreciably), and the temperature is higher. The Mach Lines, at the lower speed, will be more steeply inclined to the new surface.

Supersonic flow most commonly compresses through a shock wave; and at the leading edge of a wing, or the nose of a body, or at the mouth of a contracting duct, there is – as at this wedge – no gradual change of pressure as with subsonic flow, but a sudden rise in pressure, density, and temperature, and a sudden fall in velocity. This type of flow is called **compressive flow**. By very careful design it is possible to obtain a gradual compression by avoiding the conditions where the Mach Lines coalesce (Fig. 12.2(c)). The shock compression is, though, much more usual.

Expansive flow

There is, however, another way in which a supersonic flow can turn corners.

To understand this let us consider what happens at a convex corner, i.e. one at which, if the flow were to go straight on, it would get farther away from the surface (Fig. 12.6).

Figure 12.7 shows the result, though it must be admitted that it doesn't really indicate the reason for what happens – this unfortunately can only be given in a mathematical treatment of the subject.

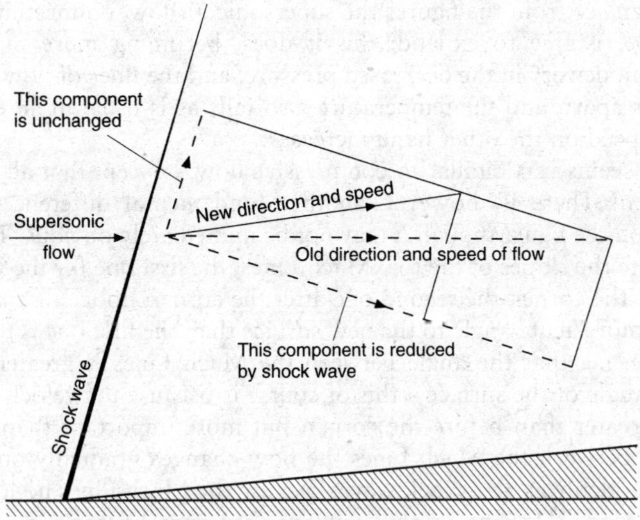

Fig 12.5 Change of direction and speed of flow

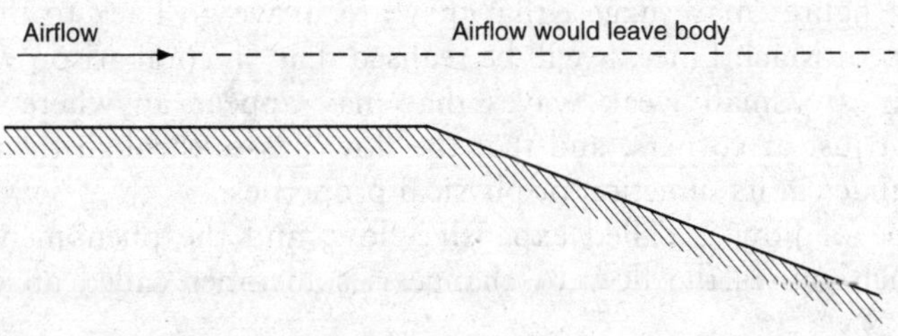

Fig 12.6 Expansive flow – convex corner

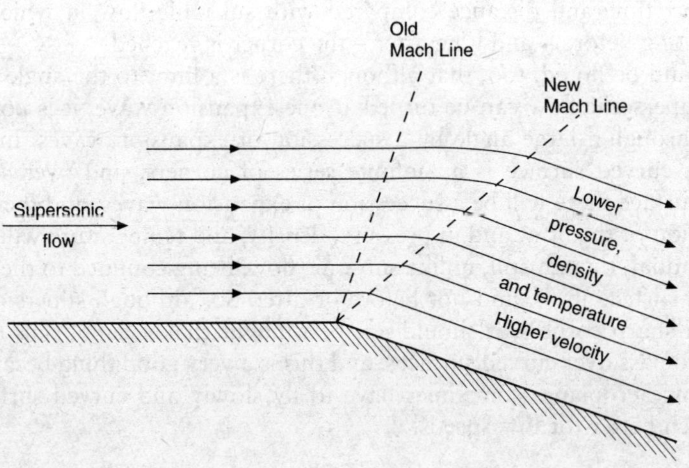

Fig 12.7 Expansive flow

As we can see from the figures the supersonic airflow, on meeting a corner of this type, is free to expand; this it does, becoming more rarefied, i.e. decreasing in density in the decreased pressure, and the lines of flow are therefore farther apart, and the temperature also falls as is usual in an expanding flow. The speed on the other hand increases.

So far it seems very similar to compressive flow – except that all the opposites happen! There is, however, another fundamental difference which is illustrated on the figure though it may not be immediately obvious. The dotted lines indicate the slopes of the two Mach Lines, the first one for the velocity of flow before the corner, the second one after the corner; notice that the second one is at a more acute angle to the new surface than the first one is to the original surface, i.e. that the angle between the Mach Lines is greater than the change of angle of the surface – this of course is because the velocity after the corner is greater than before the corner. But more important than this is to notice that between the Mach Lines the flow changes gradually on a curved path, not suddenly as at a shock wave, because the Mach Lines no longer converge as in Fig. 12.2c, but, on the contrary, they now diverge.

This gradual change of flow helps to emphasise the fact that the Mach Lines are nothing like shock waves; for there is some danger that even the dotted lines in the figures may suggest that they are. If we go back to the original explanation of Mach Lines, it will be realised that, in comparison with shock waves, they are small weak waves that may appear anywhere along the surface, not just at corners, and that the flow passes through them without sudden changes in its direction or physical properties.

This type of flow is called **expansive flow**, and the phenomenon at the corner which causes the flow to change is sometimes called an **expansion wave**.

It should be noted that although the change at an expansion wave is gradual when compared with that at a shock wave, it still takes place over a very short time and distance compared with subsonic flow in which things happen long before – and long after – the corner is reached.

It should be noted, too, that although there is a limit to the angle through which supersonic flow can be turned at one expansion wave, it is possible to turn it through a large angle by a succession of expansion waves. In fact, of course a curved surface is an infinite series of corners, and over a convex curved surface there will be a succession of expansion waves, and the changes in direction of the flow, and in pressure, density, and temperature will be even more gradual, though still, unlike subsonic flow, being confined to the passage over the surface itself, and not before or after. So, although supersonic flow **can** turn sharp corners, it should not be thought that it cannot also be persuaded to pass over curved surfaces, and this is a very good thing because even supersonic aeroplanes sometimes have to fly slowly and curved surfaces are very much better for low speeds.

Supersonic flow over an aerofoil

We are now in a position to look at the supersonic flow over an aerofoil – but what is to be the shape of our supersonic aerofoil?

Since straight lines and sharp corners seem to be at least as good as curves, the simplest aerofoil section for supersonic speeds would seem to be a flat plate inclined at a small angle of attack, and there is no doubt that if it were possible to give adequate strength to such a plate it would be the obvious answer.

If the plate were thin enough the flow would be undisturbed at zero angle of attack – and of course there would be no lift. At a small angle, on the top surface there would be an expansion wave at the leading edge and a shock wave at the trailing edge, and on the bottom surface a shock wave at the leading edge and an expansion wave at the trailing edge, the flow being as in Fig. 12.8.

On the top surface, owing to the expansion wave at the leading edge, the flow would be speeded up and there would be a decreased pressure; on the bottom surface, owing to the shock wave at the leading edge, the flow would be slowed down, and there would be an increased pressure. So there would be lift – and drag.

But a flat plate is clearly not a practical proposition, so let us have a look at a shape that is – the double-wedge.

First let us see what happens at zero angle of attack (Fig. 12.9, overleaf).

The pattern of shock waves and expansions will, of course, be the same on both top and bottom surfaces. At the nose (which corresponds exactly to the small-angled wedge of Fig. 12.4) there will be shock waves, and the consequent increases in pressure, density, and temperature, and a decrease in

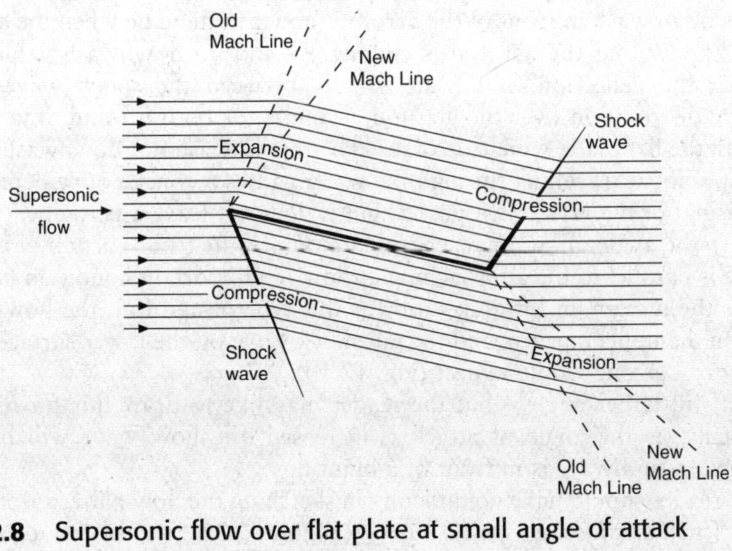

Fig 12.8 Supersonic flow over flat plate at small angle of attack

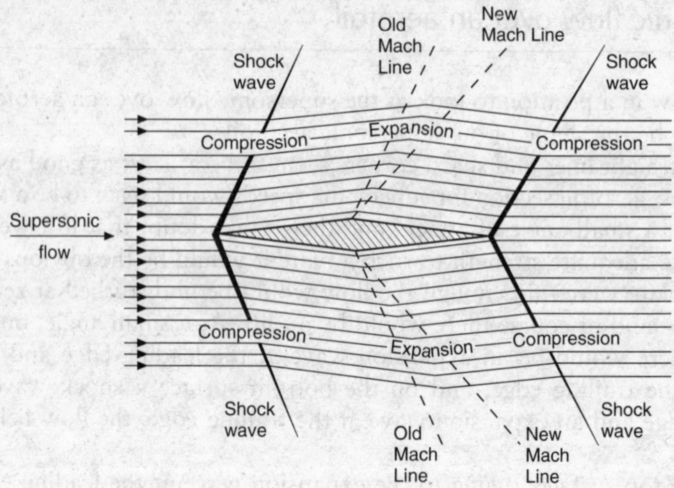

Fig 12.9 Supersonic flow over double-wedge at zero angle of attack

velocity; at the point of maximum camber there will be the expansions, and the corresponding changes over the rear portion of the aerofoil; at the trailing edge, though not perhaps quite so obvious, there will again be the wedge effect, shock waves on both surfaces, and again the changes of pressure, density, temperature, and velocity.

At very small angles of attack, for reasons that should now be quite clear, the bow shock wave on the upper surface becomes less intense and that on the lower surface more intense; the tail shock wave, on the other hand, becomes more intense on the upper surface and less intense on the lower surface. But one of the most interesting, and perhaps surprising, features of the flow is that there is no upwash in front of the aerofoil (how can there be when the airflow doesn't know that the aerofoil is coming?) – and no downwash behind the aerofoil; the deflection of the air is only between the shock waves. The pressure distribution over the aerofoil accounts for both lift and drag – as it did with the flat plate, which also caused neither upwash nor downwash. After all, a speedboat travelling through the water causes a considerable depression (and rising) of the water as it passes, but it does not leave a permanent dent!

When the angle of attack reaches that at which the front portion of the top surface is parallel to the approaching airflow (an important condition because it gives the maximum lift/drag ratio for this type of aerofoil) the bow shock wave on the upper surface, and the tail shock wave on the lower surface, both disappear – as one would expect (Fig. 12.10).

At a still larger angle – but the reader may like to draw this for himself. Eventually, as the angle of attack is increased, the bow wave will become detached, as it always is in front of a blunt nose.

The reader should have no difficulty in sketching the flow patterns for other shapes (such as those in Figs 12.20, 12.21, and 12.22), and at various angles.

Convergent–divergent nozzle

Let us now consider the flow through a contracting-expanding duct, a convergent-divergent nozzle, a de Laval nozzle – at any rate a constriction in a duct; or, as we called it in the low speed case, a venturi meter.

It is true that there are variations in the shapes of these devices according to what they are required to do, and according to the speed of flow with which they have to deal, but all have this in common – that they converge from the inlet to a throat, then diverge to the outlet.

We are familiar by now with the picture which illustrates subsonic flow through a venturi tube (Fig. 2.6); in the contracting portion the streamlines converge, the airflow accelerates, the pressure falls and, although the change in density is small (so small that we could then afford to neglect it), it is a decrease, i.e. the air is rarefied. In the expanding duct, beyond the throat, the streamlines diverge, the airflow decelerates, the pressure rises, and such small change in density as may occur is an increase, i.e. the air is compressed.

What happens if we gradually speed up the flow through the venturi tube? The reader can probably answer that from what we have discovered first about transonic flow, then about supersonic flow.

The first thing that will happen is that when the airflow, at the throat, reaches the speed of sound, a shock wave will be formed just beyond the throat just as it was on the camber of an aerofoil; so instead of the gradual rise in pressure and fall in speed beyond the throat, there will be a sudden rise in pressure, a sudden fall in speed, and the flow behind the shock wave will

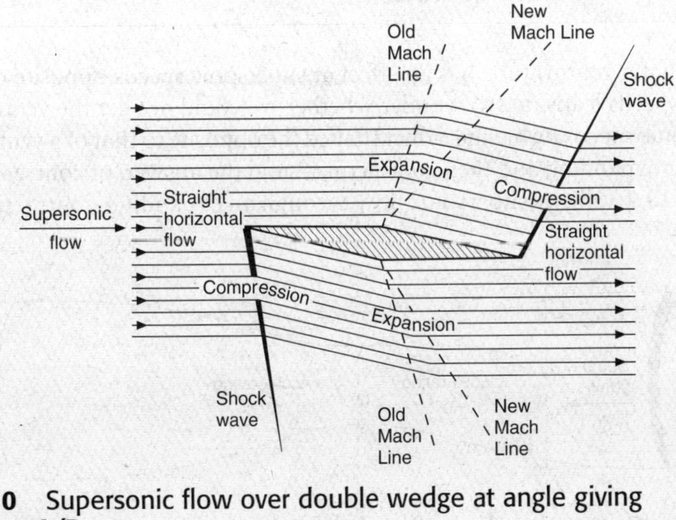

Fig 12.10 Supersonic flow over double wedge at angle giving maximum ***L/D***

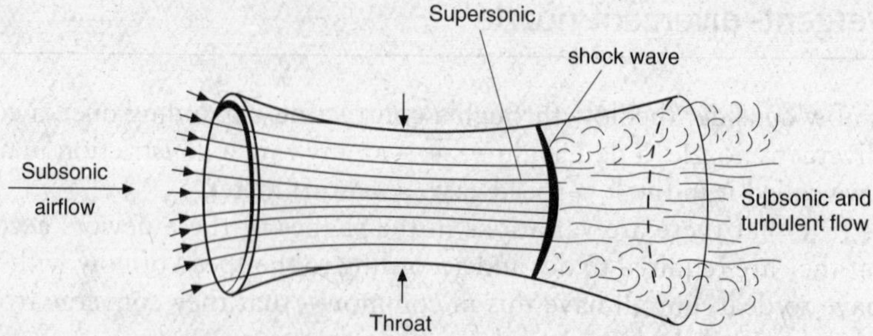

Fig 12.11 Incipient shock wave in a venturi tube

become turbulent (Fig. 12.11). The front portion of the venturi will still function reasonably like a venturi, i.e. the velocity will increase and the pressure fall, but the rear portion will no longer serve its proper purpose.

It is interesting to note that if the upstream flow is subsonic then at the throat the speed can never be greater than the speed of sound. If the flow does reach this speed it may then accelerate to supersonic speed downstream of the throat (Fig. 12.11). In this condition the duct is said to be choked. If the downstream pressure is reduced the supersonic region extends and the shock moves out of the divergent section.

If the flow in the duct ahead of the contraction is supersonic, we find that the flow behaves in the opposite way! This time the speed reduces in the convergent section and increases in the divergent section, reaching a minimum value at the throat (Fig. 12.12). The increase in speed is still, however, accompanied by a decrease in pressure and vice versa.

Expanding–contracting duct

The fact that a venturi tube has an effect at supersonic speeds opposite to that at subsonic speeds leads one to wonder whether we could not get the venturi effect at supersonic speeds by having a duct shaped the opposite to that of a venturi tube, i.e. by first expanding and then contracting – and the answer, of course, is yes.

Figure 12.13 shows supersonic flow through an expanding-contracting duct.

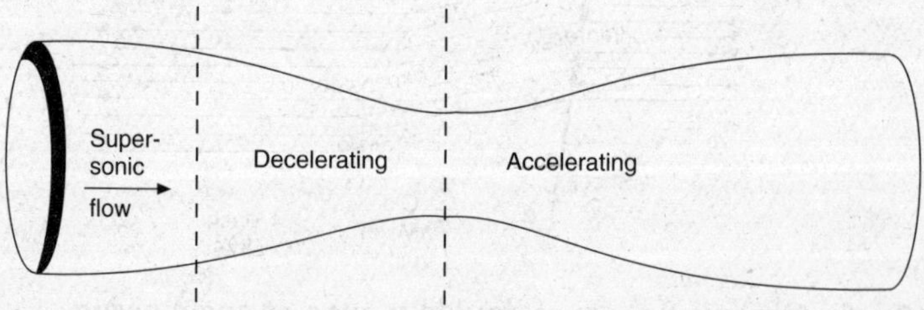

Fig 12.12 Contracting-expanding duct – supersonic upstream flow

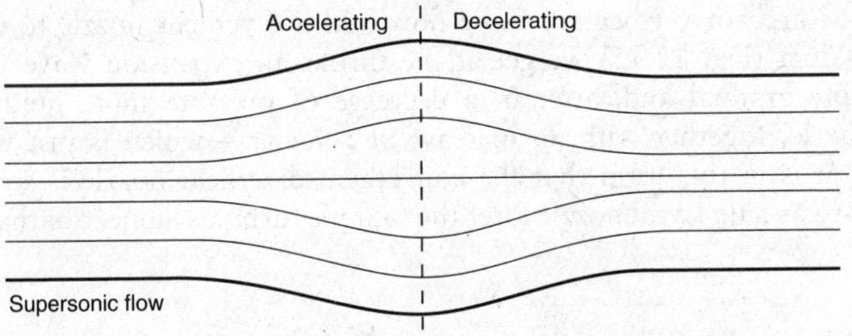

Fig 12.13 Supersonic flow through an expanding-contracting duct

The purpose of ramjets and jet engines is to provide the thrust to propel the aeroplane or missile, and it can only do this if the velocity of outflow from the engine is greater than the velocity of the aeroplane or missile through the air. The air enters the ramjet or turbojet at the inlet where it arrives with the velocity of the aeroplane; if this is above the speed of sound we can by a clever arrangement of a centre body in the inlet (Fig. 12.14) cause shock waves to be formed here and so put up the pressure which, in the case of the turbojet, is further increased by the compressor itself. The air then speeds up in the expanding duct, and the burning of the fuel adds still further to its energy. When the gases leave the jet pipe a system of shocks and expansion wave will form in the emerging jet if the pressure is not matched to that of the atmosphere at exit, resulting in losses and consequent inefficiency.

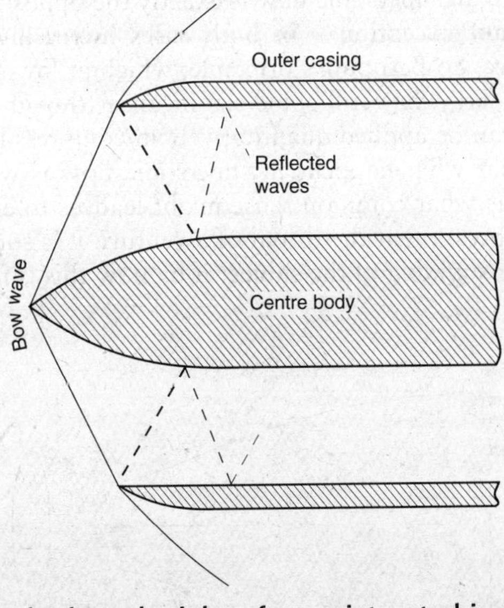

Fig 12.14 Centre body at the inlet of a ramjet or turbine
Since the angle of the bow wave will depend on the Mach Number, the centre body must be movable to be fully effective.

But we are not yet beaten. If we now add a divergent nozzle to the contracting duct (Fig. 12.15) we get at the throat an expansion wave which is reasonably gradual and, after it, a decrease of pressure more gradually to atmospheric, together with an increase of velocity – which is just what we wanted. It is in this form that the convergent-divergent nozzle is sometimes referred to as a de Laval nozzle after the famous turbine engineer of that name.

Subsonic and supersonic flow – a summary

Now, I think, we are in a position to sum up the essential differences between subsonic and supersonic flow in contracting and expanding ducts, and in similar circumstances such as over aerofoils or other bodies.

	In a Contracting Duct	*In an Expanding Duct*
Subsonic Flow	Flow accelerates Air rarefies slightly Pressure falls	Flow decelerates Air is compressed slightly Pressure rises
Supersonic Flow	Flow decelerates Air is compressed Pressure rises	Flow accelerates Air is rarefied Pressure falls

In short, everything in supersonic flow is exactly the opposite to subsonic flow with **one important exception – in both cases increasing speed goes with decreasing pressure.** So Bernoulli's principle, which at low speeds is really the conservation of energy, has still some significance (though modifications are needed before it can be applied quantitatively in compressible flow).

Notice, too, that what happens in supersonic flow is what we said in an earlier chapter was what common sense might lead us to expect – a decrease of speed and compression at the throat of a venturi. It is some measure of our learning and understanding of the subject if by now this is no longer common sense!

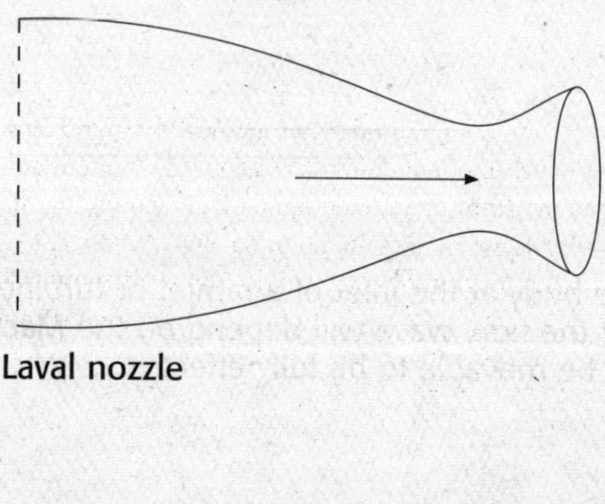

Fig 12.15 A de Laval nozzle

Boundary layer and supersonic flow

It may be noticed that in all this we have said little or nothing about the boundary layer, and it was the boundary layer that caused all the trouble in subsonic flow when it came to corners, and it was the boundary layer that was so important in transonic flow when the incipient shock wave was formed, and for which we had to think of such devices as vortex generators.

The truth is that the **boundary layer is relatively unimportant in supersonic flow**; it is **thin** and the **viscous forces within it are relatively small**. This largely accounts for the ability of supersonic flow to turn sharp corners.

Curiously enough, at the even higher speeds which will be mentioned later in this chapter, the boundary layer thickens again, and once more becomes significant. So the supersonic region is especially privileged in this respect, and in many respects the theory of the flow is simpler than over any other range of speeds.

Supersonic wing shapes – plan form

In flight at subsonic speeds the shape of the aerofoil section is in some respects more important than the plan form of the wing, but at supersonic speeds it is the plan form which is the more important (Fig. 12B, overleaf). On the other hand, the more one studies the seemingly endless variety of both aerofoil section and plan form that are not only possible but seem to have proved successful in supersonic flight, the more one is forced to the conclusion that neither shape matters very much; supersonic flow is more accommodating than subsonic flow, less fussy in what it encounters, and although, compared with subsonic flow, the lift coefficient is less, the drag coefficient greater, and the *L*/*D* ratio in consequence lower, the actual values of C_L, C_D, and *L*/*D*, and the position of the centre of pressure seem to be little affected by the shapes of either the cross-section or the plan form of the wing.

Let us consider first the plan form. It will be remembered that in the transonic region there was advantage in a considerable degree of sweepback of the leading edge because it delayed the shock stall, the increase of drag, buffeting, and so on – in other words, it raised the critical Mach Number. It is often stated that there is no advantage in sweepback after the critical Mach Number has been passed, and that straight wings are better for supersonic flight. This might be true if the **only** effect of sweepback was to delay the critical Mach Number – but actually it does more than this.

Consider, for instance, the plan shapes A, B, C, and D (Fig. 12.16, overleaf); with the possible exception of B, all these have been used on high-speed aircraft. At the apex of each are shown the Mach Lines for a Mach Number of about 1.8, and it will be noticed that the leading edges for these shapes all lie within the Mach Cone, and this in turn means that the airflow which strikes the wing has been affected by the wing before it reaches it; if, as is probable, there are also shock waves at the nose of the aeroplane, or at the apex of the wing, the whole of the leading edge of the wing will be behind these shock waves and so will encounter an airflow of speed lower than that of the aeroplane. This airflow may not be actually subsonic, but at least the resolved part of it at right angles to the leading edge, or across the chord, is likely to be. So although a swept-back wing is better than an unswept wing in the transonic region, it may retain some of its advantages even into the supersonic region – and this applies particularly to thick wings which are naturally more prone to the formation of shock waves.

Of course, if we are to keep within the Mach Cone the sweepback must increase with the Mach Number, until eventually the delta shape may be more appropriately described as an arrow-head shape (Fig. 12.17).

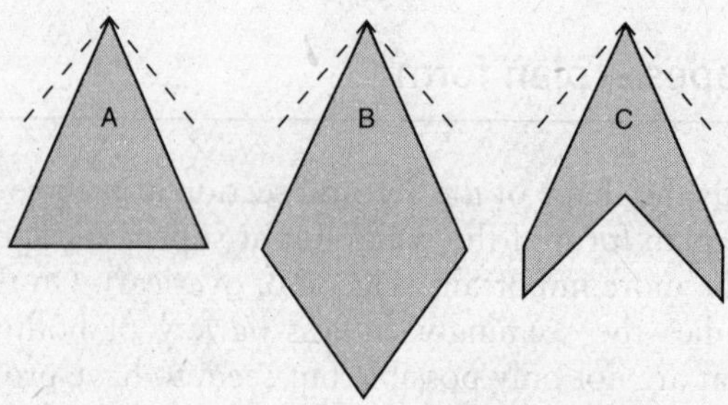

Fig 12.16 Supersonic wings – plan shapes

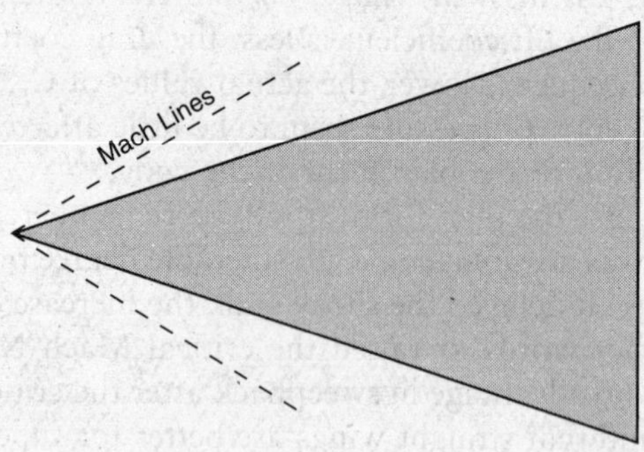

Fig 12.17 Arrow-head delta wing for high Mach Numbers

Fig 12B Supersonic configuration (opposite)
(By courtesy of British Aerospace Defence Ltd, Military Aircraft Division.)
The Typhoon Eurofighter prototype with delta wing and canard foreplane for control.

ZH588

But whenever we discuss the advantages of sweepback we must never forget its disadvantages which are largely structural; the twisting and bending stresses on a heavily swept-back wing give many headaches to the designer and mean extra weight to provide the strength. But there is also the old bogey of tip stalling and lateral control near the stalling – and landing – speed. Shapes A and B are better structurally than C and D, they are better, too, from the point of view of tip stalling; they also have an interesting, though perhaps rather concealed, advantage in that owing to the long chord the wing can be **thick** (which means a good ratio of strength to weight), yet still **slim** as regards **thickness/chord ratio** (which is what matters as regards shock drag). Have C and D then no advantages? – it would be strange if they hadn't, because they are in fact more common in practical aeroplanes than B, until recently than A also, but A is becoming increasingly popular in modern designs for supersonic transport. The advantage of C and D lies chiefly in lower drag (in spite of the point mentioned above), and so in better lift/drag ratio; they are also more suitable for the conventional fuselage and control system (if that is an advantage), and for engine installation.

One rather unexpected bonus resulting from the use of delta wings, or others with extreme taper and sweepback of more than 55° or so, comes from the stall itself; this is a leading edge stall which starts at the wing tip and progresses gradually inboard, the separation bubble is then swept back with the leading edge and shed as a trailing vortex, tightly rolled up and with a very low pressure at its core. The low pressure acts on the forward facing parts of the upper surface of the wing giving a 'form thrust' (in effect a negative drag) and a lift boost; moreover the flow in the core is stable and causes little buffeting, unlike the separation vortex on wings with sweepback of less than 50°. This is, in fact, an effective way of producing lift. Concorde used it at both subsonic and supersonic speeds. The use of fences, saw teeth and vortex generators can, at best, only give partial mitigation of the resulting stalling phenomena such as the buffeting, wing drop and pitch up.

But whatever the pros and cons of sweepback there is no doubt that there is a lot to be said for the straight rectangular wing for really high supersonic speeds (Fig. 12.18). With the small aspect ratio, and tremendously high wing loading associated with such speeds, the wings are very small anyway, and from the strength point of view a rectangular wing, or a wing that is tapered for structural reasons rather than for aerodynamic reasons, will probably win the day.

Figure 12.19 shows how only small portions of a rectangular wing at supersonic speeds (the shaded areas) can know of the existence of the tips, and these portions will tend to exhibit the normal characteristics of 3-dimensional subsonic flow, wing-tip vortices, etc., while the flow over the remainder of the wing will be straight 2-dimensional flow as if the wing was of infinite span and there were no wing tips. This leads to a rather obvious suggestion – cut off the shaded portions.

This, in fact, is sometimes evident in design, but the arguments one way and another, for sweep at leading or trailing edge (or both, or neither), for delta

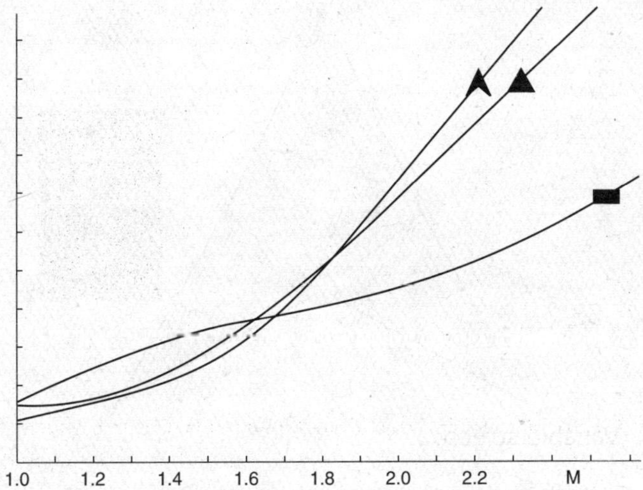

Fig 12.18 Wing shapes and drag
How the drag increases with Mach Number for straight, highly-swept and delta wing shapes.

and arrow plan forms, for tail first or wing first, for tail to be larger or smaller than the wing, even to decide which is the wing and which the tail – that these arguments are endless is clearly evident from the numerous shapes and configurations which have been tried or suggested for missiles or supersonic aeroplanes.

The fundamental difficulty, for aircraft rather than missiles, is to provide wings that are suitable not only for supersonic flight, but also for subsonic and transonic flight. After all, supersonic aeroplanes have to take off and land; and they also have to pass through the transonic region. The real answer – so far as plan form is concerned – is surely in **variable sweep** (Fig. 12.20, overleaf); advocated many years ago by an eminent British inventor but, as so often

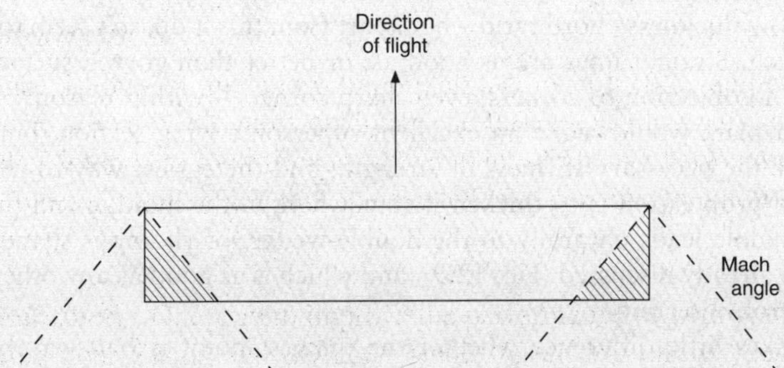

Fig 12.19 Rectangular wing at high Mach Numbers

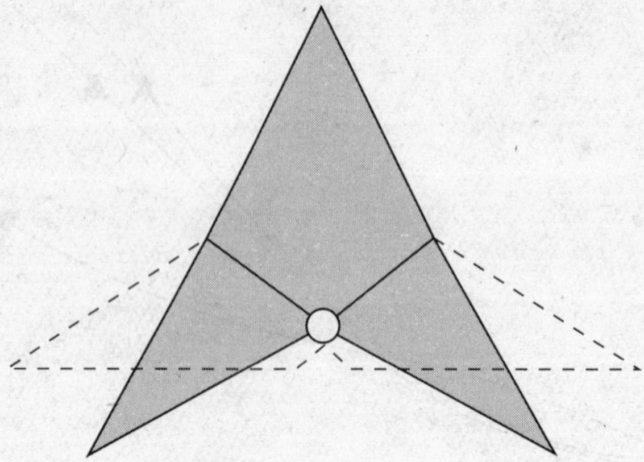

Fig 12.20 Variable sweep

happens, left to others to put into practice; the Americans had serious teething troubles with their first real effort in this direction, the General Dynamics swing-wing F–111, of which so much was expected. Perhaps as a result of this disappointment, the Boeing idea of a swing-wing supersonic transport was also abandoned. Nor, in the meantime, were the French initially successful with their Mirage G (Fig. 12C, overleaf), but accumulated experience pays off and the variable sweep concept has been adopted for a number of aircraft including the Tornado.

The design process is finely balanced though; a great number of solutions to the problem of supersonic flight abound, such as the simple long-nosed delta configuration of the Anglo-French Concorde (Fig. 12D, overleaf).

Supersonic wing shapes – aerofoil sections

What then of the shape of the aerofoil section?

The proviso here is that it must be thin – or to be more correct, that it must have a low thickness/chord ratio – but apart from this it doesn't seem to matter very much. Straight lines are as good as, or better than curved surfaces; and there is no objection to corners, even sharp corners – within reason.

A flat plate would make an excellent supersonic wing section, but would not have the necessary stiffness or strength; and the easiest way to make it a practical proposition is to thicken it somewhere in the middle, and thickness in the middle leads naturally to the double-wedge, or rhombus shape, which we have already discussed (Fig. 12.9), and which is as good as any other supersonic aerofoil section.

It makes little difference whether the thickest point is half-way back, or more or less; there is little change of drag for x/c ratios between 40 per cent and 60 per cent (Fig. 12.21), and the lift and centre of pressure positions are not affected at all. But we have always got to consider flight at subsonic speeds and, from this point of view, maximum thickness should be at 40 per cent of the chord rather than farther back; from this point of view, too, it may pay to round the corners slightly.

A variation of the double-wedge is the hexagonal shape (Fig. 12.22). This gives greater depth along the chord and so greater strength, and also makes the leading edge rather less sharp, which has advantages both as regards strength and, as will be considered later, aerodynamic heating.

A bi-convex wing is also quite good (Fig. 12.23), and this is better than the others at subsonic speeds. A bi-convex wing has about the same drag as a double-wedge with maximum thickness rather outside the best range, i.e. at about 25 or 75 per cent of the chord.

Enough has been said about supersonic aerofoil sections to make it clear why the sections in Appendix 1 are all of subsonic type; there would not be much point in giving the contour of a flat plate, or even of a double-wedge – and moreover there is little difference between the lift and drag coefficients of all reasonable shapes, and still less difference in the positions of centre of pressure.

Theory predicts a maximum value of L/D of 12.5 for a wing with a thickness/chord ratio of 4 per cent at a Mach Number above about 1.3. (Note that

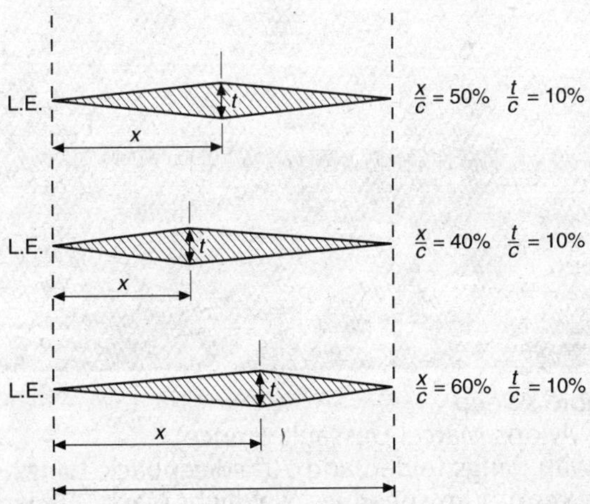

Fig 12.21 *x/c* ratio

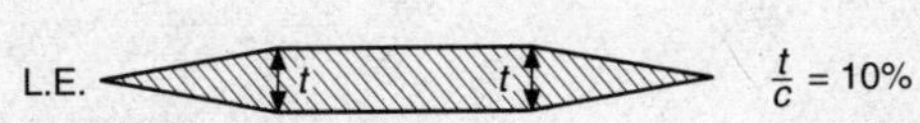

Fig 12.22 Hexagonal wing section

Fig 12.23 Bi-convex wing section

Fig 12C Variable sweep
(By courtesy of Avions Marcel Dassault, France)
The Mirage G, with wings folded and 70° sweepback, wings extended and 20° sweepback; maximum speed in level flight Mach 2.5, landing speed 110 knots; no ailerons, lateral control (wings back) by differential action of slab tail plane supplemented when wings are spread by spoilers on wings.

Fig 12D Concorde (opposite)
(By courtesy of the British Aircraft Corporation)

BRITISH AIRWAYS
BOAG

in this statement there is no reference to the shape of the wing, or where is the greatest thickness.) This value is inferior to *L/D* ratios for subsonic wing shapes (only about half), but it is reasonably economical when everything is taken into consideration. The lift coefficient is the same for all the shapes, and although it is smaller than those of subsonic aerofoils this does not matter at high speeds; where it does matter is that it means high stalling and landing speeds, which in turn mean long runways, and devices such as tail parachutes to help reduce the speed after landing. A leading-edge flap, or a permanent droop at the leading edge (sometimes called a droop-snoot), will appreciably lower the landing and stalling speed of a supersonic aerofoil section. As with plan shape the only way of making an aerofoil suitable for subsonic, transonic, and supersonic flight is to make it variable in shape; but in this case we know that it can be done because, in fact, it has long been done – so the only question is the best way of doing it.

Figure 12.24 shows the pressure distribution, and position of the centre of pressure, for three shapes at a small angle of attack. Comment would be superfluous.

Supersonic body shapes

The considerations which decide body shapes for supersonic speeds are similar to those which apply to wing sections. Bodies should be slender, but there are limits in practice owing to the need for room inside the body, for stowage, etc. All in all, the optimum fineness ratio is about 6 to 8 per cent.

Also for reasons of stowage and body capacity there are advantages in curves rather than straight lines, and in a rounded nose and tail (Fig. 12.25*a*). From the drag and speed point of view the nose should be sharp pointed, and often is; but there are disadvantages – the pilot's view is bad (hence the droop-

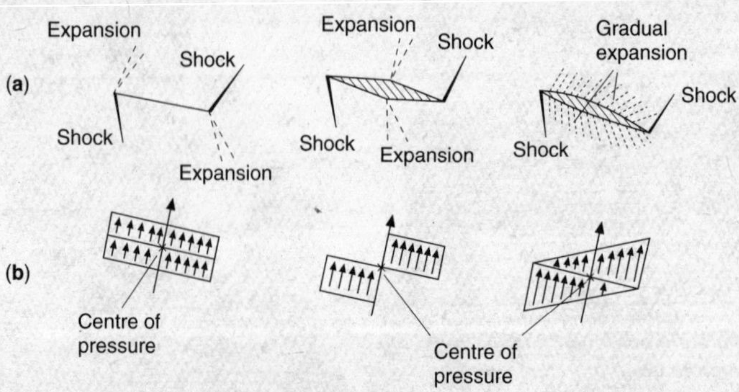

Fig 12.24 Supersonic pressure distributions
(*a*) Pattern of shocks and expansions.
(*b*) Pressure distribution and centre of pressure.

snoot as on the Concorde), the sharp point is useless for stowage, and the transmission of radar pulses is unsatisfactory. The tail portion can be cut off, like the rear of a bullet, without much loss of efficiency; and this is necessary in any case when the jet or rocket efflux is at the rear of the body (Fig. 12.25*b*).

A body or fuselage of some kind is clearly necessary if the aircraft is to carry pilot, passengers, mail, or goods, and if the wings are to be thin – but are wings really necessary at supersonic speeds? Bodies can easily be designed to give lift (whatever their shape they will give lift at a small angle of attack), but cannot the thrust be used to provide the lift? In earlier chapters we talked of flying wing; why not a flying body?

Well, of course, a rocket can be nothing more nor less than a flying body – and more will be said about rockets in the next chapter – but even rockets need guidance and, within the atmosphere at least, guidance and control are best achieved by fixed and movable surfaces. There is also, so far as aeroplanes are concerned, the not unimportant point of getting back to earth.

But the reason for mentioning this problem is something quite different. One advantage in having a wing at supersonic speeds is that the presence of the wing improves the lift on the body – there is interference between wing and body, but it is useful interference; and it is mutually useful, because the body produces an upwash which improves the lift of the wing.

There are great possibilities in the exploitation of beneficial interference at supersonic speeds, and it is something which we may hear a lot more about in the future. Another example of it has already been mentioned in connection with putting a centre body at the inlet of a ramjet or jet engine. A suggestion has even been made of beneficial biplane effects, by eliminating external shock bow waves, and using the shock between the wings to good effect as in an engine intake – perhaps making the biplane the engine. Who knows?

A form of area rule is still effective in reducing shock drag at supersonic speeds, but its application is rather more complicated. Since the shock waves and Mach Lines are now oblique, instead of being at right angles to the flow, the 'area' which must change smoothly is not that at right angles to the line of

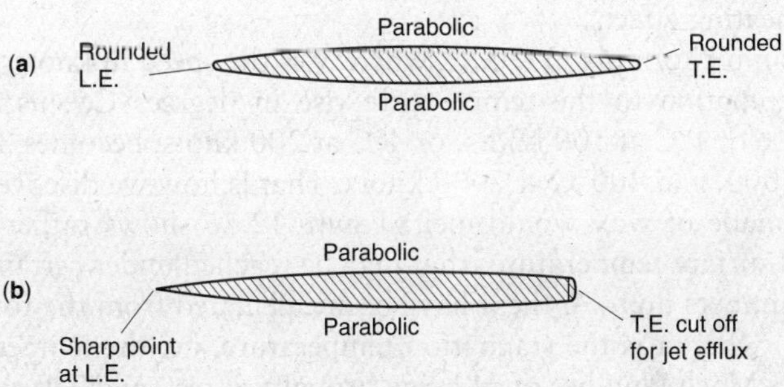

Fig 12.25 Supersonic body shapes

flight, but in planes parallel to the Mach Lines. And unfortunately the inclination of the Mach Lines depends on the Mach Number at which the aircraft is flying, so the shape of the aircraft can only be correct for a particular Mach Number.

Kinetic heating

We all know that friction increases temperature, an example of deterioration of energy from the highest to the lowest form (from mechanical energy to thermal energy), a natural process – and skin friction in the flow of fluids is no exception. We all know, too, that an increase of pressure, as in a pump, raises temperature – and the stagnation pressure on the nose of a body or wing is no exception. So when an aeroplane moves through the air it gets hot; some parts more than others, some owing to the temperature increase created by skin friction, some owing to that created by pressure.

When, then, do we first come up against this? The answer is – when we first fly! But it isn't serious? No – like many other things, it isn't serious at low speeds. It has been said that aeroplanes made of wax melt at 300 to 400 knots, those made of aluminium at 1600 to 1800, those of stainless steel at 2300 to 2400 knots. Aeroplanes are not made of wax (wind tunnel models sometimes have been), but some are made of aluminium alloy, and some of stainless steel, and of other metals (such as titanium) and their alloys, and just because of this very problem. Nor can we afford to go anywhere near the melting point; metals are weakened long before that – and what about the passengers, and crew, and freight?

Bullets and shells certainly travel at speeds where heating is significant, but within limits it doesn't matter whether bullets and shells get hot or not. Also their flight is not usually of very long duration, and it takes time for the surface to heat up. But meteors and satellites re-enter the atmosphere without any means of braking – and we know what happens to them, they get frizzled up. It is true that most manned space-craft have survived re-entry and more will be said about that in the next chapter. Let us at least consider what we can do to reduce heating effects.

A very simple formula $(V/100)^2$, where V is the speed in knots, gives a very fair approximation to the temperature rise in degrees Celsius. So what is merely a rise of 1°C at 100 knots, or 4°C at 200 knots, becomes 36°C at 600, 100°C at 1000, and 400°C at 2000 knots. That is how we discovered that the aeroplane made of wax would melt! Figure 12.26 shows rather more accurately local surface temperatures that may be reached under certain conditions at Mach numbers up to 4; these have been calculated from the formula $t/T = (1 + M^2/5)$ where t is the stagnation temperature, i.e. the temperature of air moving at a Mach Number of M being brought to rest, and T is the local temperature of the air; the figures relate to 8500 m where the local temperature is

−40°C. The temperatures shown in the graph apply to a laminar boundary layer; the temperatures are rather higher for a turbulent boundary layer. Moreover, at Mach Numbers above 2 these surface temperatures may be reached in a matter of seconds, and certainly within a minute or two, unless there is some method of insulation.

Many devices have been tried, and no doubt many more will be tried, in an effort to counter this heating problem. These devices may be classified under the following headings –

(*a*) To insulate the structure from the heat.

(*b*) To use materials which can stand the high temperatures without serious loss of strength.

(*c*) To encourage radiation from the surfaces and so reduce the temperatures.

(*d*) To circulate a cooling fluid below the surface.

(*e*) Refrigeration by any of the normal methods.

As regards materials for the aircraft structure light alloys are suitable for Mach Numbers up to 2, or even higher for short periods. Between *M*2 and *M*4 titanium alloy may be the answer, but above 3 or 3.5 stainless steel is probably better as being more readily available.

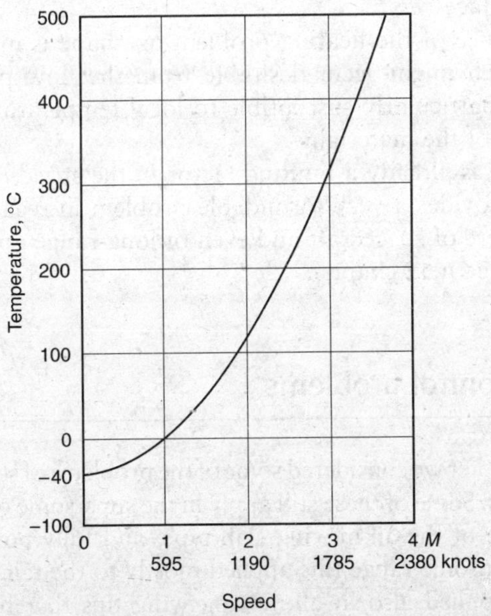

Fig 12.26 How the surface temperature rises with the Mach Number
The graph relates to a height of 28 000 ft (8500 m) where the local temperature of the surrounding air is −40°C.

It must be remembered that the crew, the equipment and the fuel must be protected as well as the structure itself, so there is no point in using materials which will stand the high temperatures, unless there is also refrigeration to keep the interior of the aircraft cool.

Perhaps the most ingenious idea is to apply the heat to a suitable working fluid (hydrogen has been suggested), and to eject the expanding fluid through a suitable nozzle, and so propel the aircraft! Ingenious and fascinating – drag produces heat, heat produces thrust to help overcome the drag. In principle it is not impossible.

An interesting aspect of surface heating is the effect of shape. It is the speed of flow adjacent to the boundary layer which is the deciding factor in the temperature rise – and to some extent, of course, the nature and thickness of the boundary layer itself – and the speed of flow depends on the shape of the body. But there is more in it than that. A rise in temperature is created owing to skin friction, and owing to the stagnation pressure, but it is also created by shock waves, and whereas the main effect of skin friction in the boundary layer is to raise the temperature of the surface, the main effect of the shock waves is to raise the temperature of the air – and that doesn't matter very much. So from the point of view of keeping down surface temperatures it is better to have wave drag than boundary layer drag. This conclusion isn't very helpful with regard to aircraft in which we try to reduce every kind of drag to a minimum, but it is a most important consideration in designing bodies for re-entry to the atmosphere from space, bodies in which we want drag, but we don't want heating of the surfaces.

Another influence of the heating problem on shape is in the avoidance of sharp edges, which might seem desirable from the flow point of view, but which would be particularly susceptible to local temperature rise and consequent weakening of the material.

Kinetic heating is already a limiting factor in the speed of certain types of aircraft, and it provides a very formidable problem in regard to the re-entry into the atmosphere of spacecraft and even of long-range missiles such as will be considered in the next chapter.

Stability and control problems

In the previous chapter we considered some of the problems of stability and control at transonic speeds. Some of these still apply in the supersonic region, and control surfaces should be of the all-moving slab type, and fully power-operated. But whereas in the transonic range this applied mostly to the tail, in the supersonic range it can be applied also in all-moving wing tips to replace conventional ailerons, and even to an all-moving fin and rudder. Since the main plane may be nearly as small as the tail plane, it, too, may be movable to give pitch; in fact in some missiles it is not easy to decide which is the main plane and which the tail plane.

But there are also new problems at supersonic speeds because the inertia forces are so great that it is practically impossible to provide the inherent natural stability of the kind that is associated with such devices as dihedral and fin area. In order to be effective against the inertia forces the surfaces would have to be so large that the cost in weight and drag would be prohibitive – and this applies particularly at great heights where there is so little air density.

Another difficulty, which is especially applicable to military aircraft, is that pilot and crew have so much to do in looking after the equipment that they must be relieved as far as possible of flying the aeroplane.

Of course we have long been familiar with the automatic pilot, but the modern conception is very different from this – nothing more nor less, in fact, than synthetic stability and automatic control. Is the pilot then necessary at all? Strictly speaking, probably not, the aircraft can be controlled from the ground like a guided missile. But pilots can still do some things that instruments cannot, they can monitor the automatic systems, tell them what to do, investigate any failure and, if necessary, take over control.

Concorde

It seems fitting to conclude a discussion on the problems of flight, from subsonic to supersonic, with some comments on the design of the late lamented Concorde, because controversy on its cost, on its sonic boom and on its commercial viability have tended to obscure the cleverness of its design. Readers who have followed the arguments put forward in this book will surely be fascinated by some of its outstanding features. It would possibly be going too far to describe one of the main features of such a costly and sophisticated piece of equipment as simplicity yet, as we shall see, there is some truth in such a description.

The now familiar ogee (or double curve) plan shape of the wing gives both a large chord (27.7 m), with its advantages, and a large span (25.6 m), with its advantages.

The large chord means that although the wing is thin, very thin, from the aerodynamic (t/c) point of view (only 3 per cent at the root and 2.15 per cent outside the engine nacelles), and so has low wave drag; yet at the same time it is deep enough (83 cm) to give the required strength and structural stiffness, usually a difficulty with slender swept-back wings. The main advantage of the large span is the reduction of vortex, or induced, drag at all speeds.

The slimness of wings and body (the Boeing 747 Jumbo Jet is longer, higher and much fatter than the Concorde), and the limitation of speed to just over Mach 2, have kept down the temperature rise and made it possible to use aluminium alloys, with which we have long been familiar, for most parts of the structure, instead of having to experiment with the more costly, and heavier, stainless steel or titanium alloys. The temperature rise in the structure is also reduced by using the fuel as a heat sink. The maximum landing weight is 1068 kN, and the maximum landing wing loading 4.786 kN/m^2, less than that of many comparable aircraft.

The ogee plan shape has another advantage in that the stalling angle is so large that it is unlikely to be reached in any ordinary condition of flight; this is because the shape leads to the formation of leading edge vortices (without any vortex generators!), and so improves the flow in the boundary layer and gives smooth changes of lift and pitching moment with angle of attack.

On the wing there are no flaps, no slots, no tabs, no spoilers, no saw teeth, no fences, nor any other devices usually required for such a large speed range as from 65.4 m/s (at 18° angle of attack) to a true speed of 649 m/s. The only moving surfaces on the wing are the six elevons (combined ailerons and elevators) which control both rolling and pitching – and very effectively too. The rudder has two sections, but is otherwise simple and conventional.

The large 'leading edge' vortices are useful when landing as lift continues to increase to large values with increasing angle of attack, the 'lift boost' and the 'form thrust' already mentioned in connection with highly-swept wings, and the large area of the delta wing gives a considerable cushioning effect when near the ground (two reasons for dispensing with flaps).

Perhaps one of the most interesting features, taking us back to one of the earliest ideas for adjusting trim is the movement of fuel between tanks, automatically between the main tanks for adjusting the centre of gravity during cruising flight and, under the pilot's control, from forward tanks to a rear tank under the fin during acceleration to supersonic flight when the aerodynamic centre moves back, and vice versa when returning to subsonic flight.

Pilots report that it is only by looking at the machmeter that they know when Concorde is going supersonic.

Flight at hypersonic speeds

In concluding this subject we must not leave an impression that once we have conquered the problems of supersonic flight, we have finished. Far from it – no sooner do we learn to get through one region of speeds than another, with new problems to solve, opens up before us.

Rather strangely, too, there does not seem to be much argument about where the change takes place in this case – above a Mach Number of 5 we talk of hypersonic speeds instead of supersonic speeds. It doesn't pay to try to see the sense of this rather extraordinary use of the English language, or perhaps we should say of Latin and Greek, which makes hypersonic superior as regards speed to supersonic; nor shall we get very far if we try to discover just why the Mach Number of 5 is so significant; actually it is a number of different factors that decides the issue, and that is far beyond the scope of this book.

To us hypersonic flight is simply supersonic flight – only more so. Here we are deep into kinetic heating effects with all the associated problems; at a Mach Number of 5 at about 61 km the temperature given by the formula is over 1000°C and by Mach 15 it has risen to over 10 000°C. The actual temperatures are found to be rather lower, but not so much lower as to give any real comfort!

Mach Lines are inclined at very acute angles; Fig. 12.27 shows the shock waves and Mach Lines over a double-wedge section at a Mach Number of 10, and Fig. 12.28 is sketched from a photograph of a bullet moving at the same Mach Number. These suggest an arrow type of aircraft as being most suitable (Fig. 12.29).

One feature of hypersonic flow is a thickening of the boundary layer and an increased importance of the nature of the flow within the boundary layer.

Aerofoil shape seems to matter even less than in supersonic flight; lift and drag coefficients tend towards a constant value as the Mach Number increases.

An interesting aspect of this part of the subject is the wide variety of experimental methods used to investigate it – arc-heated jets, gun tunnels, shock tubes, shock tunnels, hot shot tunnels, models moved by rockets or guns, and ballistic ranges; names that are all to some extent descriptive of the methods

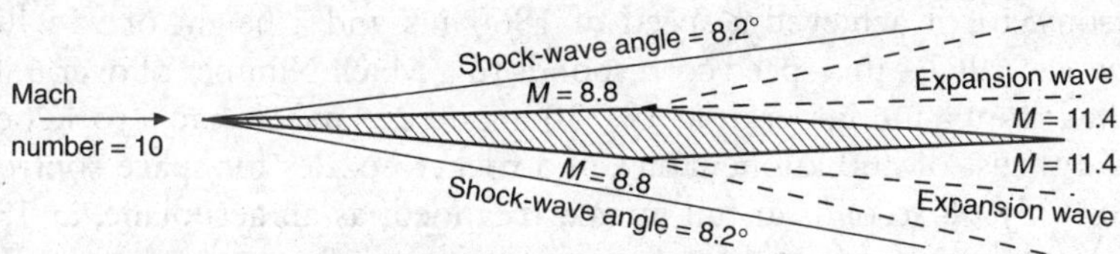

Fig 12.27 Hypersonic shock and expansion waves

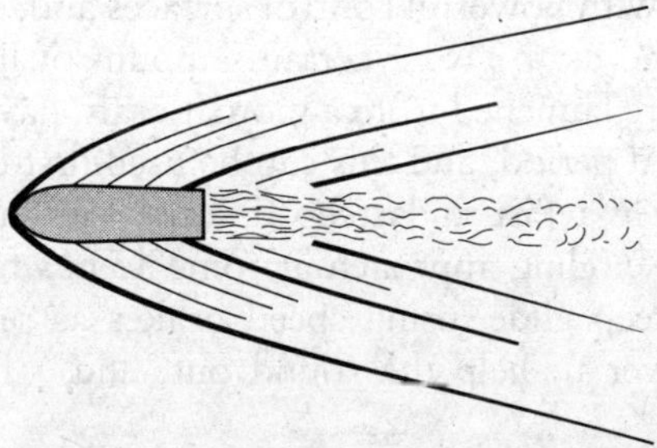

Fig 12.28 From a photograph of a bullet moving at Mach 10

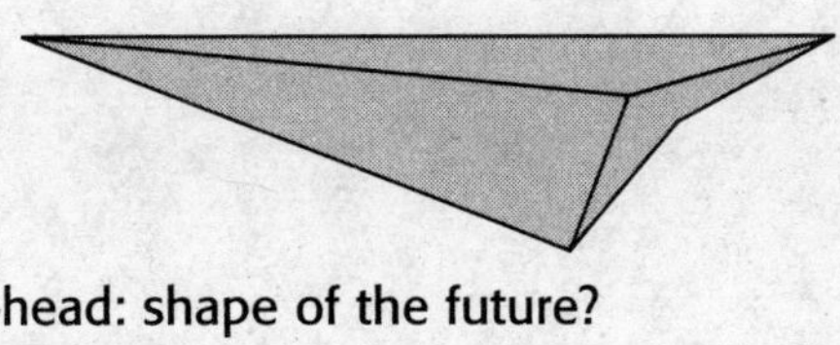

Fig 12.29 Arrow-head: shape of the future?

employed, each of which would need a chapter to itself. Mach Numbers of 15 can be achieved in still air on ballistic ranges, and even more if the projectile is fired against the airflow in a wind tunnel.

But even hypersonic flow is not the end; at Mach Numbers of 8 or 9 something entirely new begins to happen – molecules, first of oxygen, then at even higher speeds of nitrogen, dissociate, or split into atoms and ions, thus changing the very nature of the air and its physical properties (another phenomenon that is experienced by space-ships on re-entry into the atmosphere, and which affects radio communication with them). At this stage there are possibilities of the control of the flow by electro-magnetic devices.

And so it goes on. Nothing, so far, suggests an end. Figure 12.30 shows how speed records have gone up – and up – and up.

Lifting bodies

So far as aeroplanes are concerned, the North American X15 was quoted in an earlier edition of this book as the nearest approach to a piloted hypersonic aircraft – indeed, if we accept M5 as the threshold of hypersonic flight, it was hypersonic for it achieved a speed of 1860 m/s and a height of 95 940 m as long ago as 1962 – this speed corresponds to a Mach Number of over 6. It was launched from a mother craft at 13 720 m and 224 m/s; it had a rocket engine giving a thrust of 260 kilonewtons, and rocket nozzles for space control, but fuel for only 68 seconds at full power. It landed, as an aeroplane, at 134 m/s some 320 kilometres from launch.

The experimental programme with the X15 has now been superseded by a similar programme with what are called lifting bodies, with little if anything in the way of wings, with powerful control surfaces and, as their name implies, with bodies so shaped as to give a certain amount of lift at high speeds. As with the X15 these are launched from a mother craft, have rocket power available for a very limited period, and this can be used first to attain considerable height and then, after a fast and steep descent – usually without power – towards the earth, a circling approach at some 15 m/s for the last 6000 m or so, followed by a steep glide, using speed brakes as required, and finally a touch of rocket power to help the round out, and a fast landing at about 100 m/s.

These lifting bodies offer some advantages for future shuttle and sub orbital vehicles.

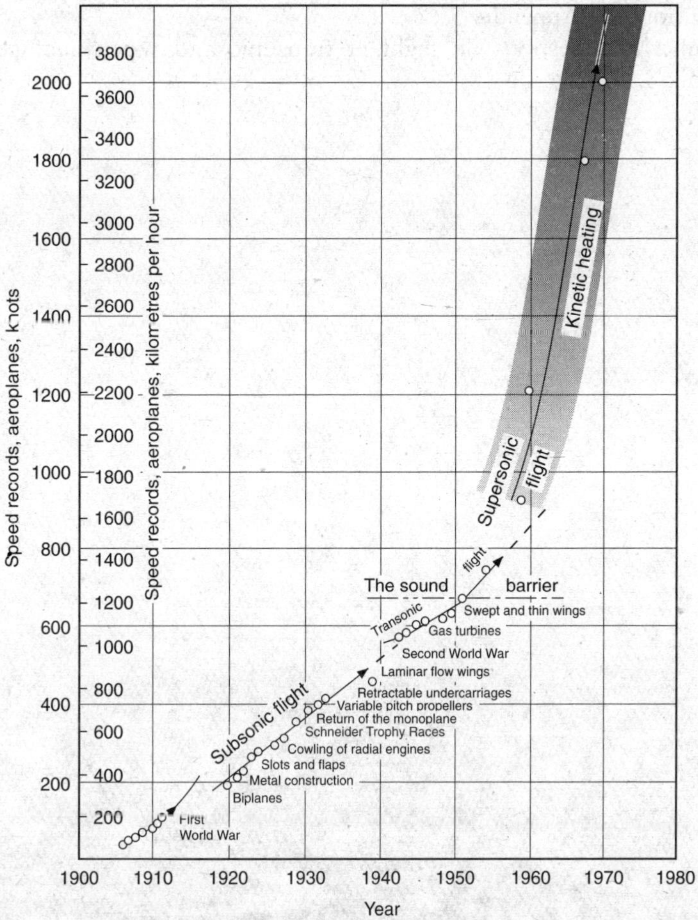

Fig 12.30 Where do we go from here?

Can you answer these?

1. What is a Mach Angle? a Mach Cone? a Mach Line?
2. Do shock waves travel at the speed of sound? If there is a difference, what effect does it have?
3. What is an expansion wave?
4. Why are sharp leading edges used on supersonic wings?
5. What influence is the heating problem having on the materials used for aircraft construction?
6. What are the main features of the wing used on the Concorde?

For solutions see Appendix 5.
For numerical questions on flight at transonic and supersonic speeds see Appendix 3.

Fig 12E Another supersonic configuration
The McDonnell Douglas F-18 Hornet which features a wing with only moderate rearward sweep on the leading edge, and a small forward sweep on the trailing edge. The sharply swept strakes also generate lift, and control the flow over the main wing.

CHAPTER 13

Space flight

Ballistics and astronautics

At first glance it may seem that space flight (Fig. 13A, overleaf) is purely concerned with ballistics and is completely divorced from aerodynamics, but as we hinted in Chapter 1 at the speeds and heights of modern flight aeroplanes behave to some extent like missiles, and missiles, given the required thrust and speed, can actually become satellites. Moreover, both missiles and artificial satellites must pass through the earth's atmosphere when they are launched, and there experience aerodynamic forces, and – what is far more important – the most hopeful means of returning them safely to earth again is to use aerodynamic forces to slow them up and control them.

In short, the subjects of aerodynamics, ballistics, and astronautics have merged into one subject, the mechanics of flight, and no apology is needed for the inclusion of this chapter in a book on that subject.

The upper atmosphere

The atmosphere with which we have been concerned in the flight of aeroplanes – i.e. the troposphere and the stratosphere – is sometimes called the lower atmosphere; the remainder is called the upper atmosphere (Fig. 13.1, overleaf).

In the lower atmosphere the temperature had dropped from an average of +15°C (288 K) at sea-level to −57°C (217 K) at the base of the stratosphere, and had then remained more or less constant. The pressure and density of the air had both dropped to a mere fraction of their values at sea-level, about 1 per cent in fact. One might almost be tempted to think that not much more could happen, but such an assumption would be very far from the truth.

There is a lot of atmosphere above 20 km – several hundred kilometres of it, we don't exactly know, it merges so gradually into space that there is really no

Fig 13A Shuttle lift off
(By courtesy of NASA)

exact limit to it – but a great deal happens in these hundreds of kilometres. The temperature, for instance, behaves in a very strange way; it may have been fairly easy to explain its drop in the troposphere, not quite so easy to explain why it should then remain constant in the stratosphere, but what about its next move? For from 217 K it proceeds to rise again – in what is called the mesosphere – to a new maximum which is nearly as high as at sea-level, perhaps 271 K; then, after a pause, down it goes again to another minimum at the top of the mesosphere. Estimates vary of just how cold it is at this height (only 80 kilometres, by the way, only the distance from London to Brighton), but all agree that it is lower than in the stratosphere, lower, that is to say, than anywhere on earth, perhaps 181 K (−92°C). But its strange behaviour doesn't stop at that and, once more after a pause at this level, as the name of the next region, the thermosphere, suggests, it proceeds to rise again, and this time it really excels itself rising steadily, inexorably to over 1200 K at 200 km, nearly 1500 K at 400 km, and still upward in the exosphere until it reaches over 1500 K at the outer fringes of the atmosphere.

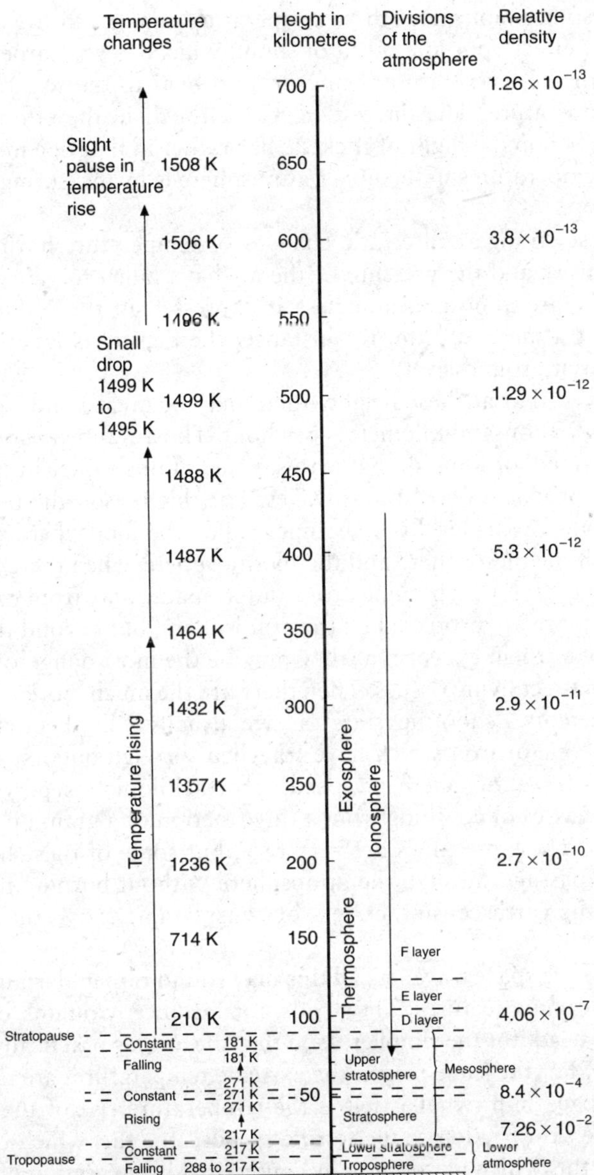

Fig 13.1 The upper atmosphere
The figures given are based on the US Standard Atmosphere, 1962, which was prepared under the sponsorship of NASA, the USAF and the US Weather Bureau.

An interesting point about these temperature changes in the upper atmosphere is their effect upon the speed of sound which, as we learned in Chapter 11, rises with the temperature, being proportional to the square root of the absolute temperature. The interest is not so much in the effects of this on shock waves, or on the flight of rockets, but rather in that one method of estimating the temperatures in the upper atmosphere is by measuring the speed of sound there.

While these strange and erratic changes of temperature have been taking place the density and the pressure of the air have fallen to values that are so low that they are almost meaningless if expressed in the ordinary units of mechanics; at a mere 100 km, for instance, the density is less than one-millionth of that at ground level.

It is believed that at these heights there may be great winds, of hundreds, perhaps even a thousand kilometres per hour. The air above about the 70 km level is 'electrified' or ionised, that is to say it contains sufficient free electrons to affect the propagation of radio waves. For this reason the portion of the atmosphere above this level is sometimes called the ionosphere, which really overlaps both the mesosphere and the thermosphere. Then there are the mysterious cosmic rays which come from outer space, and from which on the earth's surface we are protected by the atmosphere, but beyond this we know very little about them except that they may be the most dangerous hazard of all since they affect living tissues. Then there are the much more readily understandable meteors, 'shooting stars' as we usually call them, but actually particles of stone or iron which have travelled through outer space and may enter the earth's atmosphere at speeds of 100 kilometres per second, and which have masses of anything from a tiny fraction of a gram up to hundreds of kilograms. The larger ones are very rare, but some of these have actually survived the passage through the atmosphere without burning up, and have 'landed' on the earth causing craters of considerable size – these are called meteorites.

To prospective space travellers all this may sound rather alarming, but there are some redeeming features. The winds, for instance, wouldn't even 'stir the hair on one's head' for the simple reason that the air has practically no density, no substance. For the same reason the extreme temperatures are not 'felt' by a satellite or space-ship (what is felt is the temperature rise of the body itself, caused by the skin friction at the terrific speeds; it is this which burns up the meteors, it is this which has eventually caused the disintegration of many man-launched satellites on re-entering the atmosphere – but all this has little or nothing to do with the actual temperature of the atmosphere). Then, as regards the very low densities and pressures, no-one is going to venture outside the vehicle, or walk in space, or even put his head out of the window to see whether the wind stirs the hair on his head, unless he is wearing a space-suit, and we have long ago learned to pressurise vehicles because this is required even for the modest heights in the lower atmosphere. Moreover, the strong outer casing of the vehicle which is required for pressurising will in itself give protection at least from the small and common meteors, and to some extent even from the cosmic rays, the greatest unknown. So altogether the prospect is not as bad as it might at first seem to be.

The law of universal gravitation

Now let us consider the motion of bodies in this upper atmosphere, and in the space beyond.

If we throw a stone or cricket ball up into the air it goes up to a certain height, stops, and then comes down again. If we throw it vertically upwards it comes down on the same spot as that from which we threw it and, if we neglect the effects of air resistance, it returns with the same velocity downwards as that with which we threw it upwards. Moreover, again neglecting air resistance, we can easily calculate how high it will go because we know that (at first, at any rate) it loses velocity as it travels upwards at the rate of 9.81 m/s or, very roughly, 10 m/s every second, and gains it again at the same rate as it comes downwards.

But we ought to know better by now than to neglect air resistance? Yes, we certainly ought to know better, but unfortunately it wasn't only air resistance that we were neglecting in the simple examples in Chapter 1 – though, in fairness, that was our worst error, and our other omissions really were negligible in the circumstances that we were then considering. This is no longer true of the circumstances of this chapter for some of which air resistance really can be neglected, but other things that we calmly assumed most certainly cannot.

Newton would probably have been less surprised than we were when artificial satellites began to circle the globe – for these are but examples of the laws he enunciated.

When he saw the apple drop – assuming that story is true – he wondered why it did so, and eventually decided that it was because there was a mutual force of attraction between the apple and the earth; so it wasn't just a case of the apple dropping, it was the apple and the earth coming together, due to this mutual force of attraction, when there was no longer anything to hold them apart. And if the apple and the earth, why not any mass and any other mass? And so eventually, by observing the facts, and by reasoning, he came to realise that there is a force of attraction between any two masses, and that this force is proportional to the product of the masses, and inversely proportional to the square of the distance between them. This is the law of universal gravitation, perhaps the most important of all the physical laws, the law that governs the movement of bodies in space (whether they be natural or artificial), the law that Newton enunciated 300 years ago.

Escape from the earth

Consider a stone thrown vertically from the earth's surface with an initial velocity. Because the weight of the stone, caused by the gravitational attraction between the stone and the earth, opposes the stone's motion its velocity will be reduced with time (i.e. the stone has an acceleration *g* towards the earth). With a normal initial velocity gravitational acceleration will stop the stone at a certain height above the earth's surface and it will then reverse direction and accelerate towards the earth, eventually arriving at its launch point with its launch velocity but in the reverse direction (if air resistance is neglected).

For moderate initial velocity it is sufficiently accurate to assume that the weight (and hence *g*) is constant. What happens, though, if the initial velocity is large enough for a great height to be reached before the stone's direction of travel is reversed?

The weight of the stone, in accordance with the law of gravitation, is inversely proportional to the square of the distance between the masses. So supposing that the stone has a mass of 1 kilogram it will, at the surface of the earth, have a weight of 981 N. But what if it is moved away from the earth's surface altogether? What if it is thrown upwards 1 kilometre, 100, 1000, 6000 kilometres? Let us pause here for a moment because the radius of the earth is not much more than 6000 km, 6370 km in fact, so at a distance of 6370 km from the earth's surface, the force of attraction, i.e. the weight of the stone, being inversely proportional to the square of the distance from the centre of the earth – now doubled – will only be 1/4 of its weight at the earth's surface; similarly, at 12 740 km it will be 1/9, at 19 110 km only 1/16, and so on (see Fig. 13.2). Notice that in the figure the distances are given from the centre of the earth, and not from the earth's surface; for the earth is a very small thing in space, and if we are to understand the mechanics of space we must think more and more of the mass of the earth as concentrated at its centre.

The mass of the stone of course does not change, but as the weight changes so also does the acceleration (g) in proportion – this is just Newton's Second Law again. So the rate at which the stone loses speed on the outward 'flight', though starting at 9.81 m/s^2, gets less and less as the distance from the centre of the earth increases. This makes it more difficult to calculate how far the stone will go with a given starting velocity, but it has an even more interesting and important effect than this. For think of the stone returning to earth again; at great distances the rate at which it picks up speed will be very small, but the rate will increase until at the earth's surface it reaches the definite and finite value of 9.81 m/s^2. This, it will be noticed, is a maximum rate of increase, and it can be shown mathematically that even if the stone starts from what the mathematicians call infinity (which means so far away that it couldn't come from any farther!) the velocity reached will also have a definite and finite maximum value, which is in fact 11.184 km/s (about 40 250 km/h). So, if a stone is 'dropped' onto the earth from infinity, it will hit the earth at 11.184

km/s; and, by the same token, if it is thrown vertically from the earth at 11.184 km/s it will travel to infinity – and never return. This velocity is called the **escape velocity**. If it is thrown with any velocity less than this, it will return.

What happens if it is thrown from the earth at a velocity greater than the escape velocity? Or is this not possible? Yes, it is not only possible, but in a sense it has been done though not quite in this simple way. And all that happens is that it still has a velocity away from the earth when it reaches infinity – and so will go beyond infinity – but since infinity is the limit of our imagination perhaps it will be best to leave it at that. The reader may have noticed that to simplify things, we have only considered the earth's attraction on the stone, and the rest of the universe has been left out! But still the principle is illustrated.

It is important to remember that although the force of gravity, the weight of the stone, gets less and less as it travels farther and farther from the earth it never ceases altogether (at least not until the stone reaches infinity which is only another way of saying 'never'). It is often stated, quite incorrectly, that 'escape' from the earth means getting away from the pull of the earth. This we

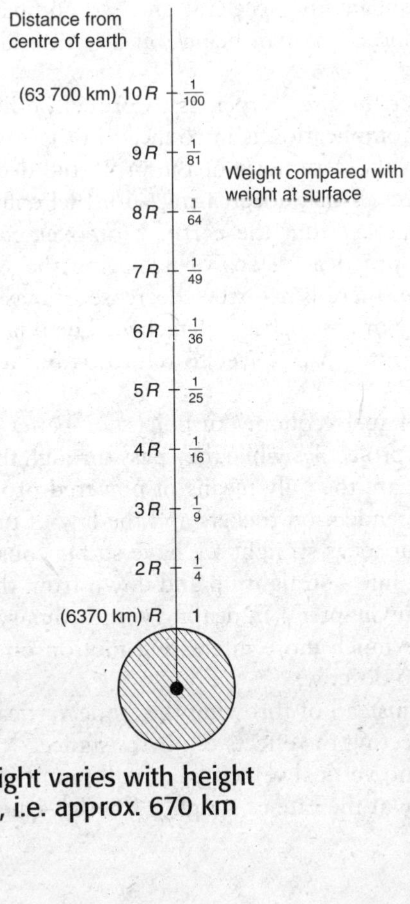

Fig 13.2 How weight varies with height
R = radius of earth, i.e. approx. 670 km

can never do, the earth 'pulls' on all other bodies wherever they are – that after all, is the universal law of gravitation. Why then do astronauts talk about 'weightlessness'? – and even demonstrate it? – we shall soon see.

In the meantime, it will be noticed that we have already introduced a new unit of velocity, the kilometre per second. Our reason for this is simply one of convenience; in this part of the subject we have to deal with very high speeds, and it is easier to remember these speeds, and even to think what they mean, as kilometres per second, than as so many thousands of knots, or metres per second. At the same time we must remember that our old friend g is still in m/s^2, so if we wish to use any of the standard formulae of mechanics we must be careful to convert the velocities into metres per second.

Projectiles and satellites

Among the many assumptions so far made one of the most impracticable has been the idea of leaving the earth's surface at speeds of about 11 km/s.

But we all know the answer to this problem now, and it lies in rocket propulsion; by this means the acceleration is so comparatively gentle that it can even be withstood by human beings, at any rate in the lying down position.

When we consider the use of rockets to propel bodies to great heights or into a space a new complication is introduced in that owing to the great rate of fuel consumption the very mass of the projectile decreases rapidly, so we have the double effect of mass decreasing with fuel consumption and weight decreasing with distance from the earth. Moreover, in multi-stage rockets, which are the only practical means of achieving the velocities required for launching into space, there is a further decrease in mass each time a stage is completed and a part of the rocket is detached. The final mass that becomes a satellite, or goes off into space never to return, is but a small fraction of the mass at take-off.

In the interests of fuel economy turbojets, or better still ramjets, may be used for the flight of projectiles while they pass through the earth's atmosphere, but in space rockets are the only means of powered propulsion, and all journeys in space are dependent on rockets and the law of universal gravitation.

In order to get our ideas straight we have so far considered the motion of missiles in a straight line – straight up and down from the earth's surface. We did the same thing in Chapter 1 in dealing with ordinary mechanics, but then we graduated to the much more interesting motion on curved paths; this is what we are going to do now.

What happens if instead of throwing the stone vertically we throw it horizontally? – still neglecting the effects of air resistance.

It will start with no vertical velocity, but will immediately begin to acquire a downward velocity at the rate of roughly $10\,m/s^2$ – meanwhile it will retain its horizontal velocity. After 1 sec it will have fallen about 5 m, after 2 sec 20 m, and so on. If its initial horizontal velocity was 100 m/s, and if it was launched from a height of 30 m, its path of travel would be something like (*a*) in Fig. 13.3, or if it was launched from twice the height, like (*b*). If its initial horizontal velocity was 200 m/s, its path of travel would be more like (*c*) or (*d*) respectively.

It will be quite clear from these figures that the distance it will travel over the ground before striking the ground depends on the height at which it is projected, and the horizontal velocity with which it is projected (the velocity being more important than the height). If it is projected at ground level it won't get any distance before hitting the ground whatever its horizontal velocity; on the other hand, if it is projected from considerable height, and at a considerable horizontal velocity, it will travel a considerable distance horizontally before reaching the ground. The path of flight, or trajectory, is a mathematical curve called a parabola.

But now we have been guilty of making yet another assumption – has the reader noticed it? In the figures, and in our reasoning, we have assumed that the earth is flat. Not just that it is free of hills and dales (these won't affect its flight path but they may obviously affect the point at which it hits the ground), but that the earth is itself a flat plane instead of being spherical or very nearly so. Such an assumption has no practical significance in the flight of stones or

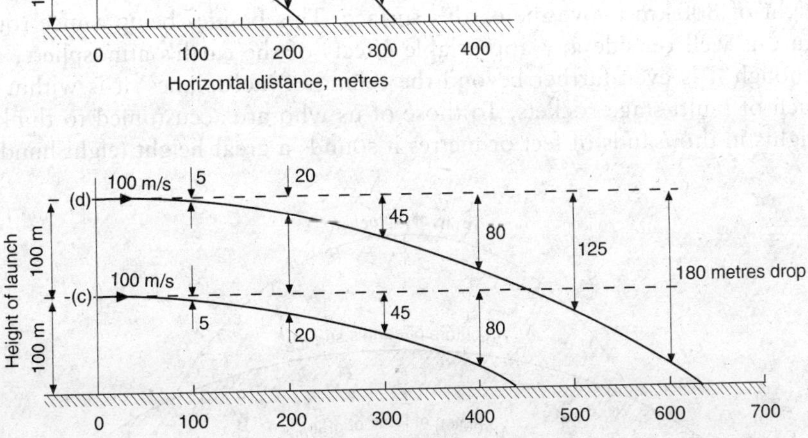

Fig 13.3 Bodies launched horizontally – paths of flight

cricket balls – but is all-important in the flight of high-speed projectiles. The curvature of the earth will affect the path of the projectile, and the distance it travels before striking the ground, for two reasons – first, because if the projectile didn't 'fall' towards the earth it would go off at a tangent and so get farther and farther away; secondly, because although the force of gravity acting on the body is always 'vertical' in the sense that it is always towards the centre of the earth, the direction of the vertical will change in space and this will change the shape of the curve – in mathematical terms it means that the trajectory is an ellipse rather than a parabola. These effects are shown in the figure (Fig. 13.4) – hopelessly exaggerated of course in the case of a stone or cricket ball, or even an ordinary shell fired from an ordinary gun, but not by any means exaggerated for modern rocket-propelled high-speed ballistic missiles and not, in this figure, even going as far as the man-launched satellite or space-ship.

It should now be clear that the greater the speed with which a projectile is launched from a given height above the earth's surface in a horizontal direction, the larger will be the curve it describes and so the greater will be the distance it travels before it hits the surface; in other words, the greater will be the range.

The actual range, of course, depends on the direction of launch as well as the speed and height but at first it is easier to consider only horizontal launches. And by 'launching' we mean that the projectile is given a velocity and then left to itself – and so becomes subject to the mechanics of ballistics. If it continues to be rocket-propelled almost anything may happen!

If we take 'launch' to have this meaning, then the launch conditions can be achieved at the end of the 'launch cycle' involving, usually, rocket propulsion between the earth's surface and the launch point where the rocket is switched off and the body left to the mercy of ballistics.

Let us assume then that we are launching a projectile horizontally from a height of 800 km above the earth's surface. This besides being a nice round figure is well outside any appreciable effects of the earth's atmosphere, and although it is even farther beyond the ceiling of 'aeroplanes' it is within the reach of multi-stage rockets. To those of us who are accustomed to think of heights in thousands of feet or metres it sounds a great height (eight hundred

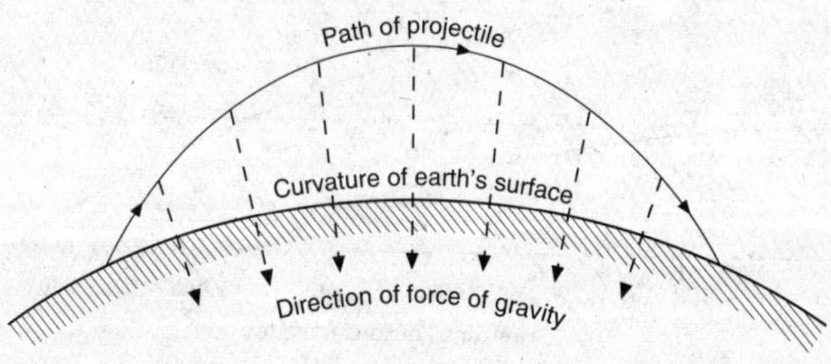

Fig 13.4 Path of projectile – curved earth

thousand metres), but in terms of space travel it is practically nothing, only one eighth of the radius of the earth away from the surface, about one four hundred and eightieth of the distance to the moon, and one hundred thousandth of the distance to the sun. At this height the weight of a body and the acceleration due to gravity are reduced by about 20 per cent of what they were at the earth's surface.

So much then for conditions at our launching platform, let us now see what happens as we increase the velocity of a horizontal launch. At first the projectile will simply get farther and farther round the earth, but always coming down to earth again at some point less than half way round (Fig. 13.5*a*, overleaf). Then at a certain velocity (about 7.16 km/s or 25 800 km/h) a most exciting thing will happen (or at least it would happen if it were not for our old enemy air resistance) – the projectile will just miss the surface on the far side of the earth and will then gain height again, and, perhaps most exciting of all, will circle the earth and come round to where it started – and will then repeat the performance – and so on. The projectile has become a satellite – it is travelling round the earth under its own steam as it were (Fig. 13.5*b*).

Unfortunately it can never happen quite like this because although it was clear of air resistance at the launch, on the far side it would have come right through the atmosphere to ground level and so would burn up owing to the heat created or, even if it could be shielded in some way from this, it would lose speed and fall to the earth.

But we have only got to increase the launching velocity a little further and the projectile will then miss the far side of the earth by an appreciable margin, and when this is say 300 km, it will miss most of the atmosphere and so continue to circle the earth on an elliptical orbit, clearing it by 800 km on one side and 300 km on the other – in any such orbit the point at the greatest distance from the centre of the earth is called the apogee, and the point nearest the centre of the earth the perigee. Our projectile is now a practical satellite – practical, but still not very probable.

Even at a height of 300 km there is some atmosphere, and there probably is even at 800 km for that matter, so this satellite will lose speed every time it dips into the atmosphere, and so will gradually lose the energy given to it at the launch and will sooner or later come down to earth on a spiral path.

But we need not be disheartened, because a further increase of launching speed will further increase the clearance on the far side, and so gradually eliminate this problem until, at a launching speed of about 7.48 km/s we reach another interesting stage at which the satellite – we can no longer call it a projectile – travels round the earth on a circular path, 800 km from the earth's surface, and there is no longer any distinction between the apogee and the perigee (Fig. 13.5*c*). The launching speed at which this occurs is called the **circular velocity.**

After this long story the reader will probably be able to guess what happens with further increase of launching speed. Yes, the circle again becomes an ellipse but the **apogee**, or farthest point, is now on the far side of the earth and

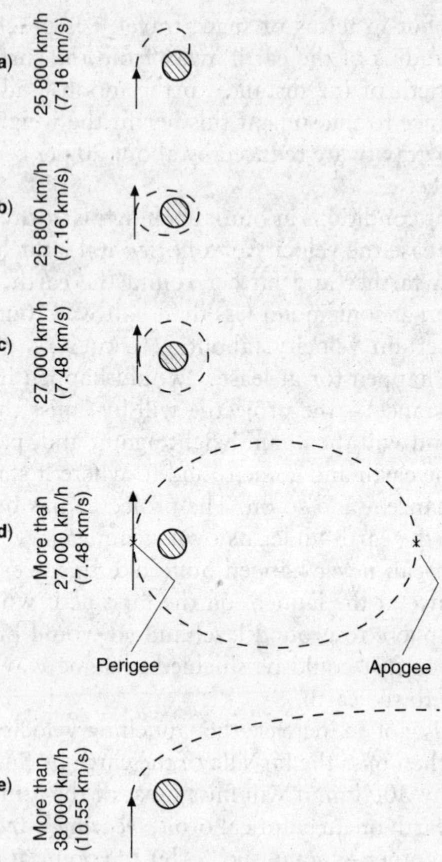

Fig 13.5 Earth satellites
Speeds refer to horizontal launches from 800 km above the earth's surface, in the interests of clarity this distance has been exaggerated in comparison with the radius of the earth.

the **perigee** is the point of launch (Fig. 13.5*d*). Still further increases of speed make the ellipse more and more elongated with the apogee getting farther and farther from the earth. It will not be difficult to understand why this type of orbit is frequently used; it practically eliminates the problems of air resistance, it is not dependent on an exact launching speed and it allows scope for travel at the apogee to great distances and so, for instance, for passing beyond the moon or other planets, or hitting them.

Is this the end of the story? Not quite. Strange as it may seem the ellipse cannot be stretched indefinitely, and at a launching speed of about 10.7 km/s – different speeds at different heights – the ellipse becomes an open curve, a parabola, and the satellite travelling on this open curve escapes from the earth for ever, and becomes a satellite of the sun (Fig. 13.5*e*). Yes, this is the escape velocity again, and it only differs numerically from the previous one because we are now launching from 800 km instead of from ground level. So the escape velocity is the highest velocity at which a body can be launched in any direction and be expected either to orbit the earth or return to it again – above that speed, speaking vulgarly, we have had it. Above that speed, too, the path of travel changes from a parabola to a different open curve called a hyperbola, but this is a subtle change for a subtle reason, and it need not worry us.

This is not quite the end of the story as at even higher speeds there comes a point where the object can escape from the solar system and carry on out into space indefinitely.

The mechanics of circular orbits

So much for the story of what happens – what is the explanation of it all? In the particular case of the circular orbit the satellite is very like a stone on the end of a string stretching from the centre of the earth to the satellite; the satellite is all the time trying to go off at a tangent but is being given an acceleration towards the centre by the centripetal force which is of course the force of gravity. So, near the earth's surface, if we neglect air resistance, the centripetal force will be the weight of the satellite, and the acceleration towards the centre will be 9.81 m/s^2. Notice that a body circling the earth is accelerating towards the centre at the same rate as a body falling straight towards the earth. So we can easily calculate the circular velocity near the earth's surface because the acceleration $= v^2/r$ (see page 14).

Now r is the radius of the earth, say 6370 km (6 370 000 m), so $v^2/r = g$, i.e. $v^2/6\,370\,000 = 9.81$

$$\therefore v^2 = 62\,490\,000$$

$$v = 7905 \text{ m/s}$$

$$= 7.9 \text{ km/s approx}$$

$$= \mathbf{28\,440\ km/h\ approx}$$

How long will the satellite take to make a complete circuit of the earth at this speed?

Circumference of earth $= 2\pi \times 6370$ km

So time of circuit at 28 440 km/h

$$= (2\pi \times 6370)/28\,440$$

$$= 1.41 \text{ hours or about}$$

1 hour 25 minutes

It will be noticed that the circular velocity we have calculated, i.e. 7.9 km/s, is higher than the circular velocity at 800 km from the earth's surface, i.e. 7.48 km/s; but there is no mystery in that and we can easily work it out for ourselves by replacing the earth's radius of 6370 km by 7170 km, and reducing g by 20 per cent, i.e. to about 7.85 m/s^2. The value of v^2, and so of v, will then be less because whereas the value of g is reduced by 20 per cent, the value of r is only increased by $12\frac{1}{2}$ per cent – this, in turn, is because the value of g depends on the force of gravity, which is inversely proportional to the square of the distance r.

And what will be the time of a complete circuit at 800 km from the earth's surface? The distance is greater, the speed less, so the time of orbit will be greater. Work it out and you will find that it is about 1 hr 40 min. Similarly at 1600 km the circular velocity is about 6.9 km/s and the time of orbit nearly 2 hours.

A distance of 35 400 km from the earth gives a particularly interesting circular orbit because the time of a complete circuit is 24 hours; so a satellite travelling at this speed – in the right direction, of course – remains over one spot on the earth; a communication satellite such as is used for transmitting TV and radio signals from one part of the earth to another.

Then at about 385 000 km the circular velocity is a mere 3700 km/h (just over 1 km/s), and the time of orbit 28 days – but on that circuit we already have a satellite that surpasses in many ways any so far launched by man – the moon.

And now we can answer an obvious question – why doesn't the moon fall on to the earth? Because it is revolving round the earth at just such velocity and radius that the centripetal force is provided by the gravitational attraction, in other words, in a sense it is 'weightless' – and this applies to all those bodies in orbit, whether circular or not, and to all the people and things inside them. Strictly speaking they are not weightless at all; in fact it is their weight, the force of attraction between them and the earth (or moon) which they are orbiting, that keeps them in orbit and prevents them from going off at a tangent. They merely seem to be weightless, and that is why a man can get out of a space-ship while in orbit, and continue in orbit himself, just like the space-ship, without any fear of 'falling' back to earth or to anywhere else. Although he may be travelling at several kilometres per second he has no sense of speed, and apparently no weight – he just floats, and has no difficulty in keeping near the space-ship which is also just floating! More correctly the man, and the space-ship, and all the other things in orbit, are falling freely, are accelerating towards the earth because of the attraction of the earth – in short because of their weight – so much for weightlessness! In the same sense the moon is falling towards the earth, though it never gets any nearer!

Notice that the reader can calculate all these circular velocities and times of orbit for himself, including that of the moon. For at all distances from the centre of the earth, the condition for a circular orbit is that the acceleration towards the centre shall be the 'g' or acceleration of gravity at that distance; this we might call g_d, and it must be equal to v^2/d.

But g_d is also proportional to the force of gravity, which is inversely proportional to the square of the distance.

Since at the earth's radius r, the acceleration is g,

g_d will be $g \times r^2/d^2$

Therefore for circular velocity at any distance d,

$$v^2/d = g \times r^2/d^2$$

i.e. $v^2 = gr^2/d$.

Figure 13.6 (overleaf) shows circular orbits at different distances from the centre of the earth.

It shows how a whole system of bodies can circle the earth, of their own free will as it were (once they have been put in orbit), and how the farther out the orbit the slower is the speed. It rather reminds one of the way in which Sir James Jeans once described the solar system as being like the traffic in Piccadilly Circus, with 'the traffic nearest the centre moving fastest, that farther out more slowly, while that at the extreme edge merely crawls – at least by comparison with the fast traffic near the centre.'* But there is an important distinction between the solar system – the work of nature – and bodies orbiting the earth – the work of man (except for the moon); in the solar system, again to quote Sir James Jeans, there is only 'one-way traffic', and the orbits of the planets round the sun are mostly circular, or very nearly so, whereas the man-made satellites orbit the earth in various directions and, as we shall soon discover, some of their paths are very far from circular.

When talking of the interesting possibilities of a 24-hour circuit we mentioned the direction of rotation. This would be all important in this case because if the satellite was rotating round the equator in the same direction as the earth's rotation it would stay over the same spot on the earth's surface but if it was travelling in the opposite direction – well, what would it do? Would it go twice round in a day? or would it merely appear to do so? or what?

But the fact that the earth is rotating will of course affect all launches, because it means that we are launching from a moving platform. The surface of the earth at the equator is travelling at a speed of rather over 1600 km/h owing to the spin of the earth on its axis, so a body launched in the same direction, i.e. towards the east, will already have the advantage of this speed and so will need 1600 km/h less extra speed to achieve circular velocity, escape velocity, or whatever it may be. Towards the west it will need 1600 km/h more extra speed. There can also be circuits of the earth in other planes altogether,

*The quotation from *The Stars in their Courses* by Sir James Jeans is given by courtesy of the Cambridge University Press.

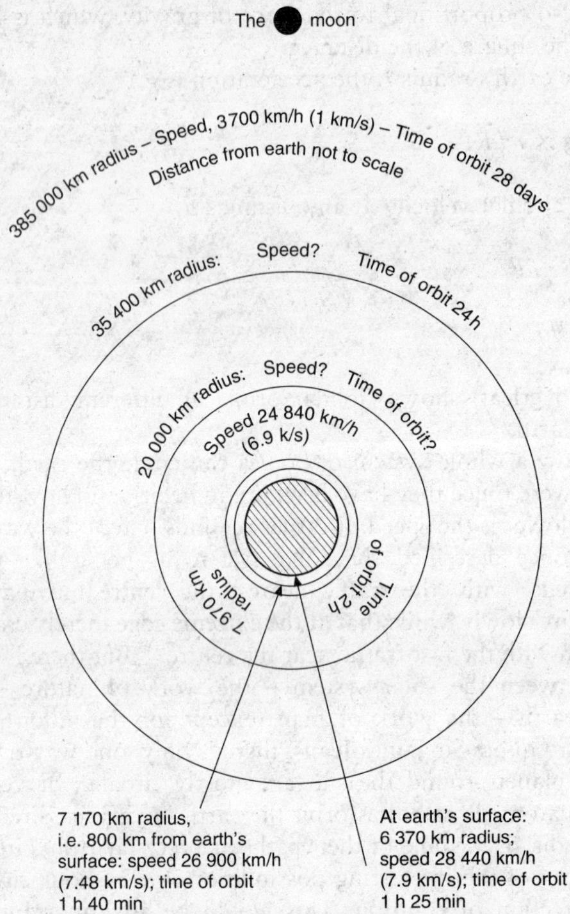

Fig 13.6 Circular orbits at different distances from centre of the earth
Note. **The speed at 35 400 km radius, and the speed and time of orbit at 20 000 km radius, have been left for the reader to work out for himself.**

e.g. over the poles, and in these cases the effect of the earth's rotation on launching and orbiting is more complicated.

It is not always realised, and it is interesting to note, that since the earth's surface at the equator is travelling at about 1600 km/h all bodies on the earth are in a sense trying to be satellites, and to go straight on instead of following the curvature of the earth. Thus there are two reasons why a body of the same mass weighs less at the equator than at the poles, first because it is farther from the centre of the earth so the true gravitational attraction is less, and secondly, because a proportion of the gravitational force has to provide the centripetal acceleration. How much is this centripetal force? Is it appreciable? Well, work it out for yourself. Take the actual velocity as 1690 km/h, the radius of the earth as 6370 km, g as 9.81 m/s^2, and you will find that the centrifugal force on a mass of 1 kg is about 0.018 N.

Escape velocity and circular velocity

As already established the velocity for a circular orbit is given by the formula –

$$v_c^2 = gr^2/d$$

where r is the radius of the earth, and d the distance from the centre of the earth.

At the earth's surface $d = r$

$$\therefore \quad v_c^2 = gr.$$

Can the escape velocity be calculated equally simply? Very nearly so. If we go back to the idea of a body being projected vertically from the earth's surface with sufficient kinetic energy to enable it to do work against the force of gravitation all the way to infinity, then if the escape velocity is denoted by the symbol v_e the

kinetic energy of a mass m will be $\frac{1}{2}mv_e^2$.

This kinetic energy must be sufficient to provide the energy needed to lift m from the earth's surface to infinity. At a particular height above the earth, the energy needed to lift m one more metre is equal to its weight which equals mg at the earth's surface, and decreases all the way to infinity when it will be zero. Since the weight changes and since the change is not a simple ratio but inversely as the square of the distance, it needs the principles of calculus to estimate the total work done, but the answer is very simple; it is the same as the weight at the earth's surface $mg \times$ the radius of the earth, i.e. mgr.

So $\quad \frac{1}{2}mv_e^2 = mgr$

or $\quad v_e^2 = 2gr$

but $\quad v_c^2 = gr$

$\therefore \quad v_e = v_c \times \sqrt{2}$

i.e. Escape velocity = $\sqrt{2}$ × Velocity of circular orbit at that radius.

Thus there is a simple relationship between all escape velocities and all circular velocities at a given radius from a mass such as that of the earth, or moon, or sun – **the escape velocity is 1.41 × the circular velocity, or 41 per cent more.**

Elliptical orbits

So much for circular orbits – now what about the elliptical orbits which, as we have already explained, are much more common for artificial satellites? They are also more common in nature, the orbit of the moon round the earth being one of the few examples of a nearly circular orbit.

Mathematically it is a little more difficult to calculate what happens during an elliptical orbit when both the velocity and the radius are constantly changing, but the reader who has followed the arguments so far should have no difficulty in understanding the principles involved. Returning to horizontal launches from a height of 800 km it will be remembered that at a speed of launch of 7.48 km/s the orbit was circular, below and above this speed it was elliptical, though at the escape velocity of 10.7 km/s it became an open curve, a parabola, and then above this a hyperbola. The real criterion is how much energy, kinetic energy, the body has when it is launched; because it is this energy which enables it to do work against the force of gravity – it is really much more like the case of the stone that was thrown vertically upwards than might at first appear. In vertical ascent the kinetic energy given to the body at the launch enables it to do work against gravity and so gradually acquire potential energy; at the highest point reached all the kinetic energy has become potential energy; then as the body falls again the potential energy is lost and kinetic energy regained until on striking the ground the original kinetic energy has all been regained (neglecting air resistance of course). The body travelling on a curved path also has to work against gravity, is also accelerating all the time, and downwards, just like the body on the vertical path, and in fact, at the same rate – once that is understood, all is clear.

The only difference then is that on curved paths the body must retain some of its kinetic energy throughout the circuit if it is to continue on its orbit. At a launching height of 800 km it has at the start both kinetic and potential energy; if the launching speed is less than the circular velocity of 7.48 km/s, but sufficient to ensure that it doesn't come down to earth before reaching the far side, then by the time the body reaches the far side of the earth it will have dropped in height, i.e. lost some of its potential energy, but by the same token will have gained some kinetic energy (just like the falling stone), and so will be travelling faster – at over 7.9 km/s in fact if it is near the earth's surface. Then as it returns to the starting point it will gain potential energy, rising again to 800 km, and lose kinetic energy, until the proportions (and values) are the same as they were at the launch. In all this we are still neglecting air resistance, and it is air resistance which in practice makes an orbit of this kind impossible; for as soon as the body travelling round the earth, and getting lower all the time, meets appreciable air density, it will experience drag, and in working against this will lose kinetic energy and so the speed necessary to take it round the far side of the earth. Thus it will fall to the ground before getting half way round or the heat generated by kinetic heating will cause it to burn up.

As the launching speed gets nearer the circular velocity, the body will keep clear of appreciable atmosphere all the way round, and the orbit becomes more practical. But until the circular velocity is reached the launching point will be the apogee, and the perigee will be on the far side of the earth, where the body will be nearer the earth's surface, and where the velocity will increase until, at the circular velocity, it remains constant all the way round.

As the launching speed is increased above the circular velocity, the body will have more kinetic energy than is necessary to keep the 800 km of height, and so it will gain height and potential energy, at the same time losing speed and kinetic energy, until it passes round the far side of the earth – now the apogee – at lower velocity and greater height. And so it will go on as the launching speed is still further increased, the ellipse becoming more and more elongated, the apogee getting farther and farther from the earth, and the residual velocity at the apogee getting less and less.

As the launching velocity approaches the escape velocity of 10.7 km/s, the body at the apogee will only just have sufficient kinetic energy to enable it to get over the top, as it were, and return again to earth. If launched at the escape velocity, or above, it won't even be able to do this, and it will go off into space on its parabola, or hyperbola as the case may be, at the mercy of the gravitational attraction of some other body, probably the sun. This new force of attraction will restore its kinetic energy and velocity as it embarks on a completely new orbit, possibly of enormous size like that of the earth round the sun (radius of orbit about 150 million kilometres).

Since nearly all practical orbits are in the form of ellipses it is interesting to consider some of the properties of these curves. The reader may know that an ellipse can be drawn on paper with a pencil and a piece of string of fixed length attached to two pins (Fig. 13.7, overleaf). In other words, an ellipse is the locus of a point moving so that the sum of the distances from these two points (the pins) is constant. These points are called the foci of the ellipse – in a circle, which is only a particular case of an ellipse, they coincide – and it will be noticed that in the figures illustrating the paths of satellites the centre of the earth is always one of the foci of the ellipse.

At launching speeds below the circular velocity the centre of the earth is the focus farthest from the launch, at the circular velocity it is of course the centre of the circular orbit, and at higher launching speeds it is the focus nearest to the launch, the other focus getting farther and farther away as the launching speed is increased. For launches at the escape velocity the distant focus has gone to infinity, and only one focus is left at the centre of the earth – a parabolic curve has only one focus. Above the escape velocity, too, the hyperbolic

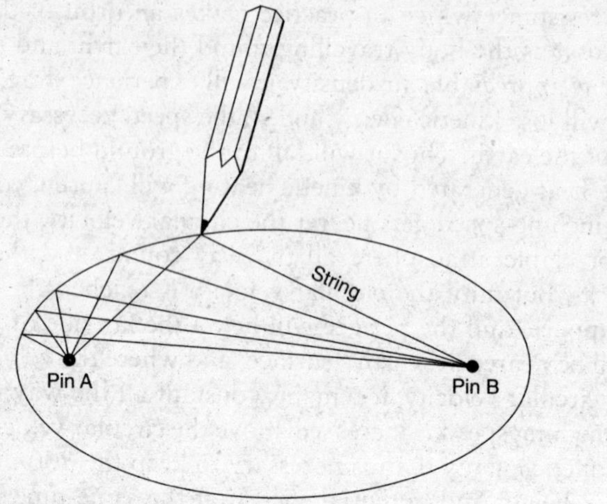

Fig 13.7 Drawing an ellipse

curve has only one focus (see Fig. 13.8); strictly speaking, a hyperbola has two foci, but this is because there are two parts of the curve, back to back as it were, and we are here only concerned with one part of the curve.

It is not of practical importance in understanding the mechanics of projectiles and satellites, but it may be of interest to know that these curves are all derived from a cone (they are the intersections of a plane and the surface of a cone) and are sometimes called 'conic sections' (Fig. 13.9).

Going to the moon?

Now let us get just a little nearer to the true state of affairs by realising at least the existence of the moon. How will this affect the stone that is thrown vertically from the earth?

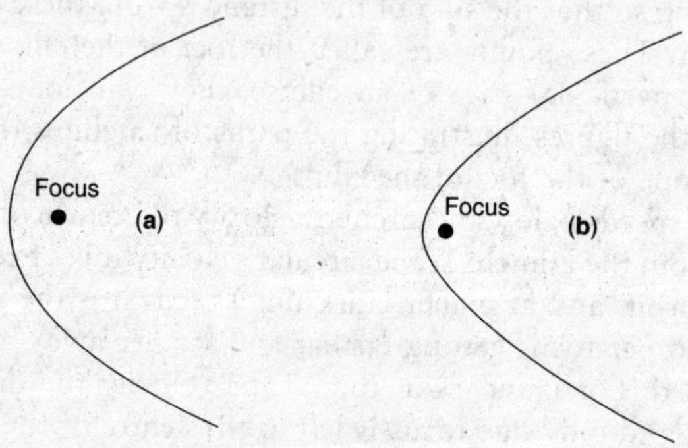

Fig 13.8 (***a***) a parabola, (***b***) a hyperbola

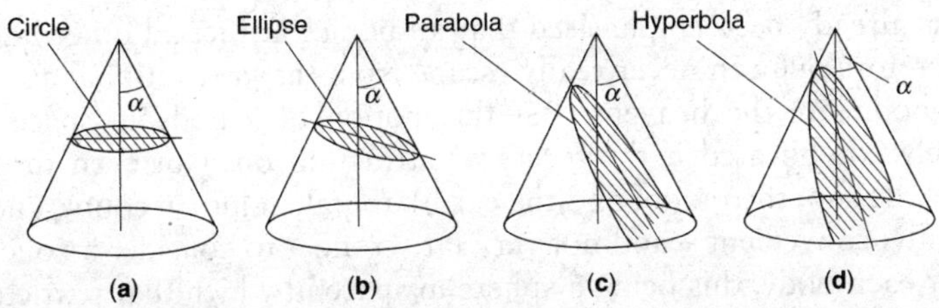

Fig 13.9 Conic sections
In (*a*) the plane cuts the cone *at right angles* to the centre line, forming a circle. In (*b*) the plane is at an acute angle to the centre line, but the angle is *greater than* α, forming an ellipse. In (*c*) the angle is *equal to* α, forming a parabola. In (*d*) the angle is *less than* α, forming a hyperbola.

Well, it will affect it quite simply, in principle at any rate, in that there will now be three masses all attracting each other – the stone, the earth and the moon.

The moon is about 385 000 km away from the earth. Just for the moment let us suppose that it remains over the point on the earth's surface from which we throw the stone, and that we increase the starting velocity of the stone until it reaches distances of 6370 km, 12 740 km, and so on. As before, the weight will decrease with the distance from the earth's surface, but now rather more so because a new factor has been introduced, the attraction between the moon and the stone. The mass of the moon is only about 1/81 of the mass of the earth, so the force of attraction at the same distance will only be a fraction of that of the earth (1/81 in fact), but as we know the moon's attraction is a very real thing, even at the earth's surface, for it is largely responsible for the tides.

We have already imagined so much that we may as well go one step further and imagine that our stone travels in a straight line between earth and moon. As it gets nearer to the moon the attraction of the earth will decrease and that of the moon increase, until a point is reached where the two attractions are the same. Since the force is inversely proportional to the square of the distance, the distance from the centre of the moon at which this occurs will be $1/\sqrt{81}$ of the distance from the centre of the earth, or approximately one ninth (Fig. 13.10, overleaf); roughly say 39 000 km from the moon and 346 000 km from the earth. So if a stone is launched with just sufficient velocity to reach this point, it will stay there – and will once more be 'weightless', this time perhaps more correctly so, though again we notice that it is just a question of the forces being balanced. In fact the balance is too delicate, and the stone will not in fact stay at this neutral position, because some other heavenly body will attract it and tip the balance, and it will fall either onto the earth or the moon. If the stone is launched with a velocity slightly greater than that required to reach this neutral point, it will still be travelling towards the moon at this point, and since as it passes the point the attraction of the moon will become greater than that of the earth the stone will pick up speed and fall on to the moon.

It has already been emphasised that all bodies attract all other bodies, and that therefore one can never really escape from the gravitational attraction of all the bodies in the heavens. But the motion of a body in space becomes extremely complicated if the forces of attraction on it of even three bodies (such as the sun, the moon, and the earth) are taken into account, and for this reason it is convenient – and not very far wrong – to consider a zone of influence for each body, this being a sphere in space in which that particular body has a greater gravitational effect than a larger body. The earth, of course, is well within the sun's zone of influence – its motion round the sun is in fact controlled by the force of attraction between it and the sun – but on the other hand, bodies near the earth's surface, although attracted by the sun, come much more under the influence of the earth's attraction than that of the sun (which is perhaps just as well, since otherwise we would all be off to the sun). The attraction of the earth remains greater than that of the sun for a distance of about a million kilometres, so the earth's zone of influence relative to the sun is a sphere of about one million kilometres radius (Fig. 13.11). Similarly the moon's zone of influence relative to the earth is a sphere of about 39 000 km radius – as we discovered in the last paragraph, though we didn't give it that name.

Well, we have described one way of getting to the moon – to go straight there in fact – but it is not quite so easy as it sounds because of the great accuracy needed both in aim and launching speed. As regards aim, we have made

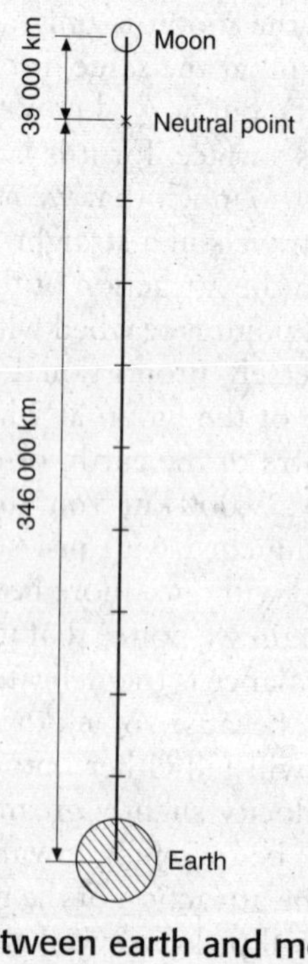

Fig 13.10 Neutral point between earth and moon

things much too easy in our imagination; in reality the earth is travelling round the sun, and is spinning on its own axis, while the moon is travelling round the earth. But even more interesting is the sensitivity to exact velocity.

The distance of 385 000 km to the moon is much less than the distance to infinity, only a minute fraction of it, and to the neutral point even less, but it requires nearly as much energy to reach this distance as it does to reach infinity because the attraction of the earth at this distance has been so reduced that nearly all the serious work in overcoming the earth's gravity has already been done; or, putting it another way, whereas 11.184 km/s is needed to send the stone to infinity, about 11.168 km/s is needed to send it to the neutral point from which it will drop on the moon. So if launched at less than 11.168 km/s it will return to earth; at about 11.168 km/s it will go to the moon; at 11.184 km/s it will go to infinity (it could hit the moon on the way); and at over 11.184 km/s – well, we have mentioned that before.

The mass of the moon being so much less than that of the earth, its escape velocity is only about 2.4 km/s, and that is the speed at which the stone would arrive on the moon if it 'fell' from infinity; it will for all practical purposes be the same if it falls the 39 000 km from the neutral point. This is about the muzzle

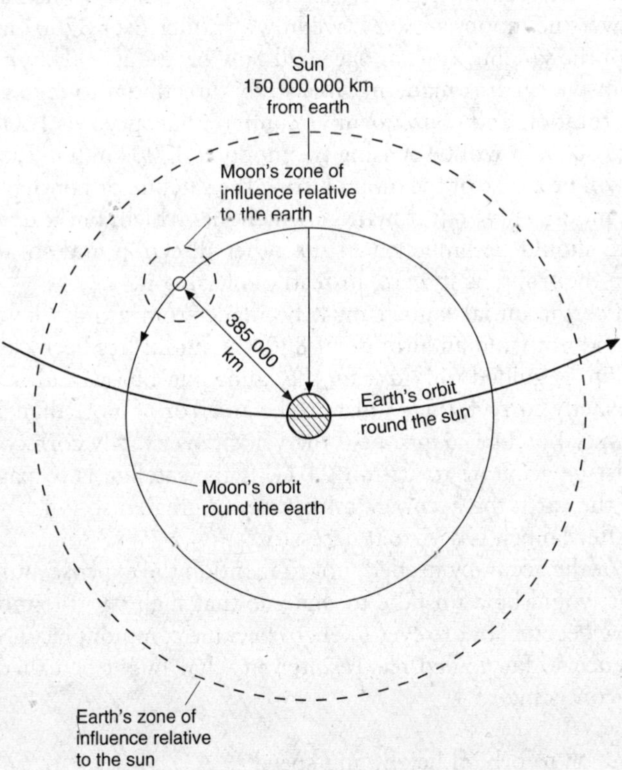

Fig 13.11 Zones of influence
Not to scale.

velocity of a shell as it leaves a long-range gun, and so the landing on the moon will not be a very soft one, and this is the minimum speed at which the stone can arrive unless there is some means of slowing it down; there is no 'atmosphere' to do this, and the only hope is to break the fall by rockets fired towards the moon.

Another way to the moon

But man has been to the moon – more than once – and has come back again; perhaps even more remarkable the Russians have sent spacecraft to the moon – without a man – have brought at least one back with samples, and have driven a moon-bug about on the surface! How has it been done? For the answer we must go back to circular and elliptical orbits. For a horizontal speed of launch of 10.46 km/s (just below the escape velocity), from a height of about 800 km, gives an elliptical orbit which will strike the moon, and at velocities of launch slightly above this, orbits will pass round both the earth and the moon (Fig. 13.12).

But the moon isn't such an easy target as all that! The shape and size of the elliptical orbit is very sensitive to the exact direction and velocity of launch, and moreover the moon is itself travelling at rather over 3700 km/h, whereas the speed of the satellite at its apogee will only be about 700 km/h. Also, if the launch from the earth is made in an easterly direction – to take advantage of the earth's rotation and consequent circumferential speed of 1600 km/h – the satellite at 700 km/h will be chasing the moon at 3700 km/h in the same direction; so it will be a case of the moon hitting the satellite rather than the satellite hitting the moon – not that it matters which hits which, but it does mean that the satellite should be launched in the other direction and so approach the moon from the front, as it were, instead of chasing it.

In practice the initial launch must be made from ground level (Fig. 13B, overleaf), and not from an altitude of 800 km, and it has been calculated that if the satellite is guided only during the launching phase, and if the angle of launch is exactly correct, there must not be an error of more than 23 m/s in the launching speed of 11 125 m/s; or if the velocity is exactly correct the angle of launch must be accurate to within 0.01°. If the satellite is to pass round the moon and the earth the accuracy must be even greater, so much so that some guidance after launch is a virtual necessity.

In view of the accuracy needed, not to mention the expense and man-power involved, it would be a mistake to imagine that flights to the moon or other planets have become, or are ever likely to become commonplace. None the less the experience so far gained has resulted in what might be called a standard procedure consisting of –

1. The launch to orbital height and speed.
2. One or more orbits of the earth.

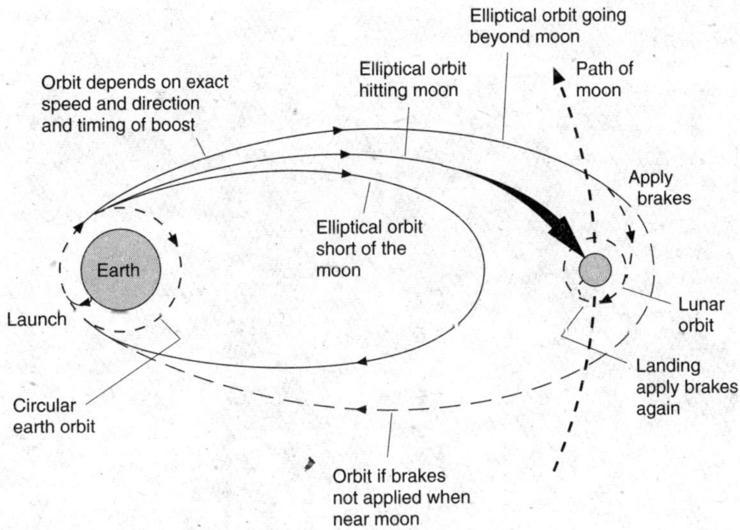

Fig 13.12 Sending spacecraft to the moon
Not to scale.

3. Rocket boost to required speed and direction for the moon.
4. Reverse burst of power to slow down, and put into orbit round the moon.
5. Separation of lunar module, and more reverse power to give a soft landing on the moon.
6. Lift-off from the moon, and into orbit to join up again with the command module.
7. Rocket boost to required speed and direction for the earth.
8. Reverse burst to slow down and put into orbit round the earth.
9. Re-entry, splash down and pick-up.

These phases have been described in detail in the Press, on radio and television, and in numerous articles and books; our purpose here is simply to indicate how the principles of mechanics apply to these various phases.

First then, the launch.

Launching a spacecraft

A projectile, whether it is launched for the purpose of escaping from the earth, or landing on the moon, or becoming a satellite, or simply travelling over the earth's surface to some other place, must first pass through the atmosphere.

Fig 13B Lift off for the moon
(By courtesy of the General Dynamics Corporation, USA)

The most important effect of this is that we cannot neglect air resistance, as we have so calmly done throughout this chapter (though with constant reminders). And the practical effect of air resistance is to reduce speeds, so the actual speeds of launching within the atmosphere must all be higher than those we have given. How much higher? From ground level, something of the order of 10 or 12 per cent, e.g. if the escape velocity is 40 250 km/h (11.134 km/s), the actual velocity of launch at ground level would have to be about 45 000 km/h (12.5 km/s), and for a circular velocity of 29 000 km/h (8.05 km/s), say 32 000 km/h (8.9 km/s). This naturally makes accuracy more difficult to achieve.

But there is a further difficulty. The launching speed cannot be attained at ground level. As has already been explained, the body on the first part of its flight is propelled by rockets; if it is required to reach great heights by multi-stage rockets. So in contrast with a shell fired from a gun there is time – and distance – in which to gather speed; and by so deciding and regulating the thrust of the rockets in relation to the mass of the projectile, and taking into account the drag due to air resistance, the acceleration can be moderated sufficiently to prevent damage to the missile itself and its mechanisms, and if passengers are to be carried, even to human beings. This moderation of the acceleration is of course an advantage, but it also makes it extremely difficult to calculate just what the speed, direction, and height of the vehicle will be when it is finally launched, i.e. when the fuel of the last launching rocket has been exhausted.

No one who has thought of this problem, even in the very elementary form such as we have attempted to explain in this book, can be anything but amazed at the accuracy that has actually been achieved in the launching of spacecraft.

As greater heights are reached there is less density of air, and so the drag decreases in spite of ever-increasing speeds. Eventually the rocket power is shut off, the last stage of the launching rocket is jettisoned, and the projectile, or spacecraft, or whatever it may be, travels on its elliptical path under the force of gravity until it begins to descend and again approaches the earth's atmosphere. The distance it travels during this ballistic phase – under its own steam, one might almost say! – will depend on the velocity it had achieved and the direction in which it was travelling when the rocket power ended. It may be hundreds or thousands of kilometres, it could be round the earth and back again, or several times round; there is no fundamental difference between a missile, a satellite and a spacecraft, they differ only in the speed, direction and height of launch.

The fact that the final launch takes place at considerable height does, at least, provide partial justification for our earlier neglect of air resistance when considering their motion. It is true that in thinking of launches at a height of 800 km we may have been guilty of going rather far though, as explained at the time, it had the advantage that we really could neglect air resistance, and so the speeds we gave for that height were reasonably correct. Typical figures for an actual launch (Fig. 13.13, overleaf) are to a height of 60 km and a speed of 6000 km/h (1.67 km/s) at the end of the first stage, 200 km and 14 500 km/h (4.03 km/s) at the end of the second stage, and 500 km and 28 000 km/h (7.78 km/s) at the end of the third stage. The take-off is vertical, the path is then inclined at say 45°, then when the velocity is sufficient there is a period of coasting or free-wheeling between the second and third stages to the required height (which will become the perigee if the missile is to be a satellite) where the path will be horizontal, then the third stage rocket boosts the velocity to that required for orbit. The more this exceeds the circular velocity, the more distant will be the apogee. The perigee of nearly all the early satellites was less than 800 km, but the apogee varied from just over 800 km for Sputnik 1 up to – well, to the moon and beyond.

The second stage, orbiting the earth, has already been considered in some detail, and there is little to add. This is the aspect of space flight of which we have had most experience, and there are now literally hundreds of 'bodies' of various shapes and sizes and masses orbiting the earth, and on a variety of orbits, and hundreds more that have finished their flights and have been burnt up on re-entering the atmosphere. There have also been several manned orbits of the earth, and space stations have been set up which can be permanently manned in 'shifts' by shuttle services, put together and enlarged up there, and used for a variety of purposes, some peaceful – others perhaps not so peaceful.

So far as going to the moon is concerned the first orbits are more or less circular and then, at the third stage, at exactly the correct part of the orbit, a burst of power is given to boost the speed and put the spacecraft on its journey to the moon. Although this journey is often represented in diagrams as a straight line it is in fact merely an elongated elliptical orbit designed to pass near the moon, so the astronauts still experience the sensation of 'weightlessness'. Mid-course and other corrections, if required, can be given by short bursts of rocket power; since there is no air resistance the thrust required to make such changes is not very great. As in all elliptical orbits the speed will decrease as the apogee is approached, but by then the spacecraft will have passed the neutral point, will be attracted by the moon and will again pick up speed, but now new problems arise and we must consider how orbits of the moon differ from those round the earth.

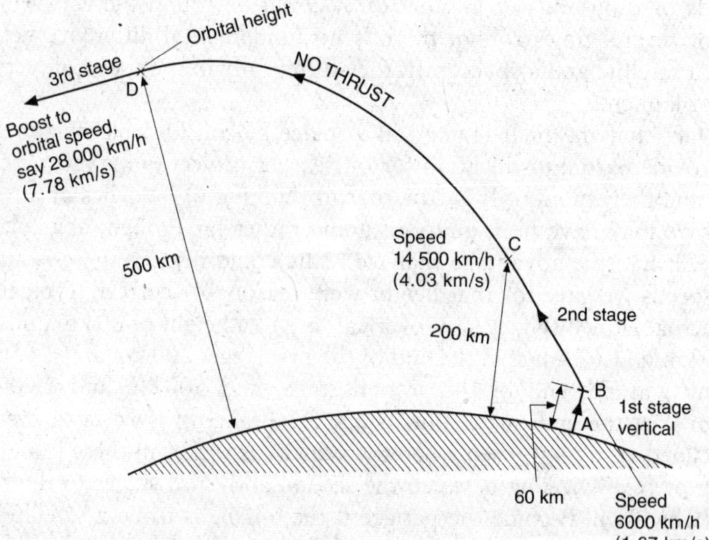

Fig 13.13 Typical flight path for launching of spacecraft
Not to scale.

Orbiting the moon

In order to understand this we must consider how the moon differs from the earth. It is, of course, much smaller, its diameter (3490 km) being rather more than 1/4 that of the earth, and its mass, which is more important from the point of view of satellite orbits, about 1/81 that of the earth. The weight of a body on the moon's surface is about one sixth of its weight on earth – if this puzzles the reader let him work it out, remembering that weight is the force of attraction which is proportional to the two masses multiplied together and inversely proportional to the square of the respective distances, i.e. the radii of the moon and the earth. The acceleration of gravity on the moon is also, of course, about one sixth of that on earth, i.e. just over 1.6 m/s^2. But the most interesting difference – and it is the result of the smaller mass of the moon and the lesser weight of bodies near the moon – is that the velocities for moon satellites, circular velocity, escape velocity, etc., are much lower than for earth satellites; the escape velocity at the surface of the moon is only about 2.4 km/s, the circular velocity being 2.4/1.41 or 1.7 km/s. Another important point is that owing to the lack of air resistance it is possible for a satellite to circle the moon very close to its surface.

But if the prospective satellite has been fired to meet the moon, the relative speed between satellite (700 km/h) and moon (3700 km/h) will be much too great, and so, unless the body actually hits the moon, it will merely go past it and escape. Thus, on first thoughts, it would seem that a body fired from the earth cannot become a moon satellite – this is true so long as there is no propulsion in the reverse direction, in other words braking; and the moon has no atmosphere to act as a brake, so the only practical means of persuading a satellite to orbit the moon, under the influence of the moon, is to provide for a rocket to act as a brake on its speed as it gets into the moon's sphere of influence (Fig. 13.12).

This, in fact, is how the spacecraft is put into orbit round the moon – a burst of power slows it down to the correct speed for the orbit required which, as already explained, can be much nearer the moon's surface than orbits of the earth (Fig. 13C, overleaf).

Thus far there has been quite a number of flights, manned and otherwise, and there have also been several soft, and some not so soft, landings of craft conveying instruments designed to send messages back to earth; but the actual landing of men on the moon has not been achieved so often that it can be considered as a matter of standard procedure. In the successful attempts so far made a small part of the spacecraft, the lunar module, has been detached at that part of the orbit which, as calculated by computer, will result in a landing at the desired point on the moon's surface. By a small burst of reverse power the lunar module is again slowed down to bring it closer to the moon, while the command module continues on its circular orbit. As the lunar module 'falls' onto the moon, fortunately not so fast as it would onto the earth, but quite fast enough to be uncomfortable, a final reverse rocket thrust is fired to enable the module to land gently on the surface – owing to the lack of air no parachutes, or any kind of air brake, are of avail in controlling the fall (Fig. 13D, overleaf).

The return flight

The moon lift-off, by another burst of rocket power, is again made just that much easier than from earth owing to the reduction in the mass of the spaceship which has thrown off the multi-stage rockets, owing too to the lesser force of gravity, the lack of air resistance, and the lower speed required for orbit; all this is just as well because the rockets that developed the tremendous thrust for lift-off from earth are no longer available, and by now much of the reserve fuel and power has been expended. Even so, a high degree of accuracy is again necessary to ensure that the lunar module gets into orbit close to the command module (though again small adjustments can be made by bursts of power), so that they can again be linked up into one spacecraft. Once they have been re-united and men, films and other souvenirs have been transferred to the command module, the lunar module can itself be discarded, and left to orbit or to hit the moon – this time probably at speed!

The next stage is a further and considerable boost to put what remains of the spacecraft out of moon orbit and on the return path to earth – once more really an elongated elliptical orbit with the apogee this time near the earth. Although this has been described as a considerable boost, the thrust required is nothing like so great as was needed to start the craft on its journey to the moon because the neutral point is now comparatively near, and once this has been passed the earth's attraction will all the time be increasing, as will the speed of the space-ship until it reaches something of the order of 10.46 km/s (more than 37 000 km/h), the speed with which it started on its journey.

Now another reverse burst is needed to slow the craft down to approximately 7.5 km/s for a similar circular orbit, or partial orbit, and in the same direction too as that used after launch.

Re-entry into the atmosphere

On the Apollo missions, the craft re-entered the atmosphere after at least a partial orbit, and after discarding the larger part of what still remained of the spacecraft leaving only the small command module, a mere 5 tonnes of the 3500 tonnes or more of the mass at launch. From re-entry to splashdown was one of the most difficult, and in some ways the crudest part of the whole procedure. Once again extreme accuracy was needed because the craft, by final

Fig 13C Orbiting the moon
(By courtesy of the Boeing Company, USA)
Scale model of a lunar orbiter; the lunar landscape (taken by Lunar Orbiter 1 on the far side of the moon) is authentic, and shows the sharp break between sunshine and shadow

use of the rocket power still available, must enter the atmosphere, at some 400 000 feet, through a 'window', as it is called, only 8 kilometres wide, and at an angle of between 5.6° and 7.2° to the top of the atmosphere – if it entered too steeply it would have burned up, if too shallowly it would have bounced off again. Not only is the angle of entry very critical, but the craft also had to be manoeuvred into such a position that it encountered maximum drag (from drag rather than skin friction) and so maximum retardation of about 6 g. Even

Fig 13D Moon landing craft
(By courtesy of the Bell Aerospace Division of Textron Inc, USA)
Astronaut Charles Conrad Jr in a lunar landing training vehicle during a simulation flight for the Apollo 12 mission

so, the speed was so high, and the skin friction so great, that the heat generated was quite alarming, the surface of the craft was burnt and scarred, and the air ionised so that radio communication between the earth and the crew was temporarily interrupted. When denser air was reached, first a drogue parachute was released, followed by at least three large parachutes, and these reduced the velocity sufficiently for any surplus fuel to be jettisoned, and finally for a reasonably soft splash-down in the sea, again with reasonable accuracy of position.

In view of the crudeness of this method of approach and landing it is not surprising that a reusable vehicle, the Space Shuttle, was developed (Fig. 13E). This has the same problems – it gets hot and needs to slow down for landing. But using wings gives another way of controlling the trajectory and means that it can land on a runway. It is really a glider – albeit an unusual one!

Flights in space

We could go on to discuss orbits round the sun and the possibility of flights to and round the various planets. It is a fascinating subject, and becomes more and more so as the possibilities become practicabilities and then historical facts. But the principles of all such flights are the same as those we have mentioned, and the author must avoid the temptation of going any further into space. The reader who is as fascinated with the subject as is the author – and who, if he is interested in the mechanics of flight at all, is not? – must seek other books, though admittedly it is not easy to find books that are at the same time reasonably simple and reasonably sensible.

Sub-orbital flight – the aeroplane – missile – satellite

Have you ever realised that the 'lift' required to keep an aeroplane in the air depends upon the direction in which it is travelling? – e.g. whether it is going with the earth or against it. In an earlier paragraph we worked out the centripetal force on a body of mass 1 kg sitting on the earth's surface at the equator – sitting still, as it seems, but in fact behaving like a stone travelling at 1690 km/h on the end of a string of 6370 km radius. The answer didn't come to much – about 0.018 N – but the principle is of extreme importance.

The corresponding value for an aeroplane of mass 10 190 kg is about 400 N, still not much perhaps, but none the less an appreciable and measurable quantity. It means that if the real force of gravity on the aeroplane is

Fig 13E The Space Shuttle
(By courtesy of NASA)

100 000 N, it would appear to weigh only 99 600 N – in fact, of course, we would call this the weight, it is the force we would have to exert to lift it.

But it is a solemn thought, though none the less a fact, that if this aeroplane were to fly against the direction of the earth's rotation, i.e. towards the west, at 1690 km/h, it would not require this centripetal force, and so would appear to weigh 100 000 N – and that is the lift the wings would have to provide. If, on the other hand, it flew towards the east at 1690 km/h, its real speed would be 3380 km/h, and the centrifugal force would be, no, not 800 N but 4 × 400, i.e. 1600 N (because the centripetal force depends on the square of the velocity), so the lift that the wings would have to provide would be 100 000 − 1600 = 98 400 N.

Similarly at a real speed of 6760 km/h (5070 km/h eastwards) the centripetal force would be 6400 N, and the necessary lift 93 600 N. At 12 800 km/h (11 200 km/h eastwards), the corresponding figures would be 25 600 N and 74 400 N; and at 25 600 km/h, 102 400 N and minus 2400 N! At approximately 29 000 km/h the centripetal force is 100 000 N and the lift required nil.

What does it all mean? Well, the reader who has followed the arguments in this chapter will surely know what it means – simply that the aeroplane travelling at 29 000 km/h near the earth's surface is travelling at the circular velocity, it doesn't need any lift from the wings, it will stay up of its own accord, it is a satellite. Nor when these velocities are reached does it make all that difference (only 1600 km/h each way) whether it travels east or west.

What it will need is colossal thrust to equal the colossal drag, which in any case will cause it to frizzle up.

But what if it flies higher – and higher – and higher? The drag for the same real speed will be less, less thrust will be needed, the circular velocity required for no lift conditions will be less, even the real force of gravity upon it will be less; it won't even create a sonic boom at ground level.

Can you answer these?

Now let us see what we know about this fascinating subject –

1. What is meant by escape velocity? What is its approximate value for the earth? Is it the same for the moon?
2. Is the escape velocity the same for a horizontal launch as for a vertical launch?
3. Distinguish between the perigee and the apogee in an elliptical orbit.
4. What is the particular significance of a satellite circling the earth at about 35 400 km from the centre of the earth?

5. What is the time of circular orbit of –
   (a) a satellite very near the earth's surface?
   (b) a satellite 1600 km from the earth's surface?
   (c) the moon?
6. Under what conditions is the path of a satellite parabolic? hyperbolic?

For solutions see Appendix 5.
For numerical examples on missiles and satellites see Appendix 3.

APPENDIX 1

Aerofoil data

The aerofoil sections, of which particulars are given in the following pages, have been chosen from among the thousands that have been tested, as being typical of the best that have been designed for particular purposes.

Although the values have been taken from standard tests, they have been modified so as to bring them as far as possible into line with each other, and simplified so as to correspond with the symbols and methods used in this book. In thus modifying the figures the aim has been to bring out the principles even at the sacrifice of some degree of accuracy. For the purpose of this book nothing is lost by this simplification and, while it is right and proper that official results should be given to the accuracy with which they can be measured, the student should remember that they are, after all, taken from experimental figures and that there is a limit not only to the accuracy of such figures, but even more so to the various corrections that have to be applied to them.

Unfortunately it is not possible to obtain results for all the sections at the same Reynolds Number, but for each the approximate Reynolds Number of the test has been given, and where alternative results are available those at the highest Reynolds' Number have been chosen.

For reasons of security it is not possible to give test results for the most modern high-speed sections; but, even if they could be given, it is doubtful whether the tests would have been made at sufficiently high Mach Numbers and Reynolds Numbers to be reliable as a guide to full-scale performance. There is no difficulty in getting lift from bi-convex or double-wedge sections used at supersonic speeds, the problem is to keep down the drag.

For the benefit of those readers who would like to sketch out the shapes of the various aerofoil sections, the co-ordinates of the upper and lower surfaces are given; the measurements are expressed as percentages of the chord, negative values being below the chord line and positive values above it.

The tables give values of C_L and C_D at various angles of attack – from negative angles to above the stalling angle – but unfortunately values of C_D are not available for aerofoils with flaps down (except for No. 8, the model of the Lightning). When a dash appears in the data columns it means that the figure is not available, or that for some reason it would be meaningless. Pitching moments are given about the leading edge, or about the quarter-chord, or about the aerodynamic centre; and an opportunity is given in the short questions that follow the data for the reader to work out one from the other. Similarly the position of the centre of pressure can be found from other data, the lift/drag ratio from C_L and C_D, and so on.

Most of the questions can be answered from the information given in Chapter 3, but the student may require a little guidance on the questions dealing with moment coefficients and centres of pressure, especially since the simplest method of reaching some of the answers is through the notation of differential calculus which has not been used in the text of the book.

For instance, to solve question (*c*) for RAF 15 (without slot):

In Chapter 3 when discussing the aerodynamic centre we arrived at the equation –

$$x/c = (C_{M.AC} - C_{M.LE})/C_L \qquad (1)$$

or

$$C_{M.AC} = C_{M.LE} + (x/c) \, . \, C_L \qquad (2)$$

By differentiating (2) with respect to C_L we get the differential equation –

$$dC_{M.AC}/dC_L = dC_{M.LE}/dC_L + x/c \qquad (3)$$

But, by definition, the moment coefficient about the aerodynamic centre does not change with the angle of attack (or with the lift coefficient), or expressed in mathematical terms –

$$dC_{M.AC}/dC_L = 0$$

Therefore

$$x/c = -dC_{M.LE}/dC_L \qquad (4)$$

By drawing a graph of $C_{M.LE}$ against C_L we can determine the slope of the curve, i.e. $dC_{M.LE}/dC_L$ for any value of C_L, and so for any angle of attack. Thus we can get x/c from (4), and substitute in (2) to get $C_{M.AC}$. To solve question (*c*) for RAF 15 (with slot), we must first find $C_{M.LE}$ from another formula in Chapter 3 –

$$\text{CP position} = -C_{M.LE}/C_L \qquad (5)$$

and then use the same method, as outlined above, to find $C_{M.AC}$. The same method can be used to solve questions (*a*) and (*b*) on the Clark YH aerofoil; and for questions (*b*) and (*c*) on NACA 0009 we can start from –

$x' = C_{M.C/4}/C_L$ as in (5) above.

Finally it should be noted that formulae (1) and (2) are approximations based on the assumption that the angle of attack is small. The answers given in Appendix 4 have been arrived at by using these simplified formulae.

The more refined formula for (2) is –

$$C_{M.AC} = C_{M.LE} + x/c\,(C_L \cos\alpha + C_D \sin\alpha)$$

and the student is advised to work out one or two of the examples at the larger angles of attack with this formula, if only to confirm that the use of the simplified formula is justified – at any rate for the angles of attack of normal flight.

When differentiating the full formula, it must be remembered that cos α, C_D and sin α all vary with C_L and so the appropriate mathematical techniques must be used; these involve drawing graphs of cos α, C_D and sin α against C_L, in order to determine $d\cos\alpha/dC_L$, dC_D/dC_L and $d\sin\alpha/dC_L$.

More extensive questions on aerofoils will be found in Appendix 3.

1. RAF 15

*1A. RAF 15 with slot

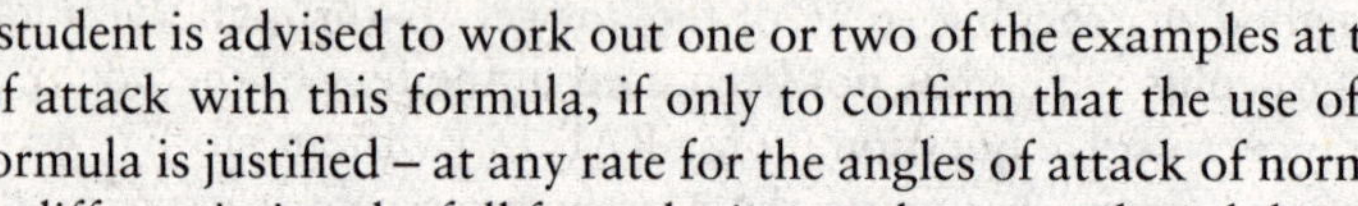

Like Clark Y, used on many early types of aircraft.
Figures relate to aspect ratio of 6.
Reynolds Number of test 3.5 million; with slot 200 000.
Slot assumed to remain open at position giving maximum lift.

(*a*) What is the max. value of *L*/*D* with slot? without slot?

(*b*) What is the stalling angle with slot? without slot?

(*c*) What is the value of $C_{M.AC}$ at +4° with slot? without slot?

(*d*) What is the value of $C_L$max/$C_D$min with slot? without slot?

(*e*) What is the value of $C_L^{\frac{3}{2}}/C_D$ (without slot) at 4° and 8°?

Distance from LE, % chord	*Upper surface*	*Lower surface*
0	1.50	1.50
1.25	3.14	0.76
2.5	3.94	0.50
5	5.00	0.18
7.5	5.37	0.02
10	6.09	0.02
15	6.67	0.18
20	6.96	0.53
30	6.94	1.02
40	6.63	1.02
50	6.13	0.71
60	5.52	0.33
70	4.79	0.06
80	3.91	0.04
90	2.81	0.21
95	2.17	0.32
100	0.94	0.94

Angle of attack	C_L	C_D	$C_{M.LE}$	$*C_L$	$*C_D$	**CP, fraction of chord*
−4°	−0.14	0.014	−0.036	−0.27	0.120	–
22°	+0.02	0.008	−0.052	−0.12	0.075	–
0°	0.14	0.008	−0.090	+0.03	0.056	–
+2°	0.32	0.012	−0.130	+0.18	0.050	0.70
+4°	0.46	0.020	−0.160	0.33	0.050	0.45
6°	0.60	0.030	−0.200	0.47	0.053	0.39
8°	0.76	0.044	−0.240	0.62	0.060	0.37
10°	0.90	0.060	−0.280	0.76	0.070	0.36
12°	1.04	0.070	−0.310	0.90	0.080	0.34
14°	1.16	0.096	−0.330	1.05	0.100	0.32
15°	1.22	0.110	−0.340	1.12	0.110	0.31
16°	1.16	0.140	−0.350	1.21	0.120	0.31
18°	1.02	0.210	−0.384	1.36	0.148	0.30
20°	0.94	0.260	−0.390	1.51	0.175	0.30
24°	–	–	–	1.70	0.234	0.30
26°	–	–	–	1.76	0.270	0.30
28°	–	–	–	1.76	0.304	0.30
30°	–	–	–	1.64	0.344	–
34°	–	–	–	1.30	0.430	–

2. Clark YH

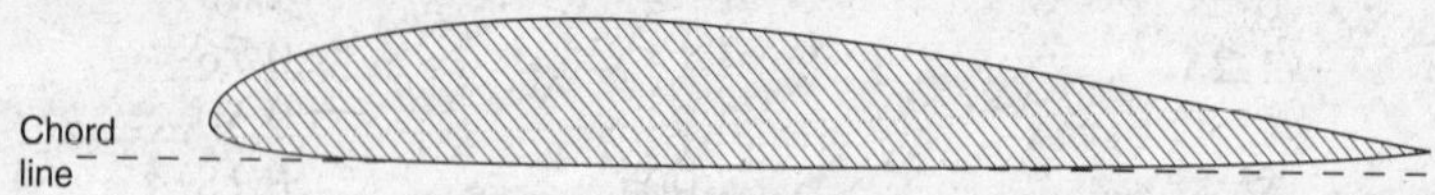

Excellent American general purpose aerofoil.
Modifications of Clark Y have been used on many types of aircraft all over the world; Clark YH was one of the first of these modifications.
Figures relate to aspect ratio of 6, and standard roughness.
Reynolds Number of test 7 million.

Distance from LE, % chord	*Upper surface*	*Lower surface*
0	3.50	3.50
1.25	5.45	1.93
2.5	6.50	1.47
5	7.90	0.93
7.5	8.85	0.63
10	9.60	0.42
15	10.68	0.15
20	11.36	0.03
30	11.70	0
40	11.40	0
50	10.51	0
60	9.15	0
70	7.42	0.06
80	5.62	0.38
90	3.84	1.02
95	2.93	1.40
100	2.05	1.85

Angle of attack	C_L	C_D	*CP, fraction of chord*	$C_{M.LE}$	*L/D*
24°	+0.09	0.010	–	+0.030	−10
22°	+0.05	0.009	0.74	−0.010	+5.2
0°	0.20	0.010	0.40	−0.046	19.3
2°	0.36	0.015	0.32	−0.072	23.2
4°	0.51	0.022	0.295	−0.116	23
6°	0.66	0.033	0.285	−0.150	20.6
8°	0.80	0.045	0.275	−0.184	17.7
10°	0.94	0.062	0.27	−0.220	15.2
12°	1.06	0.083	0.27	−0.244	13.3
14°	1.21	0.103	0.27	−0.276	11.8
16°	1.33	0.125	0.265	−0.320	11
18°	1.43	0.146	0.265	−0.352	9.9
19°	1.36	0.170	0.275	−0.356	8
20°	1.26	0.211	0.29	−0.354	7
25°	0.97	0.324	0.33	−0.354	2.9
30°	0.81	0.430	0.37	−0.352	1.9

(*a*) What is $C_{M.AC}$ at 0°, 4° and 8° for this aerofoil?

(*b*) Where is the aerodynamic centre of this aerofoil section?

(*c*) What is the stalling angle?

(*d*) What is the value of $C_L max/C_D$ min?

(*e*) What is the value of $C_L^{\frac{3}{2}}/C_D$ at 4° and 8°?

3. NACA 0009

*3A. NACA 0009 with flap

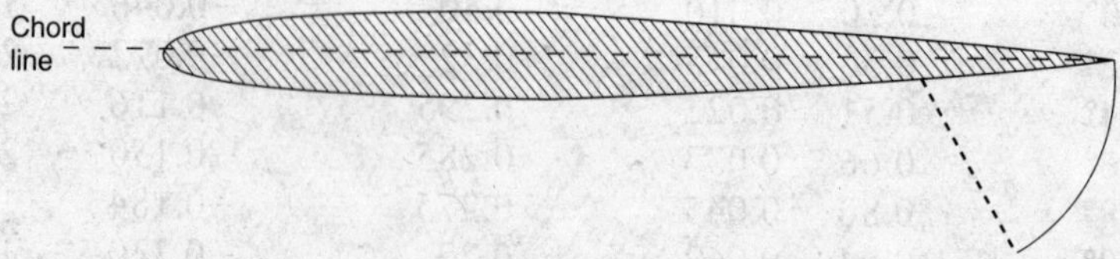

A thin symmetrical section.
All figures relate to standard roughness.
Reynolds Number of test 6 million.
Position of aerodynamic centre 0.25 of chord from LE.
*With 20 per cent split flap set at 60°.

Distance from LE, % chord	*Upper and lower surfaces % chord*
0	0
1.25	1.42
2.5	1.96
5.0	2.67
7.5	3.15
10	3.51
15	4.01
20	4.31
30	4.50
40	4.35
50	3.98
60	3.50
70	2.75
80	1.97
90	1.09
95	0.61
100	0

Angle of attack	C_L	C_D	$C_{M.C/4}$	$^*C_L$	$^*C_{M.C/4}$
28°	−0.88	0.022	0	+0.45	−0.200
26°	−0.65	0.014	0	+0.68	−0.210
24°	−0.45	0.011	0	+0.90	−0.216
22°	−0.21	0.010	0	+1.09	−0.220
0°	0	0.009	0	+1.29	−0.216
+2°	+0.21	0.010	0	+1.38	−0.218
4°	+0.43	0.011	0	+1.65	−0.222
6°	+0.64	0.014	0	+1.78	−0.225
8°	+0.85	0.018	0	+1.72	−0.230
10°	+0.90	0.021	−0.002	+1.58	−0.275
12°	+0.89	0.028	−0.004	–	–
14°	+0.87	0.036	−0.012	–	–

(*a*) What is the value of $C_{M.AC}$ (without flap)?

Is it the same at all angles, as it should be?

(*b*) What is the position of the CP (without flap) at +4°?

(*c*) What is the position of the CP (with flap) at +4°?

(*d*) What is the value of *L*/*D* (without flap) at 2°, 6°, 10°?

(*e*) What is the stalling angle (i) without flap? (ii) with flap?

4. NACA 4412

*4A. NACA 4412 with flap

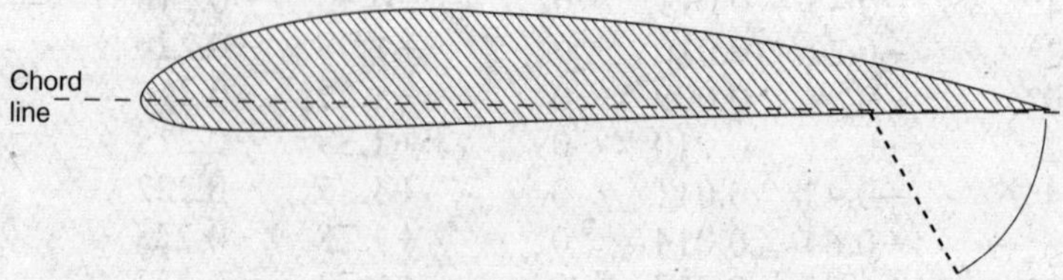

Medium thickness NACA 4-digit good all round section.
All figures relate to standard roughness.
Reynolds Number of test 6 million.
Position of aerodynamic centre 0.246 of chord from LE.
*With 20 per cent split flap set at 60°.

Distance from LE, % chord	*Upper surface*	*Lower surface*
0	0	0
1.25	2.44	−1.43
2.5	3.39	−1.95
5.0	4.73	−2.49
7.5	5.76	−2.74
10	6.59	−2.86
15	7.89	−2.88
20	8.80	−2.74
25	9.41	−2.50
30	9.76	−2.26
40	9.80	−1.80
50	9.19	−1.40
60	8.14	−1.00
70	6.69	−0.65
80	4.89	−0.39
90	2.71	−0.22
95	1.47	−0.16
100	0	0

Angle of attack	C_L	C_D	$C_{M.AC}$	$*C_L$	$*C_{M.C/4}$
−8°	−0.45	0.022	−0.097	+0.90	−0.287
−6°	−0.23	0.014	−0.092	+1.12	−0.297
−4°	−0.03	0.012	−0.092	+1.34	−0.302
−2°	+0.20	0.010	−0.092	+1.56	−0.305
0°	+0.38	0.010	−0.093	+1.75	−0.305
+2°	+0.60	0.010	−0.095	+1.95	−0.305
4°	+0.80	0.012	−0.098	+2.14	−0.305
6°	+1.00	0.014	−0.100	+2.43	−0.302
8°	+1.15	0.017	−0.100	+2.50	−0.300
10°	+1.27	0.022	−0.095	+2.65	−0.290
12°	+1.36	0.030	−0.092	+2.63	−0.275
14°	+1.35	0.042	−0.092	–	–
16°	+1.25	0.059	−0.095	–	–

(*a*) What is the value of $C_{M.C/4}$ (without flap) at 0° and 8°?

(*b*) Where is the CP (without flap) at these angles?

(*c*) What is the value of *L*/*D* (without flap) at these angles?

(*d*) What is the stalling angle (i) without flap? (ii) with 60° flap?

(*e*) What is the value of $C_L max/C_D$ min (without flap)?

5. NACA 23012

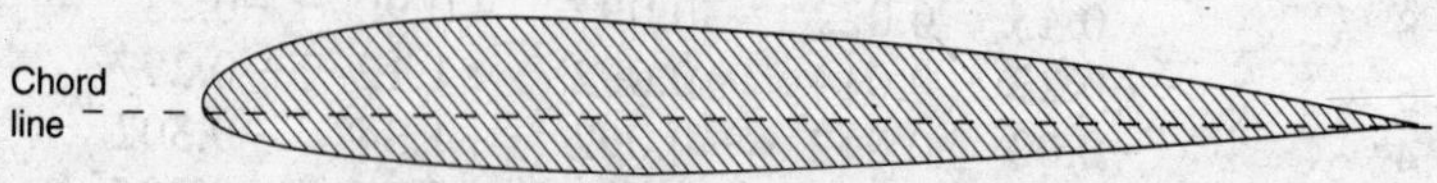

Medium thickness 5-digit section that has been much used.
Low drag; maximum camber well forward.
All figures relate to standard roughness.
Reynolds Number of test 6 million.
Position of aerodynamic centre 0.241 of chord from LE.

Distance from LE, % chord	*Upper surface*	*Lower surface*
0	0	0
1.25	2.67	−1.23
2.5	3.61	−1.71
5.0	4.91	−2.26
7.5	5.80	−2.61
10	6.43	−2.92
15	7.19	−3.50
20	7.50	−3.97
25	7.60	−4.28
30	7.55	−4.46
40	7.14	−4.48
50	6.41	−4.17
60	5.47	−3.67
70	4.36	−3.00
80	3.08	−2.16
90	1.68	−1.23
95	0.92	−0.70
100	0	0

Angle of attack	C_L	C_D	$C_{M.C/4}$	$C_{M.AC}$
−8°	−0.60	0.020	−0.018	−0.013
−6°	−0.43	0.014	−0.015	−0.013
−4°	−0.25	0.011	−0.013	−0.014
−2°	−0.08	0.010	−0.013	−0.016
0°	+0.15	0.010	−0.012	−0.016
+2°	+0.36	0.010	−0.010	−0.015
4°	+0.55	0.011	−0.008	−0.014
6°	+0.75	0.013	−0.010	−0.014
8°	+0.96	0.016	−0.013	−0.016
10°	+1.14	0.023	−0.014	−0.017
12°	+1.23	0.032	−0.012	−0.017
14°	+0.82	0.045	−0.013	–
16°	+0.77	0.065	−0.050	–

(*a*) What is the value of *L*/*D* for this aerofoil at 0°, 4°, 8°?

(*b*) Where is the CP at these angles?

(*c*) Where is the maximum thickness?

(*d*) What is the stalling angle?

(*e*) What is the value of $C_L^{\frac{3}{2}}/C_D$ at 2°, 4° and 6°?

6. NACA 23018

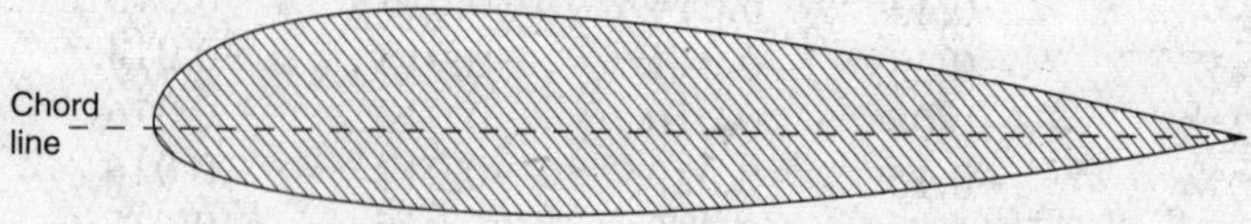

Typical thick 5-digit section of the 230 series.
All figures relate to standard roughness.
Reynolds Number of test 6 million.
Position of aerodynamic centre 0.241 of chord from LE.

Distance from LE, % chord	*Upper surface*	*Lower surface*
0	0	0
1.25	4.09	−1.83
2.5	5.29	−2.71
5.0	6.92	−3.80
7.5	8.01	−4.60
10	8.83	−5.22
15	9.86	−6.18
20	10.36	−6.86
25	10.56	−7.27
30	10.55	−7.47
40	10.04	−7.37
50	9.05	−6.81
60	7.75	−5.94
70	6.18	−4.82
80	4.40	−3.48
90	2.39	−1.94
95	1.32	−1.09
100	0	0

Angle of attack	C_L	C_D	$C_{M.C/4}$	$C_{M.AC}$
−8°	−0.62	0.016	−0.018	−0.008
−6°	−0.47	0.014	−0.010	−0.007
−4°	−0.28	0.012	−0.008	−0.007
−2°	−0.09	0.011	−0.005	−0.007
0°	+0.12	0.010	−0.002	−0.007
+2°	+0.33	0.011	−0.001	−0.007
4°	+0.53	0.012	0	−0.007
6°	+0.72	0.014	+0.002	−0.007
8°	+0.90	0.016	+0.003	−0.007
10°	+1.01	0.020	+0.004	−0.008
12°	+1.06	0.028	+0.005	−0.008
14°	+0.75	0.040	+0.002	–
16°	+0.68	0.060	−0.020	–

(*a*) What is the maximum thickness? Where is it?

(*b*) What is the value of *L*/*D* at −4°, 0°, 4°, 8°, 12°?

(*c*) Where is the CP on this aerofoil at 2°, 4° and 6°?

(*d*) What is the stalling angle?

(*e*) What is the maximum lift coefficient?

7. NACA 65_1–212

*7A. NACA 65_1–212 with flap

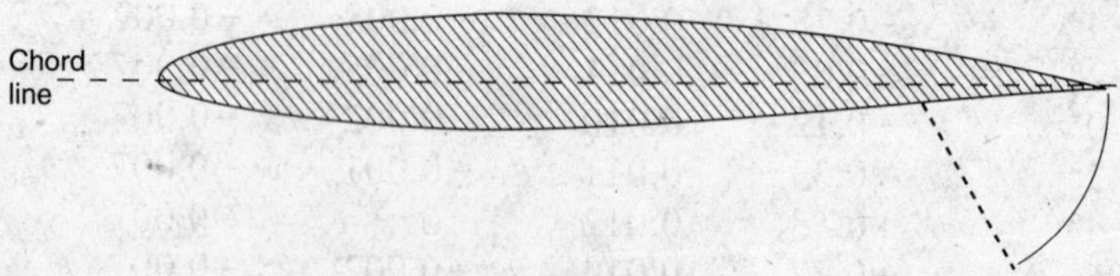

Typical of the NACA 6 series; medium thickness.
All figures relate to standard roughness.
Reynolds Number of test 6 million.
Position of aerodynamic centre 0.259 of chord from LE.
*With 20 per cent split flap set at 60°.

Distance from LE, % chord	*Upper surface*	*Lower surface*
0	0	0
0.5	0.970	−0.870
0.75	1.176	−1.036
1.25	1.491	−1.277
2.50	2.058	−1.686
5.00	2.919	−2.287
7.5	3.593	−2.745
10	4.162	−3.128
15	5.073	−3.727
20	5.770	−4.178
25	6.300	−4.510
30	6.687	−4.743
35	6.942	−4.882
40	7.068	−4.926
45	7.044	−4.854
50	6.860	−4.654
55	6.507	−4.317
60	6.014	−3.872
65	5.411	−3.351
70	4.715	−2.771
75	3.954	−2.164
80	3.140	−1.548
85	2.302	−0.956
90	1.463	−0.429
95	0.672	−0.040
100	0	0

Angle of attack	C_L	C_D	$C_{M.C/4}$	$*C_L$	$*C_{M.C/4}$
−8°	−0.68	0.020	−0.025	+0.58	−0.223
−6°	−0.50	0.015	−0.026	+0.80	−0.230
−4°	−0.33	0.013	−0.030	+1.03	−0.240
−2°	−0.10	0.010	−0.033	+1.25	−0.250
0°	+0.12	0.009	−0.035	+1.45	−0.260
+2°	+0.35	0.010	−0.037	+1.63	−0.265
4°	+0.55	0.011	−0.038	+1.80	−0.267
6°	+0.80	0.015	−0.039	+1.87	−0.264
8°	+0.95	0.023	−0.040	+1.83	−0.260
10°	+1.07	0.035	−0.040	+1.70	−0.255
12°	+1.06	0.050	−0.038	+1.48	−0.380
14°	+1.01	–	−0.035	–	–

(*a*) What is $C_{M.AC}$? Is it the same at all angles, as it should be?

(*b*) Where is the maximum thickness?

(*c*) What is the stalling angle (i) without flap? (ii) with 60° flap?

(*d*) What is the maximum value of *L*/*D*?

(*e*) What is the value of $C_L$max/$C_D$min (without flap)?

8. English Electric ASN/P1/3

*8A. English Electric ASN/P1/3 with flap

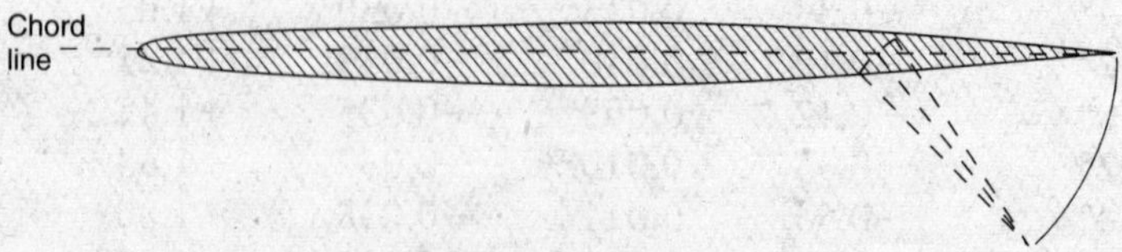

This is the symmetrical aerofoil section used on the BAC Mach 2+ Lightning. The wing was tapered and the ordinates relate to a section at 38.5 per cent of semi-span.

Reynolds Number of test 1.5 million (based on mean chord).

The values of coefficients refer to a complete model of the aircraft, not to the wing section alone.

The Lightning was a mid-wing monoplane with 60° sweepback on leading edge (see Fig. 11B).

*Model with approx 25 per cent plain flaps set at 50°.

Distance from LE, % chord	*Upper and lower surfaces % chord*
0	0
0.25	0.426
0.75	0.706
1.25	0.875
2.50	1.175
5.00	1.530
10	1.941
15	2.183
20	2.435
25	2.612
30	2.782
35	2.904
40	2.944
45	2.970
50	2.942
55	2.855
60	2.703
65	2.502
70	2.237
75	1.921
80	1.564
85	1.183
90	0.797
95	0.414
100	0.032

The values of C_M are related to a point at 0.405 of mean chord. The position of the aerodynamic centre = $0.405 + dC_M/dC_L$. The figures for the Lightning were given by courtesy of the former British Aircraft Corporation, Preston.

(*a*) What is the thickness/chord ratio of this aerofoil?

(*b*) Where is the maximum thickness?

(*c*) What is the stalling angle of the aircraft model: (i) with flaps up? (ii) with flaps down?

(*d*) What are the values of *L*/*D* of the model at 4°, 12° and 20°: (i) with flaps up? (ii) with flaps down?

(*e*) What are the positions of the aerodynamic centre of the clean aircraft at 4°, 12° and 20°?

(*Note:* The answers involve the drawing of the curve of C_M against C_L and measuring the slopes of this curve at the specific angles.)

Data for BAC Lightning model – clean aircraft

Angle of attack	C_L	C_D	C_M
0°	0	0.020	−0.017
2°	0.08	0.020	−0.013
4°	0.17	0.030	−0.008
6°	0.27	0.040	−0.006
8°	0.38	0.050	+0.005
10°	0.50	0.075	+0.010
12°	0.61	0.105	+0.016
14°	0.71	0.140	+0.026
16°	0.81	0.180	+0.040
18°	0.91	0.225	+0.055
20°	1.00	0.275	+0.070
22°	1.09	0.335	+0.088
24°	1.17	0.405	+0.108
26°	1.22	0.480	+0.124
28°	1.26	0.560	+0.132
30°	1.27	0.650	+0.140

Data for BAC Lightning model – flaps at 50°

Angle of attack	C_L	C_D	C_M
0°	0.17	0.07	−0.072
2°	0.27	0.07	−0.068
4°	0.37	0.08	−0.062
6°	0.47	0.09	−0.060
8°	0.57	0.11	−0.053
10°	0.67	0.13	−0.040
12°	0.78	0.16	−0.032
14°	0.88	0.20	−0.025
16°	0.98	0.25	−0.015
18°	1.07	0.31	+0.005
20°	1.15	0.37	+0.013
22°	1.22	0.43	+0.020
24°	1.27	0.50	+0.037
26°	1.27	0.55	+0.045
28°	1.22	0.59	+0.074
30°	1.16	0.63	+0.074

APPENDIX 2

Scale effect and Reynolds Number

This appendix is an amplification of a short note given about scale effect in Chapter 2. From the earliest days of the science of flight, even before any aeroplane had actually flown, people experimented with small models. The problem is how do we relate the behaviour of the model and the aerodynamic forces that are exerted on it to a full-size aircraft? We know that if we can measure, say, the lift force on a small model, we can work out its lift coefficient using $L = \frac{1}{2}\rho v^2 SC_L$, and from this we can calculate the lift on the full-size aircraft at any speed. However, we do not always get quite the correct answer, and sometimes we get an answer that is completely wrong. Is there something else that we should be doing? The clue to this was found by Osborne Reynolds some 150 years ago. Reynolds was not interested in aircraft aerodynamics but in the flow of liquids. In particular, he was interested in the conditions that determined whether the flow of water in a pipe was smooth and layered (laminar), which is normally associated with low-speed flow, or turbulent, which is associated with higher-speed flow. What he discovered was that the speed of flow at which the change or **transition** occurred depended on the value of the quantity:

$\frac{\rho vl}{\mu}$ which is now called a **Reynolds Number.**

In this quantity, ρ is the density of the fluid (water in his case)
v is the flow speed
l is the diameter of the pipe
μ is the viscosity of the fluid

For the problem of flow in pipes it was found that the transition from one type of flow to the other occurred at a critical value of around 2300. It was discovered that this critical value held regardless of the size of the pipe and what type of fluid was used; it even works for gases. So what does the flow of water have to do with model testing of aircraft? Well, we find that if we test our model at the same Reynolds Number as the full scale aircraft, then it will behave in the same way. For example its wing will stall at exactly the same angle of attack, and full-scale forces calculated using the lift and drag coefficients will be correct. Of course for aircraft there is no pipe diameter involved, so for the quantity l, we have to use some other characteristic length. This leads to some confusion, because different characteristic lengths have to be used for different types of model. For a wing section, the wing chord is normally used as the characteristic length, but a missile may not have a wing, so in this case we would probably use the overall missile length. It does not actually matter which length we use, as long as we are consistent between model and full-size. It is also important to say exactly what dimension one is using. All too often Reynolds Number values are quoted without this important piece of information. For the testing of a low-speed aircraft then, apparently, all we have to do is to ensure that the Reynolds Number of the model test is the same as that of the flight conditions that we are going to simulate. For example, consider an aircraft as below, flying at sea level

Flight speed	v	=	30 m/s
Wing chord		=	2 m

If we want to test a 1/10th scale model under the same sea level conditions, then the Reynolds Numbers must be the same so the speed v that we must test the model at is found from

$$\frac{\rho v \times 2/10}{\mu} = \frac{\rho\, 30 \times 2}{\mu}$$

After cancelling out the density and viscosity terms (which for this special case are the same for both model and full scale) we find that the required model test speed v is 300 m/s. This result is both surprising and unfortunate. The model actually needs to fly ten times faster than the real aircraft in order to correctly simulate the flight conditions. This is normally impractical, especially for faster aircraft, because the model would have to be flying at speeds where compressibility would totally change the flow.

We do not, of course, normally fly our models around the room; we put them in a wind tunnel and let the air flow past them. However, this does not immediately solve the problem that the relative air flow speed past a 1/nth scale model would need to be n times as fast as the full-scale aircraft. In the example above, this would mean a tunnel speed of 300 m/s, which would not only require a very strong tunnel and a huge amount of power to drive it, but would also mean that the effects of compressibility would be important. One solution is to make the tunnel very strong indeed and raise the air pressure inside. This has the effect of increasing the density, and, as will be seen from the Reynolds Number expression above, if the density is raised, then the speed can be lowered in the same proportion. Fortunately, modern large aircraft mostly cruise at very high altitude, where the density is low, so this also reduces the speed required for the model. Unfortunately, large compressed air wind tunnels are extremely expensive both to build and to run, and for this reason there are very few of them in the world; indeed the number is if anything reducing now. Making the tunnel smaller does not help, because the smaller the model, the greater the speed required. One way to overcome the problem of tunnel size is to test small critical parts of aircraft such as a wing section at large scale in a relatively small compressed air tunnel.

Compressed air tunnels do not unfortunately solve all the problems of similarity because nowadays all but light and a few specialist aircraft need to fly in the transonic region at speeds approaching the speed of sound, and fast military aircraft have to fly supersonically. Under these conditions, getting the Reynolds Number correct is less important than getting the right condition for similarity of compressibility effects. The latter entails getting the same Mach Number on the model and the full-scale aircraft. Mach Number is given simply by the ratio (speed/speed of sound). The speed of sound depends only on the square root of the absolute temperature of the air. The difference between the sea level absolute temperature on a really hot day and the temperature in the upper atmosphere is only a ratio of about 2:3, so broadly speaking, for compressible flows, matching of Mach Numbers requires us to run the tunnel air at a speed that is quite similar to that of the full-size aircraft. Notice that the size of the model does not come into this.

Trying to match both the Mach Numbers and the Reynolds numbers at the same time is very difficult. The most practical solution has been the adoption of the cryogenic tunnel, which is a variation of the compressed air tunnel, in which the air is cooled by injecting liquid nitrogen. Cooling affects the speed of sound (and hence the Mach Number), the density and the viscosity coefficient (and hence the Reynolds Number). By suitably juggling the pressure, temperature and speed, it is possible to get a simultaneous match for both the Reynolds and Mach numbers. Such tunnels are extremely expensive to construct and to run, and only a few exist in the entire world.

Needless to say, not all wind tunnel testing is carried out in such facilities. For practical purposes, we can still make reasonably accurate predictions by use of compromises. It is fortunate that matching the Reynolds Numbers is most important for very low-speed flight, so we can still make useful measurements in low-speed (about 30 to 100 m/s) wind tunnels.

The mismatch in Reynolds Number is particularly significant when it comes to the behaviour of the boundary layers. The position of transition from laminar to turbulent flow and the point at which the flow separates is strongly related to the value of the Reynolds Number, so we can expect reasonably good results if we are testing in situations where these two factors are not likely to be critical. However, when investigating stall behaviour, for example, we may need to use a large higher speed tunnel or rely more on full-scale testing.

In high flight speeds it is the matching of Mach Numbers that is important. Thus for high-speed flight, we use specially designed transonic or supersonic tunnels where we can match the Mach Numbers, but normally have to ignore the Reynolds Number. The expensive cryogenic tunnels are used only where highly accurate work is required on a major airliner or military aircraft. All of this might seem a little baffling, and indeed it requires a great deal of experience and knowledge to achieve really reliable wind tunnel results.

I am frequently asked what is the speed at which you should test a model to simulate a given full-scale speed. The answer, as you may see from the above, is that it is the speed at which the Reynolds and Mach numbers are both simultaneously the same as they would be if full scale. As we have also seen above, though, this is not normally practical unless you have a very large research budget. In practice, the answer normally is to test as fast as your tunnel will allow. The precise speed of the test is not really important, because we can determine the full-scale lift and drag etc. by using the relationships $L = \frac{1}{2}\rho v^2 SC_L$ and $D = \frac{1}{2}\rho v^2 SC_D$. The wind-tunnel data allow us to work out the lift and drag coefficients, and these can then be used to determine the values of lift and drag that would be obtained at full-scale size air density and speed. We just have to hope that the effects of Reynolds Numbers, are small.

Apart from the problems of trying to match Reynolds and Mach numbers, wind tunnels have some other drawbacks. These mainly arise from the fact that the air is constrained by the tunnel walls and cannot behave exactly as it would if there were no boundary. For example, the model forms partial blockage in the tunnel and thus it causes the flow to speed up in its proximity, thereby giving misleadingly high loads. Explaining the details of these effects is well beyond the scope of this book. What is normally done, however, is that some theoretically based corrections have to be applied. Whole books and many scientific papers have been written in the subject of wind tunnel corrections, and the science is still developing.

With all the difficulties and expense involved in wind tunnel testing it is not surprising that people have sought to find ways around it. Increasing use is now being made of computer modelling or computational fluid dynamics (CFD). After many years of development, CFD can now provide accurate predictions for many aspects of aircraft aerodynamics, but it is not such a cheap or quick solution as might have been hoped for. Also, CFD is not yet reliable for situations where important flow separations occur; it can be quite poor at predicting the drag forces from areas and components where the flow is not streamlined. At the time of writing, both wind tunnel testing and CFD are used, and there is no indication that wind tunnels are about to disappear. Experience shows that there are situations, such as in the behaviour of boundary layers, where wind tunnels work best, and others where computational methods are more appropriate. Finally, it should be mentioned that with modern telemetry and remote guidance systems it is possible to do some useful testing using flying scale models. Apart from the use of radio controlled models, there is a whole new area of testing which involves making piloted scale models, usually with relatively cheap composite material airframes.

APPENDIX 3

Numerical questions

The student may be surprised to find that in some of the examples below, we have used unfamiliar units such as knots for air speed and feet for altitude. This is quite deliberate, because flying is an international activity, and it is standard practice to use knots and feet for performance calculations. Anyone therefore who is thinking of making a career in aeronautics, whether as a pilot, an engineer, a technician or working in the area of flight management will have to get used to using these units and develop a feel for the magnitudes involved. Note that it is usually safer to convert the values to SI units before making calculations as these units are much simpler to use. Do not however forget to convert the answers back where appropriate. For convenience, we have given the necessary conversion factors below. You will soon get used to converting knots to m/s. You may not need to convert the feet to metres in all cases, because often all you need to know is what the relative density is at the given height. This can be found from Fig. 2.2 which gives the relative density against height, both in metres and feet.

In Europe, it is now common practice to use SI units for aerodynamic analysis and design, and for general scientific work, so questions of this type are in SI units.

It is recognised that in order to solve some of the examples, assumptions must be made which can hardly be justified in practice, and that these assumptions may have an appreciable effect on the answers. However, the benefit of solving these problems lies not in the numerical answers but in the considerations involved in obtaining them.

Unless otherwise specified, the following values should be used –

Density of water = 1000 kg/m^3
Specific gravity of mercury = 13.6
Specific gravity of methylated spirit = 0.78
International nautical mile = 1852 m, or approx 6076 ft
1 knot = 0.514 m/s
1 ft = 0.3048 m

Radius of earth = 6370 km
Diameter of the moon = 3490 km
Distance of the moon from the earth = 385 000 km
Aerofoil data as given in Appendix 1
C_D for flat plate at right angles = 1.2
cylinder = 0.6
streamline shape = 0.06
pitot tube = 1.00
Take the maximum length in the direction of motion for the length L in the Reynolds Number formula.
At standard sea-level conditions –
Acceleration of gravity = 9.81 m/s²
Atmospheric pressure = 101.3 kN/m², or 1013 mb, or 760 mmHg
Density of air = 1.225 kg/m³ at 1013 mb and 288°K
Speed of sound = 340 m/s = 661 knots = 1225 km/h
Dynamic viscosity of air (μ) = 17.894×10^{26} kg/ms
For low altitudes one millibar change in pressure is equivalent to 30 feet change in altitude.
International standard atmosphere as in Fig. 2.2.

Chapter 1. Mechanics

1. A car is travelling along a road at 50 km/h. If it accelerates uniformly at 1.5 m/s² –
   (*a*) What speed will it reach in 12 s?
   (*b*) How long will it take to reach 150 km/h?
2. A train starts from rest with a uniform acceleration and attains a speed of 110 km/h in 2 min. Find –
   (*a*) the acceleration;
   (*b*) the distance travelled in the first minute;
   (*c*) the distance travelled in the two minutes.
3. If a motorcycle increases its speed by 5 km/h every second, find –
   (*a*) the acceleration in m/s²;
   (*b*) the time taken to cover 0.5 km from rest.
4. During its take-off run, a light aircraft accelerates at 1.5 m/s². If it starts from rest and takes 20 s to become airborne, what is its take-off speed and what length of ground run is required?

5. A boy on a bicycle is going downhill at 16 km/h. If his brakes fail and he accelerates at 0.3 m/s^2, what speed will he attain if the hill is 400 m long?

6. Assuming that the maximum deceleration of a car when full braking is applied is 0.8 *g*, find the length of run required to pull up from (*a*) 50 km/h, (*b*) 100 km/h.

7. A rifle bullet is fired vertically upwards with a muzzle velocity of 700 m/s. Assuming no air resistance, what height will it reach? and how long will it take to reach the ground again?

8. The landing speed of a certain aircraft is 90 knots. If the maximum possible deceleration with full braking is 2 m/s^2, what length of landing run will be required?

9. A 7000 kg aeroplane touches down at 100 knots and is brought to rest, the average resistance to motion due to brakes and aerodynamic drag being 6.867 kN. To reduce the landing run by 500 m a tail parachute is fitted. If the additional equipment increases the total mass of the aircraft by 200 kg and the landing speed by 5 knots, what additional average drag must the parachute supply?

10. An athlete runs 100 m in 11 seconds. Assuming that he accelerates uniformly for 25 m and then runs the remaining 75 m at constant velocity, what is his velocity at the 100 m mark?

11. An aircraft flying straight and level at a speed of 300 knots and at a height of 8000 m above ground level drops a bomb. Neglecting the effects of air resistance, with what speed will the bomb strike the ground? (*Remember that the final velocity will have to be found by compounding the vertical and horizontal velocities.*)

12. Two masses of 10 kg each are attached to the ends of a rope, and the rope is hung over a frictionless pulley. What is the tension in the rope?

13. One of the masses in Q12 is replaced by a 15 kg mass. What will be the tension in the rope when the system is released?

14. What force is necessary to accelerate a 133 kg shell from rest to a velocity of 600 m/s in a distance of 3.5 m?

15. What thrust is necessary to accelerate an aircraft of 5900 kg mass from rest to a speed of 90 knots in a distance of 750 m?

16. Calculate the thrust required to accelerate a rocket of 1 tonne mass from rest vertically upwards to a speed of 10 km/s in 10 s (*neglect air resistance*).

17. A train of 250 tonnes mass is moving at 100 km/h. What retarding force will be required to bring it to rest in 15 seconds?

18. A 76 kg man is standing on a weighing machine which is on the floor of a lift. What will the weighing machine record when –

    (*a*) the lift is ascending with velocity increasing at 0.6 m/s^2?

    (*b*) the lift is ascending with velocity decreasing at 0.6 m/s^2?

    (*c*) the lift is descending at a constant velocity of 1.2 m/s?

19. An engine of 50 tonnes mass is coupled to a train of 400 tonnes mass. What pull in the coupling will be required to accelerate the train up a gradient of 1 in 100 from rest to 50 km/h in 2 min if the frictional resistance is 70 N per tonne?

20. An aircraft of 5000 kg mass is diving vertically downwards at a speed of 500 knots. The pilot operates the dive brakes at a height of 10 000 m and reduces the speed to 325 knots at 7000 m. If the average air resistance of the remainder of the aircraft during the deceleration is 15 kN, what average force must be exerted by the dive brakes? (*Assume that the engine is throttled back and is not producing any thrust.*)

21. A 1 tonne truck is pulled on a level track by a force of 245 N in excess of the frictional resistance. How far will it travel from rest in 30 seconds?

22. A horizontal jet of water from a nozzle 50 mm in diameter strikes a vertical wall. If the water is diverted at right angles and none splashes back, what force is exerted on the wall when the speed of the jet is 6 m/s?

23. A truck is standing on an incline of 1 in 80. If the frictional resistance is 50 N per tonne, how far will it travel in 15 s if released from rest?

24. If the air resistance of a 500 kg bomb is equal to $v^2/18$ N, where v is the velocity in m/s, what is the terminal velocity of the bomb?

25. An aircraft of 4000 kg mass has a take-off speed of 60 knots and a take-off run of 300 m in conditions of no wind. If the thrust delivered by the engine is 15 000 N, and the frictional resistance is 1000 N, what is the average aerodynamic resistance during the take-off?

26. What force is necessary to stop a 500 kg car in 10 m from a speed of 50 km/h?

27. A solid 1 kg shot is fired from a barrel of 100 kg mass with a muzzle velocity of 850 m/s. If the barrel is free to recoil against a resistance of 5 kN, how far will the barrel move back when the shot is fired?

28. A fighter aircraft of 5000 kg mass is fitted with four cannon each of which fires 600 rounds per minute with a muzzle velocity of 900 m/s. If each shot has a mass of 110 g, and if all the recoil is taken by the aircraft, find the loss in speed that the aircraft would experience in a 5 second burst of fire.

29. A 9000 kg aircraft is flying straight and level at 300 knots; what thrust is necessary to accelerate it to 450 knots in half a minute if the average air resistance of the aircraft between these speeds is 15 kN?

30. A rifle of 4 kg mass fires a 30 g bullet with a muzzle velocity of 750 m/s. Find the force a rifleman must exert on the butt of the rifle to limit the recoil to 40 mm.

31. An aircraft is fitted with brakes capable of exerting a force of 10 kN, and reversible pitch propellers capable of producing a backward thrust of 25 kN. If the aircraft has a mass of 10 000 kg and a landing speed of 110 knots, find the minimum length of runway required for the landing run. (*Neglect the effect of air resistance which will also help to decelerate the aircraft.*)

32. An aircraft carrier is steaming at 20 knots against a head wind of 30 knots. An aircraft of 9000 kg mass lands on the deck with an air speed of 100 knots; if the arrester gear must be sufficiently powerful to stop the aircraft in a distance of 25 m in these conditions, without any aid from the brakes or air resistance, find the retarding force that the gear must exert.

33. A 5000 kg aircraft touches down on the deck of an aircraft carrier with an air speed of 90 knots. If the carrier is heading into wind at 20 knots, and the wind speed is 12 knots, what kinetic energy must be destroyed by the action of the arrester gear in bringing the aircraft to rest on the deck? If the average resistance exerted by the arrester gear is 55 kN, how far does the aircraft roll along the deck before coming to rest?

34. A propeller 3 m in diameter revolves at 2250 rpm. Find the angular velocity and the linear speed of the propeller tip.

35. The piston of an aircraft engine has a stroke of 150 mm and the engine runs at 3000 rpm. Find the angular velocity of the crankshaft and the average speed of the piston.

36. Find the acceleration of the propeller tip in Q34.

37. Find the acceleration of the crankpins in the engine in Q35.

38. A stone of mass 1 kg is whirled in a horizontal circle making 60 rpm at the end of a cord 1 m long; what is the pull in the string? If it is whirled in a vertical circle, what is the pull in the string (*a*) when the stone is at the top? (*b*) when the stone is at the bottom?

39. A mass of 50 kg travelling at 7.905 km/s maintains a circular path of radius 6370 km. What is its acceleration towards the centre?

40. A sphere of mass 500 kg is travelling on a circular path of 12 800 km radius with an acceleration of 2.45 m/s^2 towards the centre. How long does it take to complete one full circle?

41. At what speed (in km/h) is a bank angle of 45° required for an aeroplane to turn on a radius of 60 m?

42. An aeroplane has a mass of 1500 kg. It is turning on a horizontal circle of radius 100 m at an air speed of 80 knots. Calculate –

    (*a*) the centripetal force exerted by the air on the aircraft,

    (*b*) the correct angle of bank,

    (*c*) the total lift normal to the wings.

43. Find the work required to lift a mass of 5 tonnes to a height of 30 m. If this is done in 2 minutes, what power is being used?

44. Find the power required to propel a 3000 kg aircraft through the air at a speed of 175 knots if the air resistance is 3924 N.

45. Find the power required to propel the same aircraft at 350 knots when the air resistance is 14.715 kN.

46. A car of mass 750 kg can climb a gradient of 1 in 12 in top gear at 40 km/h. If the frictional resistance is 100 N per tonne, find the power developed by the engine in these conditions.

47. A projectile of mass 1 kg is fired from a gun with a muzzle velocity of 850 m/s. What is its kinetic energy? What will be its velocity when the kinetic energy has fallen to 90.3125 kJ?

48. The jet velocity of a certain gas turbine is 500 m/s when the engine is stationary on the ground; if the mass flow of jet gases is 15 kg/s, find the kinetic energy wasted to the atmosphere every minute.

49. A block of wood of mass 75 kg slides down a frictionless slope on to a rough level surface. The slope is 1 in 10 and 10 m long. If the frictional resistance on the level surface is 0.981 N/kg, how far will the block travel along the level surface?

Chapter 2. Air and airflow – subsonic speeds

50. At a certain height the barometric pressure is 830 mb and the temperature 227 K. Find the density of air at this height.

51. If one fifth of the air is oxygen, what will be the mass of oxygen in 1 m^3 of air at a temperature of −33°C and a pressure of 40 kN/m^2?

52. What is the total mass of air in a room 12 m long, 8 m wide and 4 m high in standard sea-level conditions?

53. What would be the total mass of air in the room mentioned in Q52 if the temperature rose from 15°C to 25°C and the pressure dropped from 1013 m/s to 979 m/s of mercury? (*Assume that the room is not air-tight, and that therefore the air is free to enter or leave the room.*)

54. From Fig. 2.2 read the temperature, pressure and density of the air at sea-level. Taking these values and the corresponding values of temperature and pressure at (*a*) 10 000 ft and (*b*) 10 000 m, calculate the density at these two heights on the assumption that Boyle's Law and Charles' Law are true for air. Compare the calculated values with the corresponding values obtained from the relative density and density respectively given for the International Standard Atmosphere in Fig. 2.2.

55. During a gliding competition a barograph was installed in a glider to measure the altitude reached. On landing, the minimum pressure recorded by the barograph was 472 mb. Draw a graph of pressure against altitude from the values given in Fig. 2.2, and estimate the height reached by the competitor.

56. An aircraft is standing on an airfield 220 ft above sea-level on a day when the barometric pressure at ground level is 1004 mb. If the pilot sets the altimeter to read 220 ft on this day, what will it read if the barometric pressure drops to 992 mb?

57. An aircraft sets off from airfield A (126 ft above sea-level) where the ground pressure is 1010 mb and flies to B (762 foot above sea-level) where the ground pressure is 985 mb. If the pilot sets his alimeter (incorrectly) at 26 ft at A, what will it read when he lands at B?

58. The volume of the pressurised compartment of a jet aircraft is 336 m^3. If the pressurisation system has to maintain a temperature of 17°C and a cabin altitude of 10 000 ft when the aircraft is flying at 40 000 ft with a complete change of air every minute, calculate the mass of air per second which must be delivered to the pressurised compartment.

59. A light aircraft has a landing speed of 70 knots. A wind of 25 knots is blowing over the airfield. What is the ground speed of the aircraft when it touches down –

(*a*) directly into wind?

(*b*) at an angle of 30° to the wind?

(*c*) at an angle of 60° to the wind?

(*d*) directly down-wind?

60. A and B are two places 400 nautical miles apart. Find the total time taken by an aircraft flying at an air speed of 250 knots to fly from A to B and back to A –

(*a*) if there is no wind,

(*b*) if the wind is blowing at 30 knots from A towards B,

(*c*) if the wind is blowing at 30 knots at right angles to the line joining A and B.

61. A pilot must reach a destination 450 nautical miles away in one hour. He sets off at an air speed of 455 knots and after half an hour finds that he has covered only 212 nautical miles. Assuming constant wind velocity, at what air speed must he fly for the remaining half hour to reach his destination on time?

62. If the pilot in Q61 flew at an air speed of 465 knots for the first half hour, what air speed would be necessary for the remaining time to complete the journey in the hour?

63. An aircraft is taking part in a square search involving flying over the ground in the form of a square of 25 nautical miles side. If the aircraft cruises at 120 knots air speed, and there is a wind of 30 knots down one of the sides of the square, calculate the time of flight for each of the four sides.

64. A flat plate of area 0.25 m^2 is placed in a 30.8 m/s airstream at right angles to the direction of the airflow. Calculate the air resistance of the plate in these conditions.

65. What would be the resistance of the flat plate of Q64 at 61.6 m/s.

66. What will be the resistance of a sphere of radius 75 mm moving through air at 30.8 m/s? ($C_D = 0.55$)

67. What would be the resistance of the same sphere moving at the same speed through water?

68. A 1/8th scale model of a streamlined body, when tested in a water tank at 5 m/s, had a resistance of 0.6 N. Neglecting any 'scale effect', what would be the resistance of the full-size body at 75 m/s in air?

69. A wind of 7.7 m/s causes a pressure of 50 N/m^2 on a flat plate at right angles to it. What wind velocity would produce a total force of 1 kN on 3 m^2 of a similar plate?

70. Of two exactly similar parts of an aeroplane, one is situated in the slipstream from the propeller and the other is outside the slipstream. If the velocity of the slipstream is 1.4 times the velocity of the aeroplane and if the resistance of the part outside the slipstream is 100 N, what will be the resistance of the corresponding part within the slipstream?

71. The air resistance of the fuselage of an aircraft is 13.35 kN at ground level at an air speed of 90 m/s. What will be the resistance of this fuselage at a true air speed of 113.1 m/s at 20 000 ft, assuming that the density of air at this height is half the value at ground level?

72. If the undercarriage of an aircraft has a frontal area of 0.45 m^2, and a resistance of 475 N at a speed of 42.3 m/s, what is the value of the drag coefficient?

73. A rough egg-shaped body with a circular cross-section 75 mm in diameter is tested in a wind tunnel at 51.4 m/s and the air resistance is found to be 1.8 N. What is the value of the drag coefficient?

74. The drag of a loop aerial on an aircraft was found to be 400 N at a speed of 113.1 m/s. In order to reduce this, a fairing of 0.1 m^2 cross-section and drag coefficient of 0.11 was fitted to the aerial. By how much did this reduce the drag at this speed?

75. A 1/5th scale model of an aeroplane is tested in a wind tunnel at a speed of 25 m/s, and the drag is found to be 56 N. What will be the drag of the full-size machine at 61.7 m/s? (*Neglect any 'scale effect', and assume that the density of the air is the same in each case.*)

76. A 1/10th scale model of an aeroplane is tested in a wind tunnel, and the air resistance is 65 N at a speed of 100 m/s. What would be the resistance of an 1/8th scale model at 130 m/s? (*Assume that the air density is the same for both tests.*)

77. A 1/10th scale model of a hull of a flying boat is tested in a water tank and has a resistance of 135 N at a speed of 12 m/s. What would be the resistance of the full-size hull in water at a speed of 23.1 m/s?

78. A streamlined shape with a cross-sectional area of 0.01 m^2 is tested in a compressed air tunnel at a speed of 34 m/s and a pressure of 25 atmospheres. If the resistance is 18 N, what is the value of the drag coefficient?

79. What would be the resistance of the same body in water at a speed of 3 m/s?

80. A 1/5th scale model has a resistance of 19.5 N when tested in a wind tunnel. What would be the resistance of a half-scale model at half the speed in air five times as dense?

81. If the drag coefficient of the flaps used on an aircraft is 0.92, what would be the drag of these flaps at 100 knots if their area totalled 3 m^2?

82. An aeroplane is to be modified, and in order to estimate the effect of the modification the drag of two models, one of the original and one of the proposed modified type, is measured in a wind tunnel.

    The model of the original is 1/20th scale and, when tested in air of density 1.225 kg/m^3 at 33.4 m/s, the drag is 62 N.

    The model of the modified type is 1/16th scale and, when tested in air of density 1.007 kg/m^3 at 30.8 m/s, the drag is 48 N.

By what percentage will the modification increase or decrease the drag coefficient of the aeroplane?

83. The resistance of a part of an aeroplane is 640 N when the aeroplane is flying at 150 knots near sea-level. What will be the resistance of this part at a height of 20 000 ft if the 'indicated' air speed is the same, i.e. 150 knots? (*Don't forget to change knots to m/s!*)

84. What would be the resistance of this part at 20 000 ft if the 'true' air speed were 150 knots?

85. If the static atmospheric pressure is 101.3 kN/m^2, and the air density is 1.225 kg/m^3, what will be the pressure on the pitot side of the diaphragm in an air speed indicator when the forward speed of the aircraft is 51.4 m/s?

86. An aircraft is flying at a true air speed of 138.8 m/s at a height of 20 000 ft, where the air density is 0.653 kg/m^3 and the pressure is 466 mb.

    What are the pressures transmitted to the air speed indicator via (*a*) the static tube and (*b*) the pitot tube?

    What will be the indicated air speed if the density assumed in the calibration of the instrument was 1.225 kg/m^3?

87. If an aircraft stalls in straight and level flight at an indicated speed of 100 knots at sea-level, at what true air speed will it stall at

    (*a*) 20 000 ft? (*b*) 40 000 ft?

88. At what indicated air speed will it stall at –

    (*a*) 20 000 ft?

    (*b*) 40 000 ft?

89. An air speed indicator is being calibrated with a U-tube containing mercury. Calculate the speed that corresponds to a height difference in the liquid levels in the two limbs of 40 mm.

90. An aircraft flying at 10 000 ft runs into severe icing which blocks up the static tube, but leaves the pitot tube open. The aircraft descends and approaches to land at sea-level with the static tube still blocked by ice. If the pilot approaches at a true air speed of 60 knots, what speed will be indicated on the air speed indicator?

91. An aircraft is flying at sea-level at a true air speed of 77.1 m/s. Calculate the static pressure at –

    (*a*) the stagnation point,

    (*b*) a point on the wing surface where the local flow velocity is double the free stream velocity.

Chapter 3. Aerofoils – subsonic speeds

92. The table shows the lift coefficient of a flat plate at angles of attack from 0° to 90°. A flat plate of 6 m span and 1 m chord is tested in an airstream of velocity 30 m/s (equivalent airspeed). Plot a graph showing how the lift of such a plate varies as its angle to the airflow is increased from 0° to 90°.

    (*a*) What is the maximum lift obtained?

    (*b*) What would be the maximum lift of an aerofoil with the lite curve given in Fig. 3.13 of the same area under similar conditions?

Angle of attack	0°	5°	10°	15°	20°	30°	40°	50°	60°	70°	80°	90°
Lift coefficient	0	0.36	0.68	0.80	0.78	0.80	0.76	0.68	0.56	0.38	0.20	0

93. If an aeroplane of mass 950 kg has a wing area of 20 m^2, what is the wing loading in N/m^2?

94. A pressure plotting experiment is carried out in a wind tunnel on a model aerofoil of chord 350 mm, and large aspect ratio. Methylated spirit is used in the manometer. The table overleaf shows the distances of the holes *a*, *b*, *c*, *d*, etc., from the leading edge, and also the corresponding pressures recorded at these holes in millimetres of methylated spirit, the negative values representing pressures below the static pressure in the tunnel. The air speed was 45 m/s and the angle of attack 4°. Find the lift coefficient for the aerofoil at this angle of attack.

    Note. Strictly speaking, from the data given, it is impossible to find the lift coefficient because we do not know the surface friction forces on the aerofoil, which will also make a small contribution to the life force (defined to be at right angles to the free stream direction). This can be safely neglected when calculating lift, but not when calculating drag.

	Distance from leading edge (mm)	*Pressure, mm of methylated spirit*
Upper surface –		
Hole *a*	5	−228
Hole *b*	20	−203
Hole *c*	41	−195
Hole *d*	74	−153
Hole *e*	103	−112
Hole *f*	153	−76
Hole *g*	216	−64
Hole *h*	292	−25
Lower surface –		
Hole *k*	15	+97
Hole *l*	46	+99
Hole *m*	89	+86
Hole *n*	153	+56
Hole *o*	228	+25
Hole *p*	305	+8

The student is advised to work this question out because it will help in understanding several aspects of the subject. Proceed as follows –

Draw the chord line to some suitable scale, preferably on squared paper, marking off the position of each hole a', b', c', etc., as shown in the figure. (There is no need to incline the chord at 4°; exactly the same result will be obtained, rather more simply, if it is drawn horizontal.)

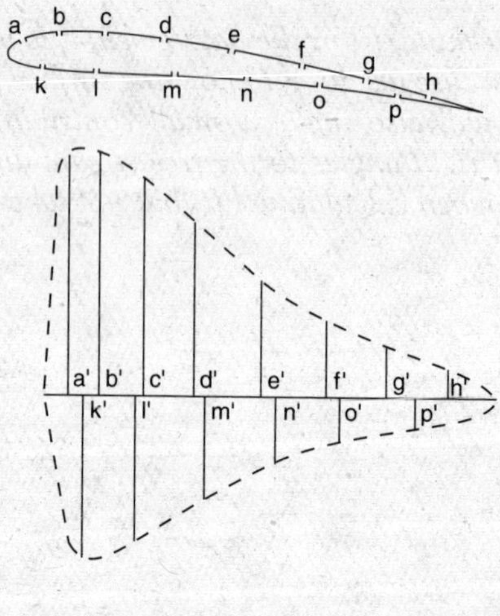

At each point set off a vertical line to some suitable scale to represent the pressure at the corresponding hole; the line should be upwards if the pressure is below atmospheric, and downwards if above atmospheric.

Draw a pressure distribution curve through the ends of these lines.

By one of the mathematical methods, or by counting the squares, or by using a planimeter, find the total area enclosed by this curve.

Divide the area by the length of the diagram, thus finding the average height; this represents the average pressure difference between the bottom and top surfaces of the aerofoil – at right angles to the chord line – in millimetres of methylated spirit. (Readers who have worked on engines, will recognise the similarity of this method to that of finding 'mean effective pressure' from an indicator diagram.)

Convert the average pressure into N/m² of wing area (S).

Then total force (normal to chord) = average pressure × S.

This is called the normal force.

Just as lift $= C_L \cdot \frac{1}{2}\rho V^2 \cdot S$, *so normal force* $= C_Z \cdot \frac{1}{2}\rho V^2 \cdot S$, *where* C_Z *is called the normal force coefficient.*

Equating average pressure × S to $C_Z \cdot \frac{1}{2}\rho V^2 \cdot S$, *the wing area (S) will cancel out; taking* ρ *as* 1.225 kg/m³ *and V as* 45 m/s *we can find* C_Z. *At this low angle of attack we can take* C_L *as approximately equal to* C_Z

95. Taking the values of the static pressures given in the table in Q94, find the speed of the airflow at each of the holes and construct velocity distribution diagrams for the upper and lower surfaces.

Note. Bernoulli's Principle (static pressure + dynamic pressure = constant) can be applied directly. Over the top surface the static pressure decreases, so the dynamic pressure and thus the speed of the airflow increases. On the under surface, the static pressure increases and the speed of the airflow decreases. The velocity distribution diagrams should be plotted with the aerofoil chord as abscissa (horizontal) and the velocity as ordinate (vertical), and the diagrams will be similar in shape to the pressure plotting diagrams. Assume the density of the air remains constant at 1.255 kg/m³.

96. When the angle of attack of a certain aerofoil is 12°, the direction of the resultant lies between the perpendicular to the chord line and the perpendicular to the airflow, being inclined at 8° to the latter (see figure). If the total force is 700 N, find its component in the direction OC.

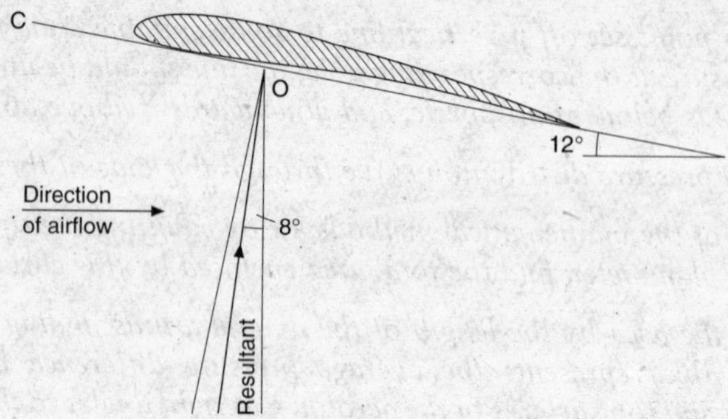

97. What are the lift and drag on the aerofoil in Q96?

98. From the tables given in Appendix 1 draw curves of lift coefficient against angle of attack for –

(*a*) The 'general purpose' biplane aerofoil section, RAF 15.

(*b*) The 'thin symmetrical' section, NACA 0009.

(*c*) The 'glider and sailplane' section, NACA 23012.

(*d*) The 'Lightning model' section, ASN/P1/3.

From your curves write down the maximum lift coefficients of these four aerofoil sections. (*These curves should be drawn to the same scale so that comparisons can be made between the various sections. The student will find it useful to keep these curves, and also those in the following questions, as they will be valuable in solving other problems.*)

99. Draw curves of drag coefficient against angle of attack for the four sections mentioned in Q98.

100. Draw curves of lift/drag ratio against angle of attack for the four sections mentioned in Q98.

101. Draw graphs to show how the centre of pressure moves as the angle of attack is altered on the following sections –

(*a*) Clark YH.

(*b*) NACA 0009.

(*c*) NACA 4412.

(*d*) NACA 23018.

Which of these four sections gives the least movement of the centre of pressure over the ordinary angles of flight?

102. A model aerofoil section (span 0.3 m, chord 50 mm) is tested in a wind tunnel at a velocity of 30.8 m/s. The maximum lift obtained is 11 N. Find the value of the maximum lift coefficient.

103. From the curves drawn for Q98 find the lift of a RAF 15 wing of area 25 square metres at 75 knots, the angle of attack being 4°.

104. What would be the lift of the wing in Q103 if it had been of NACA 4412 section?

105. A sailplane of total mass 300 kg is fitted with a NACA 23012 section wing of area 20 square metres. From the graphs drawn for Q98 and 100, find the indicated air speed at which it should be flown so that the wings are operating at the maximum lift/drag ratio.

106. Draw curves of lift coefficient against angle of attack for the following –

(*a*) NACA 4412.

(*b*) NACA 4412 with 20% split flap set at 60°.

(*c*) Lightning Model – clean aircraft.

(*d*) Lightning Model with 25% flap set at 50°.

From these curves write down the values of the maximum lift coefficients, and the angles of attack at which they occur.

107. In a test on an aerofoil section the moment coefficient about the leading edge was found to be −0.005 at zero lift. At 6° angle of attack the moment coefficient about the leading edge was −0.240, and the lift and drag coefficients were 0.90 and 0.045 respectively. Find the position of the aerodynamic centre of this aerofoil section.

108. What is the position of the centre of pressure of the aerofoil section of Q107 at 6° angle of attack?

109. The aerodynamic centre of a certain aerofoil section is at 0.256 c. At an angle of attack of 8° the moment coefficient about the aerodynamic centre is −0.100 and the lift and drag coefficients are 1.05 and 0.035 respectively. What is the moment coefficient about the quarter-chord point at this angle of attack?

110. What is the position of the centre of pressure of the aerofoil section of Q109 at 8° angle of attack?

111. An elliptical planform wing has a span of 12 m and a chord of 2 m. What is the induced drag coefficient when the lift coefficient is 0.8?

112. What is the induced drag at 100 knots of a monoplane of mass 2400 kg having a wing span of 16 m?

113. An aircraft has a mass of 7200 kg and a wing span of 12 m. If the wing loading is 1.968 kN/m^2, what is the induced drag when flying at speeds of 200 knots and 300 knots at sea-level?

114. A monoplane has a mass of 15 000 kg, and has a span of 10 m. What is the induced drag at –

(*a*) 120 knots,

(*b*) 240 knots, at sea-level? (*assume straight and level flight*)

115. What would be the induced drag of the aircraft in the previous question at 40 000 ft at an indicated air speed of 240 knots?

116. An aircraft of mass 15 000 kg is a monoplane of 20 m span. Calculate the induced drag at sea-level at speeds ranging from 80 to 400 knots. Plot the results in the form of a graph of speed (abscissa) against induced drag (ordinate). (*The student will find it useful to keep this graph as it will be referred to in later questions.*)

117. Calculate the power (in kilowatts) required to overcome the induced drag of the aircraft at the speeds used in the previous question, and plot the results in the form of a graph of speed (abscissa) against power (ordinate).

118. The sailplane in Q105 has an aspect ratio of 8. Find the percentage reduction in induced drag that would result from increasing the aspect ratio to 12.

119. The following table gives the values of all the drag except induced drag (i.e. of form drag + skin friction) of the aircraft in Q116 at various speeds at sea-level –

Speed	80	120	160	200	240	280	320	360	400 knots
Form drag + skin friction	2.4	5.4	9.6	15.0	21.6	29.4	38.4	48.6	60.0 kN

Plot these values on the graph drawn in Q116, and construct a graph showing the variation of the total drag with air speed.

What percentage of the total drag is induced drag at 120, 200, 280 and 360 knots?

120. At what speed is the induced drag exactly half the total drag? (*Note. This speed is the speed for minimum total drag. Keep these graphs; they will be referred to in later questions.*)

Chapter 4. Thrust

121. If a propeller, on a stationary mounting, blows back 36 kg of air per second at a speed of 45 knots, what thrust does it produce?

122. An aircraft engine is being tested on the ground before take-off. If the propeller has a diameter of 3 m, and the velocity of the slipstream is 80 knots, what thrust is being produced? (*In calculating the mass flow of air past the propeller take the average velocity of the air well in front of and behind the propeller, i.e. 40 knots.*)

123. What thrust will be produced by the propeller in the previous question when the aircraft is flying at 160 knots at 5000 ft, if the velocity of the slipstream relative to the aircraft is now 180 knots?

124. A twin-engined aircraft, with propellers of 3 m diameter, is flying at 10 000 ft at an air speed of 350 knots. If the aircraft has a mass of 10 000 kg, and has a lift/drag ratio of 5 to 1 in these conditions, what thrust is being produced and what is the speed of the slipstream relative to the aircraft?

125. The jet velocity of a gas turbine on the ground is 400 m/s; if the mass flow is 18 kg per second, what thrust does it produce?

126. An aircraft powered by two gas turbines is flying at 600 knots. If the jet velocity is 440 m/s, and the mass flow is 66 kg/s for each engine, what is the total thrust being developed?

127. What is the total power being developed by the gas turbines in Q126?

128. If a rocket can burn and eject 100 kg of fuel per second, what jet velocity is required to give a thrust of 50 kN?

129. A rocket, the total mass of which is 25 kg, contains 10 kg of fuel. If all the fuel is burnt in 2 seconds, and is ejected with a velocity of 500 m/s, what thrust is produced? What would be the initial acceleration if the rocket were fired off vertically upwards?

130. An aircraft using a rocket-assisted take-off system on an aircraft carrier has a mass of 4500 kg complete with rockets. It has a jet engine capable of producing 15 kN of thrust, and each rocket will produce 1566 N of thrust for 8 seconds. If the carrier steams at 20 knots into a head wind of 24 knots, how many rockets will have to be fired to get the aircraft airborne in a distance of 60 m along the deck if the take-off speed is 80 knots, and the average air and wheel resistance during take-off is 8.4 kN?

131. The pitch of a propeller is 2.5 m. If the slip is 15% when running at 1200 rpm, what is the speed of the aeroplane to which it is fitted?

132. The maximum speed of a light aeroplane is 80 knots when the engine is revolving at 3000 rpm; if the pitch of the propeller is 1.1 m, what is the percentage slip?

133. What is the torque of an engine which develops 1500 kW at 2400 rpm?

134. If the engine of the previous question is fitted with a propeller which has an efficiency of 83% at an advance per revolution of 3 m, what will be the thrust of the propeller?

135. When a certain aeroplane travels horizontally at an air speed of 300 knots, the engine develops 800 kW. If the propeller efficiency at this speed is 87%, find the thrust of the propeller.

136. The efficiency of a propeller is 78%. At what forward speed will it provide a thrust of 3 kN when driven by an engine of 210 kW power?

137. A propeller is revolving at 2600 rpm on an aircraft travelling at 280 knots. The thrust is 4 kN and the torque 2500 N-m. What is the efficiency of the propeller?

138. The diameter of a propeller is 3 m. The blade angle at a distance of 1 m from the axis is 23°. What is the geometric pitch of the propeller?

139. The geometric pitch of a propeller of 3 m diameter is to be 3.6 m, and should be constant throughout the blade. Find the blade angles at distances 0.8 m, 1 m, and 1.2 m respectively from the axis of the propeller.

Chapter 5. Level flight

140. The mass of an aeroplane is 2000 kg. At a certain speed in straight and level flight the ratio of lift to drag of the complete aircraft is 7.5 to 1. If there is no force on the tail plane, what are the values of the lift, thrust and drag?

141. In a flying boat the line of thrust is 1.6 m above the line of drag. The mass of the boat is 25 000 kg. The lift/drag ratio of the complete aircraft is 5 to 1 in straight and level flight. If there is to be no force on the tail, how far must the centre of pressure of the wings be in front of the centre of gravity?

142. An aeroplane of 10 000 kg mass is designed with the line of thrust 0.9 m above the line of drag. In normal flight the drag is 18.2 kN and the centre of pressure on the main plane is 150 mm behind the centre of gravity. If the centre of pressure on the tail plane is 10 m behind the centre of gravity, what is the load on the tail plane?

143. In a certain aeroplane the line of thrust is 100 mm below the line of drag. The mass of the aeroplane is 1500 kg, and the drag is 2.3 kN. If the aircraft is to be balanced in flight without any load on the tail, what must be the position of the centre of pressure relative to the centre of gravity?

144. In a certain aeroplane, which has a mass of 2000 kg, the centre of lift and the centre of gravity are in the same vertical straight line when in normal cruising flight. If the thrust is 4.5 kN, and is 180 mm below the centre of drag, what force must there be on the tail plane which is 6 m behind the centre of gravity?

145. What force would be required on the tail plane of the aeroplane in Q144 if the centre of lift had been 25 mm behind the centre of gravity?

146. A jet aircraft with a mass of 6000 kg has its line of thrust 150 mm below the line of drag. When travelling at high speed, the thrust is 18.0 kN and the centre of pressure is 0.5 m behind the centre of gravity. What is the load on the tail plane which is 8.0 m behind the centre of gravity?

147. An aircraft with a mass of 5500 kg is flying straight and level at its maximum speed. The thrust line is horizontal and is 0.3 m above the drag line which passes through the centre of gravity. If the drag is 12.0 kN, and the centre of pressure is 0.6 m behind the centre of gravity, find the load on the tail plane which is 5.5 m behind and on a level with the centre of gravity.

148. When the aircraft in the previous question is flying at its minimum speed, the thrust line is inclined at 25° to the horizontal. If the centre of pressure is now a horizontal distance of 0.1 m in front of the centre of gravity, what is the vertical load on the tail plane assuming that the drag (now 10 kN) still acts through the centre of gravity?

149. An aircraft with a mass of 14 000 kg is fitted with a wing of 60 m^2 area and of Clark YH section. Taking the values of the lift coefficient for this section from Appendix 1, calculate the indicated air speeds corresponding to the angles of attack from −2° to +25°, draw these graphs –

(*a*) Indicated air speed (abscissa) *v.* Angle of attack (ordinate).

(*b*) Indicated air speed (abscissa) *v.* Lift coefficient (ordinate).

At what speed must this aircraft fly to ensure that the wings are operating at the angle of attack giving the maximum lift/drag ratio? (*Note that this speed will be the aerodynamic range speed for the aircraft.*)

150. If the aircraft of Q149 uses 1500 kg of fuel during a flight, what should be the speed at the end of the flight to keep the wings operating at the angle of attack which gives maximum lift/drag ratio?

151. Draw a graph of the values of $C_L^{\frac{3}{2}}/C_D$ against angle of attack for the aerofoil section Clark YH. What angle of attack gives the maximum value of $C_L^{\frac{3}{2}}/C_D$? Refer to the graphs drawn in Q149 and determine the air speed that corresponds to this angle of attack for the aircraft in question. Compare this air speed with four-fifths of the range speed as found in Q149.

152. The aircraft in Q116 and Q119 is fitted with a 1000 kg load which is stowed internally. What is the new speed for maximum range? Compare the result with that obtained in Q120. (*Note. The maximum range speed is that which gives the minimum total drag of the aircraft. The curve of total drag is rather flat in this region, so it is not easy to estimate the minimum position accurately. A better method is to find the speed at which the induced drag is equal to all the other drag, i.e. form drag + skin friction, as this condition will also give the minimum total drag. All that is necessary in this question is to calculate the induced drag at the various speeds for this new load, and plot the results on the graph drawn for Q119. The intersection of the new induced drag curve and the old form drag 1 skin friction curve will give the new range speed.*)

153. The aircraft of Q116 is a twin-engined aircraft, and on a certain flight one engine fails. The pilot continues flying with the propeller of the dead engine wind-milling. This increases the form drag + skin friction to the following values –

Speed	80	120	160	200	240	280	320	360	400	knots
Form drag + skin friction	3.0	6.7	12.0	18.7	27.0	36.8	48.1	60.8	75.2	kN

What is the speed for maximum range under these conditions? (*See note under Q152*)

154. The aircraft in Q152 is fitted with a 1000 kg torpedo instead of the previous 1000 kg load; the torpedo, however, has to be carried externally and this increases the form drag + skin friction to the following values –

Speed	80	120	160	200	240	280	320	360	400	knots
Form drag + skin friction	2.7	6.1	10.9	16.9	24.4	33.1	43.3	54.8	68.0	kN

What is the speed for maximum range under these conditions?

155. Taking the values of the total drag at sea-level found in Q119, determine the power needed to drive the aircraft through the air at speeds from 80 to 400 knots. Plot the results and find the speed at which minimum power is required, i.e. the speed for maximum endurance. Compare this speed with the range speed found in Q176.

156. From the graph drawn for the previous question, read off the power required at sea-level to fly the aircraft at its maximum endurance speed, and calculate the power required for the same indicated air speed at 5000 ft, 10 000 ft and 15 000 ft above sea-level. (*Note. Remember that power depends on true air speed.*)

157. The following table gives the fuel consumption of a twin-engined training aircraft at various speeds at sea-level. Draw graphs of air speed against kg per hour, and air speed against air nautical miles per kg, and determine the speeds for maximum range and maximum endurance.

True air speed	80	90	100	110	120	130	140	150	160	knots
kg/hour	162	152	151	158	172	190	214	241	275	

158. Determine the speeds for maximum range over the ground for the aircraft of the previous question when flying –

(*a*) against a head wind of 40 knots,

(*b*) with a tail wind of 40 knots.

159. If the aircraft in Q119 is to be propelled by jet engines instead of piston/propeller engines, what will be –

(*a*) the speed for maximum range?

(*b*) the speed for maximum endurance?

160. The fuel consumption of a jet-propelled aircraft at 220 knots indicated air speed, at various altitudes, is as follows –

Sea-level	1240 kg/h
10 000 ft	1046 kg/h
20 000 ft	990 kg/h
30 000 ft	880 kg/h

Calculate the air nautical miles per kilogram for each altitude, and plot a graph showing the variation of air nautical miles per kilogram with altitude.

Chapter 6. Gliding and landing

161. When there is no wind, a certain aeroplane can glide (engine off) a horizontal distance of $1\frac{1}{2}$ nautical miles for every 1000 ft of height.

What gliding angle does this represent?

162. What will be the gliding angle, and what will be the horizontal distance travelled per 1000 ft of height, if the aeroplane in the previous question glides against a head wind of 20 knots? The air speed during the glide can be taken as 60 knots and the wind direction as horizontal.

163. An aeroplane glides with the engine off at an air speed of 80 knots, and is found to lose height at the rate of 1500 ft/min. What is the angle of glide? (*Assume conditions of no wind.*)

164. From the result of the previous question, what is the value of the lift/drag for this aeroplane on this glide?

165. At angles of attack of 1°, 4° and 10° the values of the lift/drag ratio of a certain aeroplane when gliding with the engine off are 3.5, 8 and 4.5 respectively. What horizontal distance should a pilot be able to cover from a height of 10 000 ft if he glides at each of these angles of attack?

166. An aeroplane of 5000 kg mass is flying at 20 000 ft. It glides to 10 000 ft at an angle of attack giving a lift/drag ratio of 10 to 1. What is the horizontal distance covered?

167. A model of a sailplane is tested in a wind tunnel, and the following values of C_L and C_D are found at the angles of attack stated:

Angle of attack	0°	2°	4°	6°	8°	10°	12°
C_L	0.09	0.23	0.39	0.53	0.68	0.83	0.98
C_D	0.009	0.010	0.016	0.023	0.036	0.049	0.064

Neglecting any scale effect, what is the flattest possible gliding angle that should be obtainable with the full-size sailplane? and at what angle of attack should the sailplane be flown to cover the greatest horizontal distance? (*Assume conditions of no wind.*)

168. What angle of attack should be used with the sailplane of the previous question to give the minimum rate of descent? and what is the gliding angle under these conditions? (*Assume no wind.*)

169. A sailplane with an all-up weight of 2452 N has a lift/drag ratio of 24 to 1 when gliding for range at 40 knots. Calculate the angle of glide, and the sinking speed in ft/s.

170. When flying with a heavier pilot, the all-up weight of the sailplane of the previous question is 2600 N. At what speed must he fly to cover the same range (in still air)? and what will be the sinking speed?

171. When flying for endurance, the sailplane of Q169 has a lift/drag ratio of 21.5 and a speed of 34 knots. Calculate the angle of glide and the sinking speed, and compare these with the values found in Q169.

172. Find the minimum landing speed of an aeroplane of mass 500 kg and a wing area of 18.6 m^2. The maximum lift coefficient of the aerofoil section is 1.0.

173. When an aeroplane is fitted with an aerofoil having a maximum lift coefficient of 1.08, the minimum flying speed is 42 knots. If, by the use of slots, the maximum lift coefficient can be increased to 1.60, what will be the minimum flying speed when fitted with slots?

174. It is required that a light aeroplane of mass 400 kg should land at 40 knots. What wing area will be required if the following aerofoil sections are used –

(*a*) Clark YH.

(*b*) NACA 4412.

(*c*) NACA 23012.

(*d*) NACA 4412 with flap.

175. From curves for NACA 23018 aerofoil read the values of the lift coefficient at angles of attack of 2°, 6° and 12°. If an aeroplane of mass 1000 kg is fitted with this aerofoil, what air speed will be necessary for horizontal flight at each of these angles of attack, assuming that the effective wing area is 20 square metres?

176. The total weight of an aeroplane is 14 715 N, and its wing area is 40 m^2. The maximum lift coefficient of the aerofoil is 1.08. What will be the maximum speed of flight –

(*a*) at sea-level?

(*b*) at 10 000 ft?

177. The total loaded mass of an aeroplane is 1100 kg. When NACA 65_1-212 aerofoil section is used the minimum landing speed is 56 knots. With a view to decreasing this landing speed, the following alterations are considered –

(*a*) the fitting of slots which will increase the maximum lift coefficient by 40 per cent,

(*b*) the fitting of flaps which will increase the maximum lift coefficient by 80 per cent, (*c*) a 20 per cent increase in wing area.

It is estimated that the increase in total mass necessitated by these alterations would be (*a*) 25 kg, (*b*) 30 kg and (*c*) 50 kg. What will be the resulting reduction in landing speed which can be achieved by each of the three methods?

178. An aeroplane of mass 13 500 kg has a wing loading of 2.75 kN/m^2. At 8° angle of attack the lift coefficient is 0.61. What is the speed necessary to maintain horizontal flight at this angle of attack at sea-level?

179. An aircraft is to be fitted with a NACA 23018 aerofoil section, and flaps which increase the maximum lift coefficient by 60 per cent. If the landing speed must not be more than 85 knots, what is the highest possible value of the wing loading?

180. An aircraft of 20 000 kg mass is to be fitted with a NACA 4412 aerofoil section with a 20 per cent split flap. If the area of the wing is to be 84 m^2, what will be the landing speed –

(*a*) with flaps lowered to 60°?

(*b*) without flap?

Chapter 7. Performance

181. An aeroplane of mass 2700 kg has a wing area of 30 m^2 of NACA 65_1-212 aerofoil section. During the take-off the pilot allows the aeroplane to run along the ground with its tail up until it reaches a certain speed; he then raises the elevators and so increases the angle of attack of the main planes to 12° which just enables the aeroplane to leave the ground. What is the required speed?

182. What would have been the speed required for the aeroplane of the previous question if the pilot had allowed it to continue running along the ground until it took off at an angle of attack of 4°?

183. An experimental aircraft of total mass 45 000 kg has four normal engines, and four auxiliary lift engines to provide a short take-off when required. The average thrust produced by each of the normal engines during take-off is 12.5 kN, and the take-off run required in still air is 2000 m. If the average resistance to motion is 11.0 kN, what is the take-off speed?

184. When using the auxiliary lift engines, the aircraft of the previous question can take off in 1000 m at the same angle of attack in still air. What thrust must each lift engine produce?

185. An aeroplane of 3000 kg mass is climbing on a path inclined at 12° to the horizontal. Assuming the thrust to be parallel to the path of flight, what is its value if the drag of the aircraft is 5.0 kN?

186. An aeroplane of 12 000 kg mass climbs at an angle of 10° to the horizontal with a speed of 110 knots along its line of flight. If the drag at this speed is 36.0 kN, find –

(*a*) the power used in overcoming drag.

(*b*) the power used in overcoming the force of gravity.

Hence find the total power required for the climb.

187. A jet aircraft with a wing loading of 2.4 kN/m², and a mass of 4500 kg, has a maximum thrust of 30 kN at sea-level. If the drag coefficient at a speed of 270 knots is 0.04, what will be –

(*a*) the maximum possible rate of climb?

(*b*) the greatest angle of climb?

at this speed.

188. An aircraft of 5000 kg mass is powered by an engine capable of producing 1500 kW power. Calculate the maximum angle of climb at an air speed of 130 knots if the efficiency of the propeller is 80 per cent, and the drag at this speed is 6.7 kN.

189. At sea-level the total drag of an aircraft of mass 5500 kg is 5.0 kN at a speed of 160 knots. Calculate the rate of climb and angle of climb at an indicated air speed of 160 knots at 10 000 ft, if the power available is 980 kW and the relative air density is 0.739.

190. For the aircraft of the previous question calculate the rate of climb and angle of climb at the same angle of attack, if the mass of the aircraft is increased to 7000 kg by the addition of internal load, and the power available at 10 000 ft remains unchanged.

191. A jet aircraft weighing 58 860 N has a climbing speed of 250 knots. If the rate of climb is 9000 ft/min, and the drag of the aircraft in this condition is 8.2 kN, find the thrust being delivered by the engines.

192. The following table gives particulars relating to a certain aeroplane of 1050 kg mass –

Speed of level flight	45	50	55	60	65	70	75	80	85	90	95	100	knots
Power available from propeller	135	170	205	225	240	250	255	255	250	240	230	220	kW
Power required for level flight	250	115	93	90	100	120	150	180	215	255	300	350	kW

Estimate

(a) the minimum speed of level flight,

(b) the maximum speed of level flight,

(c) the best airspeed for climbing purposes,

(d) the power available for climbing at this airspeed and

(e) the maximum vertical rate of climb.

193. By throttling down the engine of the aeroplane of the previous question the power available is reduced by 30 per cent throughout the whole range of speed. Find the maximum and minimum speeds for level flight under these conditions.

194. An aeroplane of 1500 kg mass has a minimum landing speed of 50 knots, and the minimum power for level flight is 60 kW at 80 knots. If 250 kg of extra load is added, find the new minimum landing speed, and the new minimum power and speed for level flight, i.e. at the same angle of attack as before.

195. The following table gives particulars of a certain aeroplane of 6000 kg mass which is designed for either piston engine/propeller or jet propulsion –

Speed of level flight in knots	50	70	90	110	130	160	200	240	280	320	360	400	440
Power required for level flight in kW	2000	750	580	540	540	640	610	1340	2000	2870	4000	5500	7500
Power available from propellers in kW	900	1120	1300	1480	1640	1870	2130	2340	2550	2700	2800	2800	2650
Power available from jets in kW	660	950	1200	1500	1750	2170	2720	3260	3800	4340	4900	5450	6000

Estimate the maximum and minimum speeds for level flight when the aircraft is propelled by the piston engine/propeller combination.

196. Estimate the maximum and minimum speeds for level flight when the aircraft in the previous question is propelled by the jet engines, and compare the speeds with those found in the previous question.

197. Estimate the best climbing speed and rate of climb at this speed for the aircraft in Q195 when propelled by the piston engine/propeller combination.

198. Estimate the best climbing speed and rate of climb at this speed when the aircraft in Q195 is propelled by the jet engines. Compare these answers with those obtained in the previous question.

199. When the mass of the aircraft of Q195 is increased by carrying an extra 900 kg of fuel in overload tanks, the power required for level flight is increased to the following:

Speed of level flight in knots	60	70	90	110	130	160	200	240	280	320	360	400	440
Power required for level flight in kW	1600	1000	760	670	650	700	960	1385	2045	2910	4040	5535	7535

Estimate the maximum and minimum speeds for level flight in this condition when propelled by the piston engine/propeller combination.

200. Estimate the maximum and minimum speeds for level flight of the aircraft in the previous question when propelled by the jet engines.

201. The lift produced by the wings of an aircraft travelling at 250 knots at sea-level is 70 kN. At what speed must the aircraft travel at 45 000 ft to produce the same lift at the same angle of attack of the wings?

202. What will be the indicated air speed of the aircraft of the previous question at 45 000 ft?

203. If the aircraft of Q201 has a lift/drag ratio of 10 to 1 when travelling at an indicated air speed of 250 knots, calculate the power required to propel the aircraft at this indicated air speed –

(*a*) at sea-level,

(*b*) at 45 000 ft.

204. When flying at its endurance speed of 140 knots, an aircraft of 7250 kg mass has a drag of 13.4 kN. Calculate the power required to propel the aircraft at altitudes from sea-level to 45 000 ft (at 5000 ft intervals) and construct a graph of –

Altitude (abscissa) against Power Required (ordinate).

Estimate the altitude at which the power required is double that at sea-level.

205. The following table gives values of the maximum power available from the propeller fitted to the aircraft of the previous question when flying at the endurance speed at various altitudes –

Altitude (ft)	0	5000	10 000	15 000	20 000	25 000	30 000	35 000	40 000	45 000
Power available (kW)	2050	2095	2140	2180	2240	2085	1965	1840	1740	1630

Note that the rated altitude of the engine is 20 000 ft. Estimate –

(*a*) the absolute ceiling of the aircraft,

(*b*) the service ceiling (the altitude at which the maximum rate of climb is reduced to 100 ft/min).

206. If the engine fitted to the aircraft of Q204 uses fuel at the rate of 0.275 kg/kW hr, calculate the endurance of the aircraft –

(*a*) at sea level, and (*b*) at 30 000 ft,

assuming that the aircraft has 1000 kg of fuel available for level flight, and that the propeller efficiency is 80 per cent at both altitudes.

207. An aircraft weighing 13 500 kgf has a range speed of 175 knots indicated. At this speed the lift/drag ratio is 8 to 1, and the propeller efficiency is 82 per cent. If the fuel consumption is 0.27 kg/kW hr, calculate the air nautical miles covered per kilogram of fuel at –

(*a*) sea-level,

(*b*) 20 000 ft,

(*c*) 40 000 ft.

208. If the aircraft of the previous question is propelled by jet engines at the same indicated speed, and if the fuel consumption is then 57 kg/kN of thrust hr, calculate the air nautical miles per kilogram of fuel at –

(*a*) sea-level,

(*b*) 20 000 ft,

(*c*) 40 000 ft.

Compare the results with those obtained in the previous question.

Chapter 8. Manoeuvres

209. Find the correct angle of bank for an aeroplane of 1200 kg mass taking a corner of radius 60 m at 75 knots.

210. What will be the total lift in the wings of the aeroplane in the previous question while taking this corner at the correct angle of bank?

211. An aircraft with a mass of 1000 kg does a steady turn at 55 knots and an angle of bank of 45°. Calculate (*a*) the acceleration, (*b*) the force required to produce the acceleration, and (*c*) the wing loading on the aircraft during the turn if the wing area is 14.14 m^2.

212. If the aerofoil section used on the aeroplane of Q42 is NACA 23012, and if the total wing area is 25 metres, calculate –

    (*a*) the angle of attack required for normal horizontal flight at the same speed of 80 knots, and

    (*b*) the angle of attack required to produce the necessary lift when turning the corner at this speed.

213. An aeroplane of 1750 kg mass makes a horizontal turn at an angle of bank of 25°. If the speed in the turn is 85 knots, what is the radius of the turn?

214. Calculate the radius of turn and angle of bank of an aircraft doing a Rate 1 turn (180° per minute) at 550 knots. What is the acceleration in multiples of *g*? And if the mass of the aircraft is 5000 kg, what is the lift force on the wings during the turn?

215. Calculate the loading on an aircraft in a correctly banked turn at angles of bank of 60°, 75°, 83°, 84°. (*Note. The loading is expressed as a factor found by dividing the lift force on the wings by the force on the wings in straight and level flight, i.e. the weight.*)

216. Calculate the accelerations in multiples of *g* for each of the turns mentioned in the previous question.

217. If the stalling speed of an aircraft is 60 knots in straight and level flight, what is the stalling speed in correctly banked turns at angles of bank of 45°, 60°, 75°, 83°, 84°?

218. Calculate the rate of turn of an aeroplane in degrees per minute if the acceleration during the turn is 4 *g* at speeds of 100 knots, 150 knots, 300 knots, 500 knots.

219. If the maximum load factor a certain aircraft can sustain without structural failure is 8, what is the maximum angle of bank it can use in a correctly banked turn?

220. If the stalling speed of the aircraft of the previous question is 80 knots in straight and level flight, what will the stalling speed be in the turn at the maximum permissible angle of bank?

221. What will be the radius of the turn of the aircraft of Q219 when it is flying on the stall at its maximum permissible angle of bank?

222. What angles of bank are required for Rate 1 turns (180° per minute) at 100 knots, 200 knots, 300 knots, 400 knots?

223. What are the loadings on the aircraft in the turns in Q222?

224. If an aircraft stalls at 110 knots in straight and level flight, what will be the load factor on the aircraft if it stalls in a turn at 246 knots?

225. What will be the radius of turn when the aircraft of the previous question stalls at 246 knots?

226. An aeroplane of 1500 kg mass performs a loop. If it is assumed that the top of the loop is in the form of a circle of radius 80 m, what must be the speed at the highest point in order that the loads on the aeroplane may be the same as those of normal horizontal flight? (*If the centripetal force on the aeroplane at the top of the loop is just equal to the weight, then there will be no lift; but if the centripetal force is double the weight, then the lift will be the same as in normal flight, and the pilot will be sitting on his seat with the usual force, but upwards!*)

227. If the terminal velocity of the aeroplane of Q226, in a vertical nose dive with engine off, is 365 knots, what is its drag when travelling at this speed?

228. A spherical ball of 1 kgf weight and diameter 75 mm is dropped from an aeroplane. What will be its terminal velocity in air of density 0.909 kg/m^3? (*Take C_D for the sphere as* 0.8.)

229. If, without any appreciable increase in weight, the ball is faired to a streamline shape, with C_D of 0.05, what will be the terminal velocity in air of the same density?

230. An aircraft of 6000 kg mass completes a vertical loop at a constant speed of 425 knots, with a height range from top to bottom of the loop of 3000 m. What is the acceleration and, if the area of the wing is 40 square metres, what is the wing loading (*a*) at the top, (*b*) at the bottom of the loop? (*Assume that the loop is in the form of a perfect circle, which is very unlikely in practice.*)

231. An aircraft is in a vertical terminal velocity dive at 400 knots. If, in pulling out of the dive, it follows the arc of a circle, what will be the acceleration if the height lost is 5000 ft? What is the maximum loading during the pull-out?

232. What will be the acceleration and maximum loading of the aircraft in Q231 if the loss in height in the pull-out is only 2500 ft?

Chapter 9. Stability and control

Note. Numerical questions on stability and control are too complex to be included in the scope of this book.

Chapter 10. A trial flight

233. The landing speed of a light aircraft at sea-level is 50 knots. At what speed will it land on an airfield situated at an altitude of 5000 ft?

234. The normal stalling speed of an aircraft is 55 knots. At what ground speed would it stall if it were flying at low level (*a*) into wind, (*b*) down wind, if the wind speed was 20 knots?

235. A pilot is flying a low-level cross country exercise at a speed of 100 knots in a wind of 20 knots. He wishes to pass vertically over a point on the ground on completion of a Rate 1 (i.e. 180° heading change in 1 minute) turn of 90° from an into-wind to a cross-wind direction. How far from the point must the pilot start the turn?

236. A monoplane of span 30 m and aspect ratio 7 has a profile drag coefficient of 0.008, and a lift coefficient of 0.17, when flying at 400 knots at sea-level. Calculate the total drag of the monoplane under these conditions.

237. Find the power required from an engine to drive a propeller which is 80% efficient when it is producing 3.6 kN of thrust at 120 knots.

238. When a certain aeroplane is in horizontal flight, the thrust and drag lie along the same line which is 150 mm above the centre of gravity. The extreme positions of the centre of pressure for level flight are 25 mm in front of and 200 mm behind the centre of gravity. If the tail plane is 5.5 m behind the centre of gravity, what will be the load on the tail plane in each case? The mass of the aeroplane is 3500 kg, and the thrust required for both conditions is 7 kN.

239. The maximum value of the lift/drag ratio for a certain aeroplane is 5.5 to 1. Find the flattest possible gliding angle (in degrees) with the engine off.

240. The total loaded mass of a two-seater aeroplane is 1700 kg, and the corresponding minimum landing speed is 45 knots. What will be the minimum landing speed if it is flown as a single-seater, the reduction in mass being 80 kg?

241. The normal climbing speed of an aircraft of 2500 kg mass is 110 knots. At this speed 210 kW is required to overcome drag. With 15° of flap lowered, the climbing speed is 85 knots, and 220 kW is required to overcome the drag. If the maximum power available from the propeller is 375 kW, what is the angle of climb and rate of climb in each case?

242. If the stalling speed of an aeroplane in normal flight is 42 knots, what will be its stalling speed when executing a correctly banked turn at 45° angle of bank?

243. An aircraft completes a loop at a constant speed of 200 knots in 15 seconds. If the loop can be considered as a vertical circle, calculate the radius of the loop, and the maximum and minimum loadings during the manoeuvre.

Chapters 11 and 12. Flight at transonic and supersonic speeds

244. If the speed of sound is proportional to the square root of the absolute temperature, calculate the speed of sound at −50°C, +50°C and +100°C.

245. An aircraft has a critical Mach Number of 0.85. If the pilot cannot control the aircraft at higher Mach Numbers than this, what is the maximum permissible speed of the aircraft (*a*) at sea-level? (*b*) at 40 000 ft?

246. What are the maximum indicated air speeds for the aircraft of the previous question (*a*) at sea-level? (*b*) at 40 000 ft?

247. If the temperature at sea-level on a certain day rises to 25°C, what will be the maximum speed of the aircraft of Q245 at sea-level on that day?

248. A certain aircraft has a critical Mach Number of 0.80, and a structural limitation which prevents it being flown at an indicated air speed greater than 438 knots. At what altitude are these two speeds equal? (*This question is best solved graphically.*)

249. The following table gives the values of the drag coefficient of a thin aerofoil at various Mach Numbers –

Mach Number	0.6	0.7	0.8	0.9	1.0	1.1	1.2	1.3	1.4
Drag Coefficient	0.01	0.011	0.019	0.05	0.068	0.063	0.052	0.044	0.037
Mach Number	1.5	1.6	1.7	1.8	1.9	2.0			
Drag Coefficient	0.032	0.028	0.026	0.022	0.021				

Calculate the drag of the aerofoil at sea-level at these Mach Numbers taking the area of the aerofoil as 40 square metres, and plot a graph of Drag (ordinate) *v.* Mach Number (abscissa).

250. Calculate the power necessary to propel the aerofoil of the previous question at speeds corresponding to the Mach Numbers given, and plot a graph of Power (ordinate) against Mach Number (abscissa). Compare the shape of this curve with that obtained for Q249.

251. Construct a graph similar to that in Fig. 11.9 for an aircraft which stalls at 115 knots at sea level, and has a critical Mach Number of 0.85. Estimate the maximum height at which it can fly, assuming that it cannot fly at Mach Numbers above the critical Mach Number.

252. The diagram shows the variation of the specific excess power (P_s) in m/s for a supersonic fighter-type aircraft. Estimate –

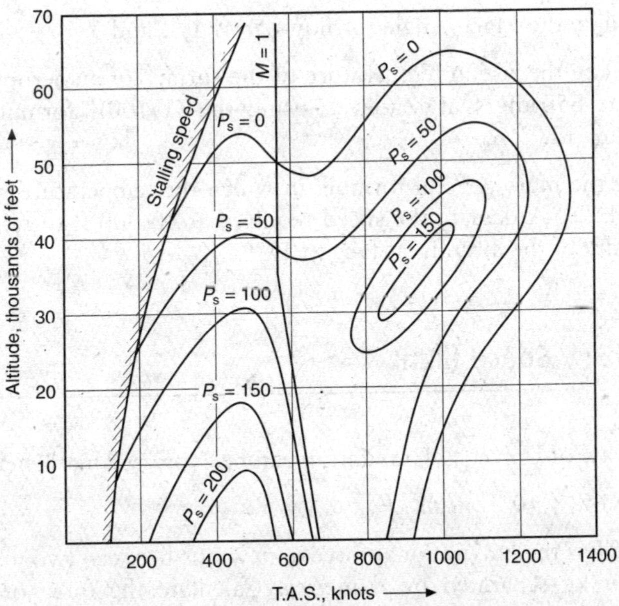

Specific excess power (P_s) in m/s for supersonic aircraft

(*a*) The speed range in Mach Nos. at 50 000 ft altitude.

(*b*) The speed range in Mach Nos. at 60 000 ft altitude.

(*c*) The highest altitude at which the aircraft can fly at $M = 2$.

(*Note. The student may like to consider how the pilot would get to 60 000 ft when flying at* M = 1 *at* 50 000 ft.)

253. The following table gives values of the drag coefficient for a supersonic aircraft together with the specific fuel consumption (kg/kN Thrust h) of the engines fitted to the aircraft for three cruising altitudes –

Mach Number	1.7	1.8	1.9	2.0	2.1	2.2
C_D	0.037	0.034	0.032	0.031	0.031	0.031
SFC at 50 000 ft	54	54	54	55	56	57
SFC at 55 000 ft	66	66	67	68	69	70
SFC at 60 000 ft	85	86	86	87	88	89

Determine the speed and altitude at which the aircraft should be flown for optimum range performance in still air.

254. Calculate the Mach Angle for flows at M 1, 2 and 3.

255. Calculate the rise in temperature of the surface of an aeroplane travelling at 650 knots at sea-level, using the $(V/100)^2$ formula given in Chapter 12.

256. Using the more general formula in which the temperature rise is given by $(M^2T)/5$, calculate the speed required at 40 000 ft to raise the temperature of the aircraft surface to 15°C.

Chapter 13. Space flight

257. The Law of Universal Gravitation can be expressed mathematically as –

$$F = 6.67 \times 10^{-11}\ m_1 m_2 / d^2$$

where F is the gravitational force in newtons between two masses of m_1 and m_2 kg separated by d metres. Calculate the force of attraction between (*a*) the earth and the moon, and (*b*) the earth and the planet Venus (*mass 0.81 that of the earth*).

258. Calculate the acceleration due to gravity at a distance of 12 000 km from the centre of the earth.

259. Calculate the acceleration of the moon towards the earth.

260. A communications satellite is positioned on a 'stationary orbit' over a geographical point on the equator. Calculate the radius of its orbit and its speed.

261. A satellite of 100 kg mass is circling the earth on a radius of 18 000 km from the centre of the earth. Calculate –

(*a*) the force exerted on the satellite by the earth, and

(*b*) the time taken for one complete orbit.

262. A satellite of 250 kg mass is required to orbit the moon at a distance of 1750 km from the centre of the moon; at what speed must it travel round its orbit? (*Take the mass of the moon as* 0.0123 *times the mass of the earth.*)

263. To what value must the speed of a 6000 kg space vehicle be reduced so that it will orbit 100 km above the surface of the moon?

264. Calculate the thrust to be produced by the retro-rocket, and the time for which it must operate, to put the space vehicle of the previous question into its desired orbit of the moon if its transit speed is 10.5 km/s and the deceleration is to be 50 m/s^2.

265. Calculate the thrust required to lift a 1500 kg space vehicle off the surface of the moon and accelerate it vertically upwards to a speed of 1.5 km/s in 30 s.

266. A satellite is circling the equator at a distance of 800 km from the surface of the earth, and in an east to west direction. If it crosses the Greenwich meridian at noon, when will it be vertically over the 180° line of longitude?

267. The radius of the planet Venus is about 6115 km, and its mass is about 0.81 times that of the earth. Calculate the acceleration due to gravity at the surface of Venus.

268. Find the escape velocity at a distance of 2000 km from the earth's surface.

269. The radius of Mars is about 3376 km, and its mass about 0.1 times that of the earth. What will be the velocity of a satellite circling Mars at a distance of 10 000 km from the centre of the planet?

270. Venus and Mars are approximately 42 000 000 and 56 000 000 km respectively from the earth. What are the radii of their zones of influence relative to the earth?

271. Using the data of the previous questions calculate the escape velocities at the surface of –

(*a*) Venus,

(*b*) Mars.

Appendix 1. Aerofoil data

Five questions (*a*), (*b*), (*c*), (*d*), (*e*) are given at the end of the data for each aerofoil section.

Appendix 2. Scale effect

272. The viscosity of air varies with the temperature according to the formula –

$$\frac{\mu_1}{m_2} = \left(\frac{T_1}{T_2}\right)^{\frac{3}{4}}$$

where μ_1 and $\mu_2$are the viscosities at the absolute temperatures T_1 and T_2 respectively. Find the viscosity of air at −25°C, −5°C, +5°C, and +25°C.

273. An aircraft 12 m in length cruises at 77.1 m/s at sea-level. Find its Reynolds Number under these conditions.

274. The average length of the chord of a wing of a certain aircraft is 3.05 m. Taking this as the length *L*, calculate the Reynolds Number when flying at sea-level at 102.8 m/s.

275. A 1/10th scale model of the aircraft of the previous question is tested in a wind tunnel at 67 m/s, at a temperature of 288 K, the density of air in the tunnel being 1.225 kg/m^3. What is the Reynolds Number of the test?

276. What will be the Reynolds Number if the test of the previous question is conducted in a compressed air tunnel in which the pressure is 1013 kN/m^2 and the temperature 298 K?

277. What must be the pressure in the compressed air tunnel of the previous question if the 1/10th scale model test is to have the same Reynolds Number as the full-scale aircraft in Q280, assuming that the maximum speed of the tunnel under these conditions is 75 m/s and the temperature 298 K?

278. By what percentage will the Reynolds Number be raised or lowered if the temperature in the compressed air tunnel in the previous question is reduced to 5°C, assuming that no more air is pumped into the tunnel during the cooling process?

279. Find the Reynolds Number of a test conducted in a decompressed air tunnel on an aerofoil of 0.46 m chord at a Mach Number of 0.85, a temperature of 15°C and a pressure of 20.26 kN/m^2.

APPENDIX 4

Answers to numerical questions

Chapter 1

1. (*a*) 114.8 km/h, (*b*) 18.5 s
2. (*a*) 0.25 m/s^2, (*b*) 458 m, (*c*) 1833 m
3. (*a*) 1.39 m/s^2, (*b*) 26.8 s
4. 58 knots, 300 m
5. 58 km/h
6. (*a*) 12.3 m, (*b*) 49.2 m
7. 24.97 km, 2 min 22.7 s
8. 536 m
9. 5.503 kN
10. 11.36 m/s
11. 425.1 m/s
12. 98.1 N
13. 117.7 N
14. 6.84×10^6 N
15. 8.432 kN
16. 1009.8 kN
17. 462.96 kN
18. (*a*) 80.65 kgf, (*b*) 71.35 kgf, (*c*) 76 kgf
19. 113.5 kN
20. 16.62 kN
21. 110.25 m
22. 70.7 N
23. 8.17 m
24. 297.1 m/s
25. 7648 N
26. 4823 N
27. 722.5 mm
28. 7.7 knots
29. 38.15 kN
30. 1582 N
31. 457.5 m
32. 119.1 kN
33. 2226 kJ, 40.47 m
34. 236 rad/s, 353.6 m/s

35. 314 rad/s, 15 m/s
36. 83 350 m/s^2
37. 7402 m/s^2
38. 39.48 N, (*a*) 29.67 N, (*b*) 49.29 N
39. 9.81 m/s^2
40. 3 h 58 min 50 s
41. 87.34 km/h
42. (*a*) 25.41 kN, (*b*) 59°55′, (*c*) 29.36 kN
43. 1471.5 kJ, 12.26 kW
44. 353.3 kW
45. 2650 kW
46. 7.65 kW
47. 361.25 kJ, 425 m/s
48. 112 500 kJ
49. 10 m

Chapter 2

50. 1.044 kg/m^3
51. 0.116 kg
52. 470.4 kg
53. 439.7 kg
54. (*a*) 0.906 kg/m^3 (0.905), (*b*) 0.414 kg/m^3 (0.414)
55. 19 700 ft
56. 580 ft
57. 776 ft
58. 4.687 kg/s
59. (*a*) 45 kt, (*b*) 48.35 kt, (*c*) 57.5 kt, (*d*) 95 kt
60. (*a*) 3 h 12 min, (*b*) 3 h 15 min, (*c*) 3 h 13 min
61. 507 knots
62. 497 knots
63. 17 min, 13 min, 10 min, 13 min
64. 175 N
65. 700 N
66. 5.7 N
67. 4.632 kN
68. 10.6 N
69. 20 m/s
70. 196 N
71. 10.549 kN
72. 0.804
73. 0.251
74. 314 N
75. 8.537 kN
76. 172 N
77. 50.24 kN
78. 0.091
79. 4.1 N
80. 152 N
81. 4.474 kN
82. Decrease by 29.3%
83. 640 N

84. 341 N
85. 102.9 kN/m^2
86. (*a*) 46.6 kN/m^2, (*b*) 52.9 kN/m^2, 101.3 m/s
87. (*a*) 137 kt, (*b*) 201 kt
88. (*a*) 100 kt, (*b*) 100 kt
89. 181 knots
90. 408 knots
91. (*a*) 104.9 kN/m^2, (*b*) 90.3 kN/m^2

Chapter 3

92. (*a*) 2.7 kN, (*b*) 6.3 kN
93. 466 N/m^2
94. 0.84
95. Diagram
96. 48.8 N
97. 693 N, 97 N
98. (*a*) 1.22, (*b*) 0.90, (*c*) 1.23, (*d*) 1.27
99. (*a*) 0.0074, (*b*) 0.009, (*c*) 0.010, (*d*) 0.019
100. (*a*) 26.8 at 2.2°, (*b*) 47.3 at 7.5°, (*c*) 60.0 at 8°, (*d*) 7.6 at 8°
101. NACA 0009
102. 1.26
103. 10.486 kN
104. 18.236 kN
105. 31 knots
106. (*a*) 1.37 at 13°, (*b*) 2.66 at 10.5°, (*c*) 1.27 at 30°, (*d*) 1.28 at 25°
107. 0.261 *c*
108. 0.267 *c*
109. −0.106
110. 0.352 *c*
111. 0.034
112. 425 N
113. 1.700 kN, 756 N
114. (*a*) 29.516 kN, (*b*) 7.379 kN
115. 7.379 kN
116. Diagram
117. Diagram
118. 33⅓%
119. 57.7%, 15.1%, 4.4%, 1.7%
120. 130 knots

Chapter 4

121. 833 N
122. 7.336 kN
123. 6.719 kN
124. 19.62 kN, 367 knots
125. 7.2 kN
126. 17.336 kN
127. 5351 kW
128. 500 m/s
129. 2.5 kN, 90.19 m/s^2
130. Four

131. 83 knots
132. 25.2%
133. 5996 N-m
134. 10.375 kN
135. 4.510 kN
136. 106 knots
137. 84.6%
138. 2.67 m
139. 35°36′, 29°48′, 25°31′

Chapter 5

140. 19.62 kN, 2.616 kN, 2.616 kN
141. 0.32 m
142. 3.157 kN downwards
143. 15.6 mm behind C.G.
144. 135 N upwards
145. 53.5 N upwards
146. 4.284 kN downwards
147. 7.341 kN downwards
148. 318 N upwards
149. 180 knots
150. 170 knots
151. $5\frac{1}{2}$°, 151 knots
152. 134 knots
153. 122 knots
154. 126 knots
155. 100 knots
156. 720 kW, 776 kW, 838 kW, 908 kW
157. 116 knots, 96 knots
158. (*a*) 130 knots, (*b*) 110 knots
159. (*a*) 173 knots, (*b*) 130 knots
160. 0.177, 0.245, 0.304, 0.408 a.n.m. per kg

Chapter 6

161. 6°16′
162. 9°22′, 1847 m (0.9974 n.miles)
163. 10°40′
164. 5.3 to 1
165. 5.76, 13.17, 7.41 n.miles
166. 16.46 n.miles
167. 2°21′, 4°
168. 6°, 2°29′
169. 2°21′, 2.77 ft/s
170. 41.2 knots, 2.45 ft/s
171. 2°40′, 2.67 ft/s
172. 40.3 knots
173. 34.5 knots
174. (*a*) 10.58 m^2, (*b*) 11.12 m^2, (*c*) 12.30 m^2, (*d*) 5.71 m^2
175. 96 kt, 65 kt, 53 kt

176. (*a*) 46 kt, (*b*) 53 kt
177. (*a*) 8 kt, (*b*) 14 kt, (*c*) 4 kt
178. 167 knots
179. 1.986 kN/m^2
180. (*a*) 74 kt, (*b*) 103 kt

Chapter 7

181. 71.7 knots
182. 99.5 knots
183. 114 knots
184. 55.2 kN
185. 11.12 kN
186. (*a*) 2037 kW, (*b*) 1157 kW, 3194 kW
187. (*a*) 13 197 ft/min, (*b*) 28°51′
188. 13°15′
189. 1829 ft/min, 5°34′
190. 839 ft/min, 2°16′
191. 29.124 kN
192. 47.5 kt, 89 kt, 64 kt, 142 kW, 2714 ft/min
193. 79.5 kt, 49.5 kt
194. 54 kt, 75.6 kW, 86 kt
195. 312 kt, 59 kt
196. 398 kt, 64 kt
197. 175 kg, 4181 ft/min
198. 245 kt, 6438 ft/min
199. 309 kt, 67 kt
200. 397 kt, 72 kt
201. 568 knots
202. 250 knots
203. (*a*) 900 kW, (*b*) 2045 kW
204. 39 700 ft
205. (*a*) 37 000 ft, (*b*) 36 250 ft
206. (*a*) 3 h 1 min, (*b*) 1 h 51 min
207. (*a*) 0.36, (*b*) 0.36, (*c*) 0.36 a.n.m. per kg
208. (*a*) 0.19, (*b*) 0.25, (*c*) 0.37 a.n.m. per kg

Chapter 8

209. 68°26′
210. 32 kN
211. (*a*) 9.81 m/s^2, (*b*) 9.81 kN, (*c*) 981 N/m^2
212. (*a*) 4°, (*b*) 10°
213. 418 m
214. 5402 m, 56°30′, 1.51 *g*, 88.87 kN
215. 2.0, 3.86, 8.2, 9.53
216. 1.73, 3.73, 8.14, 9.51
217. 71 kt, 85 kt, 118 kt, 172 kt, 185 kt
218. 2621°/min, 174°/min, 874°/min, 524°/min
219. 82°49′
220. 226 knots

221. 174 m

222. 15°21′, 28°47′, 39°29′, 47°42′

223. 1.04, 1.14, 1.30, 1.49

224. 5

225. 333 m

226. 77 knots

227. 14.7 kN

228. 78 m/s

229. 312 m/s

230. 31.87 m/s², (*a*) 3.31 kN/m², (*b*) 6.25 kN/m²

231. 27.8 m/s², 3.83

232. 55.6 m/s², 6.7

Chapter 10

233. 54 knots

234. (*a*) 35 knots, (*b*) 75 knots

235. 1191 m

236. 31.012 kN

237. 277.8 kW

238. 156.1 N up, 1.296 kN down

239. 10°18′

240. 43.9 knots

241. 6°50′, 1324 ft/min, 8°19′, 1244 ft/min

242. 49.9 knots

243. 245.5 m, 5.4, 3.4

Chapters 11 and 12

244. 582 kt, 700 kt, 752 kt

245. (*a*) 562 kt, (*b*) 488 kt

246. (*a*) 562 kt, (*b*) 242 kt

247. 572 knots

248. 10 000 ft

249. Diagram

250. Diagram

251. 73 500 ft

252. (*a*) M 0.66 to 1.0 and M 1.24 to 2.33, (*b*) M 1.54 to 2.11, (*c*) 63 000 ft

253. M 1.8 at 55 000 ft

254. 90°, 30°, 19°28′

255. 42°C

256. 734 knots

Chapter 13

257. (*a*) 1.984×10^{26} N, (*b*) 1.094×10^{18} N

258. 2.76 m/s²

259. 0.0027 m/s²

260. 41 850 km, 11 100 km/h

261. (*a*) 123 N, (*b*) 6 h 41 min

262. 1.673 km/s

263. 1.718 km/s

264. 300 kN for 2 min 55.64 s

265. 77.421 kN

266. 1247 hours G.M.T.

267. 8.62 m/s^2

268. 9.753 km/s

269. 1.995 km/s

270. Venus 19 900 000 km,
Mars 13 450 000 km

271. (*a*) 10.27 km/s, (*b*) 4.856 km/s

Appendix 1

RAF 15

(*a*) (i) 11.3, (ii) 26.8

(*b*) (i) 27°, (ii) 15°

(*c*) (i) −0.085, (ii) −0.051

(*d*) (i) 36, (ii) 174

(*e*) (i) 15.60, (ii) 15.06

CLARK YH

(*a*) (i) 0.001, (ii) 0.004, (iii) 0.005

(*b*) 0.236 *c*

(*c*) 18.2°

(*d*) 159

(*e*) (i) 16.55, (ii) 15.90

NACA 0009

(*a*) 0, Yes (from −8° to +8°)

(*b*) 0.25 *c*

(*c*) 0.385 *c*

(*d*) (i) 21.0, (ii) 45.7, (iii) 40.9

(*e*) (i) 9.6°, (ii) 6°

NACA 4412

(*a*) (i) −0.091, (ii) −0.095

(*b*) (i) 0.489 *c*, (ii) 0.333 *c*

(*c*) (i) 38.0, (ii) 67.6

(*d*) (i) 13°, (ii) 10°

(*e*) 137

NACA 23012

(*a*) (i) 15, (ii) 50, (iii) 60

(*b*) (i) 0.33 *c*, (ii) 0.265 *c*, (iii) 0.264 *c*

(*c*) 0.30 *c*

(*d*) 12°

(*e*) (i) 21.60, (ii) 37.08, (iii) 49.96

NACA 23018

(*a*) 0.18 *c* at 0.30 *c*

(*b*) (i) −23.3, (ii) 12.0, (iii) 44.2,
(iv) 56.3, (v) 40.8

(*c*) (i) 0.253 *c*, (ii) 0.25 *c*,
(iii) 0.253 *c*

(*d*) 12°

(*e*) 1.06

NACA 65_1–212

(*a*) −0.032 ± 0.002 between −8° and +8° angle of attack

(*b*) 0.40 *c*

(*c*) (i) 10.5°, (ii) 6°

(*d*) 53.3

(*e*) 120

ASN/P1/3

(*a*) 0.0594

(*b*) 0.45 *c*

(*c*) (i) 30°, (ii) 25°

(*d*) (i) 5.7, 5.8, 3.6,
(ii) 4.6, 4.9, 3.1

(*e*) (i) 0.442 *c*, (ii) 0.468 *c*,
(iii) 0.507 *c*

Appendix 2

272. 15.99×10^{-6} N-s/m^2
16.95×10^{-6} N-s/m^2
17.43×10^{-6} N-s/m^2
18.36×10^{-6} N-s/m^2

273. 63.39×10^6

274. 21.48×10^6

275. 1.399×10^6

276. 13.63×10^6

277. 1426 kN/m^2

278. Raised by 5.3%

279. 1.82×10^6

Tail end
Tailplane of the Antonov An-225 Mriya.

APPENDIX 5

Answers to non-numerical questions

Chapter 1

1. The lift is decelerating in the downwards direction, so the acceleration is upwards.

2. (*a*) Pressure is a scalar quantity: it has no direction; pressure is measured by the force that it would produce on an area.

   (*b*) A moment is the product of a force and a distance (the moment arm). Momentum is the product of a mass and a velocity.

   (*c*) Power is the rate of doing work: force × distance/time.

3. To pull a body up an inclined plane only requires that the force be equal and opposite to the component of the weight acting in the direction of the slope of the plane.

   The same work is done in each case; on the inclined plane, the force is smaller, but the distance moved is correspondingly larger.

4. Mass is a measure of the quantity of matter in a body. Weight is the force produced by the gravitational attraction between the body and a heavenly body (the earth, unless otherwise specified).

5. To accelerate the aircraft requires that the thrust exceeds the drag, but when a steady speed is reached, it will be maintained as long as the drag equals the thrust.

6. Yes, the aircraft is accelerating.

7. Yes, the centre of gravity of a ring is outside the material of the ring. The centre of gravity of a piece of bent wire will also not normally lie on the wire.

8. (*a*) Yes.

   (*b*) No, it will normally be accelerating.

9. (*a*) Less.

   (*b*) Same.

   (*c*) More.

   (*d*) More.

   (*e*) Same.

10. Yes, otherwise they would be in equilibrium, and no movement would take place, so it would be stalemate.

11. (*a*) Yes.

    (*b*) Yes.

12. The flag will hang down limply, because there will be no relative motion between the balloon and the air.

Chapter 2

1. The pressure altimeter only tells you what height the external pressure would be given by in an International Standard Atmosphere.
2. Air density is a measure of the mass of air in a given volume.
3. The pressure.
4. The altimeter can be set so that it reads zero height when the pressure reaches the ground-level value at a specified location, usually the local airfield ground-level; the aircraft should then not hit the runway before the indicated height reaches zero. Alternatively, while en route, it can be set to read zero at the sea-level pressure, so that the pilot knows how high he is above features given on a map.
5. The temperature, density and pressure will all be lower.
6. A streamlined shape is one where the flow is able to follow the contours without separating.
7. See the definitions in the chapter.
8. This is the dynamic pressure.
9. This is the error on the air speed indicator reading caused by the static hole not being located at a position where the pressure is exactly equal to the free-stream static pressure.

10. The troposphere is the lowest part of the atmosphere, where the temperature varies almost linearly with height. The stratosphere is above the troposphere, and is the part where the temperature remains nearly constant with height.

11. Subsonic means that the speed of the object relative to the air is less than the local speed of sound. Supersonic means that it is greater.

12. The symbol q stands for dynamic pressure.

Chapter 3

1. As the angle of attack increases, a lift force develops due to the difference in pressure between upper and lower surfaces. This is mainly caused by a reduction in pressure on the upper surface. When the angle of attack increases too far the flow separates from the upper surface, this leads to a reduction in lift as the aerofoil stalls.

2. The centre of pressure is the point through which the aerodynamic force can be considered to act on the aerofoil.

3. Because the forces on an aerofoil are primarily dependent on the dynamic pressure and the wing area, if the lift or drag are divided by the product of these quantities, a force coefficient is obtained which depends mainly on angle of attack. This makes it easier to translate the results from, for example, an experiment to the full-scale aircraft. It must be remembered, though, that Reynolds number and Mach number will also have an effect (see Appendix 2 and Chapter 11).

4. The aerodynamic centre of an aerofoil section is the point about which the moment coefficient remains constant with changes in angle of attack.

5. The stalling angle of an aerofoil is the angle of attack at which the flow separates from the top surface, leading to a reduction in lift. The exact angle at which this happens is open to discussion – see the text! Apart from Reynolds number and Mach number effects (see question 3) this happens at a particular angle of attack independently of the air speed.

6. The aspect ratio of a wing is equal to the square of its span divided by its area. It is important because it influences the strength of the trailing vortices for a given lift, and hence the drag produced by these vortices.

Chapter 4

1. A ramjet has no turbine or compressor. Air is compressed purely by aerodynamic effects before it enters the combustion chamber. It is only of practical use at supersonic speeds where use can be made of shock wave compression.

2. The blade angle is the angle at which the propeller blade is set in relation to the plane of rotation. The angle reduces as the tip is approached because the speed at the tip is higher than it is near the spinner. The angle at which the blade is set must therefore be reduced as the tip is approached, to maintain the same local angle of attack at each section of the blade.

3. The advance per revolution is the actual forward distance travelled in one propeller revolution. The geometric pitch is equal to $2\rho\tau \tan \theta$and is equivalent to the advance of a screw of 'blade' angle θ through a solid object. The experimental mean pitch is the advance per revolution at zero thrust.

4. Slip is the difference between the actual advance per revolution and the experimental mean pitch.

5. The angle of attack of a propeller blade is reduced by an increase in the aircraft air speed. To compensate for this it is desirable to use a higher blade angle (coarser pitch) as air speed increases.

6. Tip speed is important because it is at the tip that the local air speed is highest. Therefore the speed of sound may be reached at the tip long before the forward speed of the aircraft approaches this value. This may lead to the formation of shock waves over the tip with a marked loss of efficiency.

7. The solidity influences the ability of a propeller of given diameter and rotational speed to absorb power. It may be increased by increasing the number of blades or the chord of the blades.

8. Outside the atmosphere propulsion can be achieved by any system which ejects matter from the craft. It is necessary to employ a high velocity of projection so that adequate thrust can be obtained without ejecting too much mass. If the required ejection velocity is produced by combustion then the oxidant must be carried in the craft since no air is available. This is effectively rocket propulsion.

Chapter 5

1. The four forces are lift, weight, thrust, and drag.
2. For equilibrium the sum of the forces must be zero and they must produce no moment.
3. The weight of the aircraft will change as fuel is used or stores are dropped and this will also change the position of the centre of gravity. The lift will change to compensate for the weight change in level flight, or to provide additional lift for manoeuvres. The centre of lift will also change, particularly if the change in lift is produced by a change in angle of attack rather than speed. Changes in lift will be accompanied by changes in drag which in turn must be accompanied by an equal change in thrust to maintain equilibrium. Changes in the line of action of the latter forces are likely to be fairly small for conventional aircraft.
4. The lift of the aircraft depends on angle of attack and speed. The speed can be changed by changing the angle of attack to give the same overall lift as the speed is altered. Generally this will be accompanied by some change in drag so some adjustment will be needed to the throttle to compensate. The speed range can be further extended by the use of high-lift devices to increase the maximum lift coefficient that can be obtained from the wing.
5. The relationship will be the same for indicated air speed but not for true air speed (see p. 163). At height a higher true air speed will be needed to give the same lift at a given angle of attack.
6. As the weight is increased a higher angle of attack will be needed at a given air speed.
7. (*a*) As far as the airframe is concerned the minimum drag will be approximately independent of altitude. The aircraft should then be flown at the best altitude for maximum engine efficiency, i.e. the height at which the throttle setting is fully open for the most economical fuel–air ratio. At greater heights the engine efficiency will be lower and the range consequently less.

   (*b*) For maximum endurance with a piston engine we need to fly at minimum power as far as the airframe is concerned. Because the true air speed increases with height at the optimum angle of attack for minimum power, it is best to fly at low altitude for maximum endurance.

   (*c*) Because the best angle of attack, and hence speed, for minimum power required by the airframe is different than that for minimum drag.

8. (*a*) Because the efficiency of a jet engine increases with speed a compromise between the best operating speed for the airframe and the engine is required. The aircraft must therefore be flown at a speed greater than that for minimum airframe drag.

   (*b*) True air speed for a given angle of attack increases with height and temperature falls. Both of these improve engine efficiency, therefore it is best to cruise high for maximum range in a jet aircraft.

   (*c*) Because the fuel flow in a jet aircraft is approximately proportional to thrust, the aircraft should be flown at the minimum drag speed for best endurance. There is some advantage in flying high from the point of view of improved engine efficiency.

Chapter 6

1. If the aircraft is operating at the minimum angle of glide then it is operating at its minimum drag for the weight. If the pilot pulls the nose up then the drag will increase and the glide angle steepen.
2. Lowering flaps during the glide will generally steepen the glide angle because the best lift to drag ratio is likely to be in the flaps-up configuration.
3. The load carried will not greatly alter the minimum glide angle, because the maximum lift to drag ratio, at which the minimum glide angle occurs, depends on angle of attack. To get the same lift at this angle of attack, however, the gliding speed must increase at heavier weights.
4. No! The flattest glide occurs at minimum drag. For maximum time in the air we need to operate at the minimum power condition so that the potential energy due to height is lost at the lowest possible rate.
5. (*a*) The true airspeed at stall will be higher the greater the altitude. However, the speed indicated on the air speed indicator will be approximately the same irrespective of height because the instrument works by sensing the dynamic pressure and no correction is made in its calibration for the change in density with height.

   (*b*) The stalling angle will not change with height.
6. An engine-assisted approach allows the glide angle to be increased if necessary by a reduction in the throttle setting. If a 'go around' is required the engine can be set to full power quicker.
7. The lowering of flaps may change the position of the centre of pressure of the wing and produce a pitching moment (generally nose-down) which must be overcome by the tailplane.

8. A couple of things to get you started. After that you are on your own! Happy landings.

   You need to fly at minimum glide angle, i.e. at minimum drag. Are the things you are carrying mounted outside the aeroplane and increasing the drag? Before you release them, is there a headwind? To reduce the effect of this you need a high gliding speed and so a lot of weight. If there is a tail wind you need a low speed to increase the effect of wind on range. Now it's up to you.

Chapter 7

1. When the aircraft is climbing its path is inclined to the horizontal. Because the lift is measured at right angles to the flight path, only the component of weight which is also normal to the flight path must be balanced by the lift. The other component must be supplied by the thrust.

   If you still don't believe it, imagine an aircraft with an engine thrust greater than the weight. It can 'stand on its tail' and climb vertically with no lift at all.

2. For a jet aircraft maximum true air speed does not vary greatly over a wide altitude range. It is usually limited by compressibility effects (see Chapters 11 and 12). For a piston engine the relationship can be complicated by the question of supercharging, but will fall off at high altitude; engine power is limited and does not increase with forward speed like a jet engine. Because of the reducing density, the best true air speed as far as the airframe is concerned increases with altitude and the power available from the engine is insufficient to cope.

   The minimum true air speed will increase with altitude, although the indicated minimum speed will be nearly constant.

3. Ceiling is the maximum height that can be reached. Service ceiling leaves something in hand for manoeuvres. It is usually specified as the height at which the rate of climb falls to a specified level.

4. No! It is the attitude of the aircraft that matters and this will require a change in true air speed at different heights.

5. An increase in weight will mean a reduction in range and endurance. There will also be an increase in minimum speed and a reduction in maximum rate of climb and ceiling.

Chapter 8

1. The degrees of freedom are –

    (*a*) Three translational degrees of freedom along the three aircraft axes.

    (*b*) Three rotational degrees of freedom along the same axes.

2. The radius of turn may be limited by –

    (*a*) Wing stalling because of increased lift required in the turn.

    (*b*) Engine power because the increased lift increases the drag and required power.

    (*c*) The structural strength of the wing being exceeded because of increased lift.

3. This is a bit of a trick question – be careful!

    Because the aircraft complete the circuit in the same time, the one at the greater radius is going faster. If you check the acceleration towards the centre for a constant circuit time, you will find it is proportional to the radius of the turn. The aircraft at the higher radius will therefore need to bank more than the one on the inside.

4. As the aircraft turns and climbs the upward motion reduces the angle of attack on the wings. Because the inner wing is turning on a smaller radius than the outer wing it suffers a greater reduction in angle of attack and a consequent loss of lift compared to the outer wing. This tends to increase the bank which must be held off by the ailerons. The reverse is clearly true in a gliding turn.

5. In a spin one wing is stalled with the aircraft describing a spiral path downwards. The variation of lift with angle of attack at the stall is such that the aircraft can become locked into this attitude and will not recover naturally.

Chapter 9

1. If the directional stability is very large and lateral stability small, then the aircraft will be prone to spiral instability.

2. If the lateral stability is large and the directional stability small, then spiral instability will not be a problem. However, another problem may be encountered in which there is a motion consisting of a combined rolling and yawing oscillation (known as Dutch roll).

3. In manually operated controls the aerodynamic forces acting on the control surface may make them very heavy to operate, especially at high speed. The use of suitable aerodynamic balancing can offset some of the hinge moment and make the control lighter to operate.

4. Aerodynamic balance is used to reduce the aerodynamic hinge moment on the control. Mass balancing is used to alter the inertial characteristic of the control to change its dynamic behaviour and prevent flutter of the control surface.

5 At high angles of attack, particularly, the down-going aileron produces a substantial increase in drag and hence yawing moment away from the intended turn.

   If the angle of attack is too high it may even be that the downward deflection of the aileron is sufficient to reduce the lift rather than increasing it because of stall.

6. The yawing moment problem can be tackled by suitable design of the ailerons to ensure that the up-going aileron also produces a significant drag increment. A bigger rudder may also help.

   The use of a spoiler instead of, or supplementing, the up-going aileron will produce the desired drag increment and can be used to drop that wing significantly with respect to the other, even with no aileron control on either wing.

7. The use of spoilers at low speed is explained above. They can also be a useful aid during the ground run on landing. At high speed they may be better than an aileron for roll control because they avoid imposing wing twist which is caused by the deflection of ailerons near the tips – a problem which is particularly severe on swept wings. Tip ailerons at transonic Mach numbers may also cause undesirable local changes in flow due to compressibility effects.

Chapter 11

1. The speed of sound in water is roughly four times the speed of sound in air.

2. The speed of sound reduces with height up to the stratosphere. It is proportional to the square root of the absolute air temperature, which reduces with height until the stratosphere is reached, where it remains approximately constant.

3. The shock forms first at the region where the local flow first becomes supersonic. This is usually on the upper surface of the wing.

4. The buffet boundary of a transonic aircraft is the Mach number at which an interaction between shock waves formed on the wing and the local boundary layer causes an unstable separation which causes a buffeting on the aircraft.

5. Mach number is air speed divided by the speed of sound. It is possible to talk of the Mach number at which an aircraft operates, or flight Mach number, as its true air speed divided by the speed of sound in the surrounding atmosphere. It is also possible to refer to a local Mach number relating to the flow over some particular part of the aircraft. This is the speed of the local flow divided by the local speed of sound. The local speed of sound will be different in different parts of the airflow because of temperature differences.

   The critical Mach number can be defined as the flight Mach number at which local supersonic flow first appears somewhere on the aircraft.

   A Mach meter is an instrument for measuring the flight Mach number of an aircraft.

6. A shock wave causes a rapid increase in pressure as the air passes through it. On a wing it is therefore likely to reduce the extent of the suction peak on the upper surface. It may further modify the pressure distribution by causing flow separation.

Chapter 12

1. A Mach line is the trace of a weak pressure wave produced by a body in a supersonic airstream. Its angle to the flow, the Mach angle, depends only on the Mach number. A Mach cone is a three-dimensional surface consisting of Mach lines. For a small body in a supersonic airstream this surface is conical.

2. Shock waves travel at a speed greater than the speed of sound. This means that a shock wave can form upstream of a body travelling at supersonic speed.

3. An expansion wave in supersonic flow is a region where the speed increases while the pressure, density and temperature decrease.

4. Sharp leading edges are used on supersonic wings to reduce the drag due to shock waves (the wave drag).

5. At cruising Mach numbers above about 2, the structure may be subjected to aerodynamic heating to a degree which makes the use of conventional aluminium alloys impossible. In this case alternative materials such as titanium must be used at critical parts of the structure where heating is

severe. In the Concorde the problem is solved by employing the fuel as a heat sink to reduce the local structural temperatures.

6. In order to achieve low wave drag at supersonic speeds a slender wing (i.e. one which is long compared to its span) must be used. For efficient subsonic cruise a high aspect ratio wing is needed. This leads to the Concorde's highly swept 'ogee' planform. Because the high sweep leads to flow separation, a sharp leading edge is used so the wing has separated flow over the top surface at nearly all flight conditions. This separation, however, is in the form of two well-controlled vortices above the wing which contribute to the lift and do not produce the buffet and increase in drag associated with the normal separation process leading to a conventional wing stall.

Chapter 13

1. Escape velocity is the velocity which is required by a body, with no assistance from propulsion, if it is to avoid being returned to the earth by gravitational attraction. The escape velocity required for a body starting on the earth's surface, and ignoring such important things as air resistance, can be calculated to be approximately 11.2 km/s. The escape velocity from the moon is less because the moon's mass, and hence gravitational attraction, is less than that of the earth.

2. Again, ignoring air resistance the escape velocity is the same for both horizontal and vertical launch, since the critical factor is the amount of kinetic energy possessed by the body. If we are clever, though, we can get part of the escape velocity on launch from the rotation of the earth if we launch near the equator and in the direction of rotation.

3. For an elliptical orbit round the earth the perigee is the point on the orbit nearest the earth and the apogee is the point furthest away.

4. This is the height at which the orbital period of the satellite is the same as the period of the earth's rotation. If the plane of the orbit is in the plane of the equator the satellite therefore remains stationary with respect to the earth's surface (called the geostationary orbit). This is clearly a help with communication satellites.

5. (*a*) 1 hr 25 min.

   (*b*) 2 hr approx.

   (*c*) 28 days.

6. A satellite launched at the escape velocity has a parabolic path. Above this speed it is hyperbolic.

Index

absolute pressure 61
acceleration 3, 9–10, 232–4
centripetal 12–16, 233–4
and gravity 6–7
uniform 9
advance per revolution 134–7
aerobatics 247–53
in adverse weather 255
aerodynamic centre 87–90
aerofoils 67–119, 412–30
aerodynamic centre 87–90
angle of attack 70, 72, 77, 78
aspect ratio 99, 101
camber 93–4, 97, 109
centre of gravity 86–7, 92
centre of pressure 74, 75–8, 80, 85–7, 91–2
chord line 70, 330, 360
circulation theory 106–8
coefficients 80
depth 92, 93–4
design process 96–8
fences 338
high lift 111–12, 191–3
icing problems 118–19
laminar flow 94–5, 324–5
nomenclature 96–8
pitching moment 78–80, 85–7
plan area 79
power requirements 91
pressure distribution 72–4
pressure plotting 72
reference point 86, 92
slots 109–14, 191–2, 196–7
spars 92
stagnation point 74
for supersonic speeds 351–2, 362–6
total resultant force 75
transition point 95
variable camber 109
and velocity 78
vortices 100, 106
see also flaps; lift; wing shapes
Aeronautical Society of Great Britain 318
ailerons 197, 243, 274, 277, 281–2
aileron drag 281
differential ailerons 283
frise 283
and high speed 285, 333–4
slot–cum–aileron control 284
air *see* atmosphere
air brakes 116, 142, 192, 210
and gliding 186
air resistance 11, 381
see also drag
air speed 35–6
and altitude 166, 223–4
and the angle of attack 163–6
and endurance flying 221
and engine power 165–6, 222
and landing 187, 189–90, 191, 194–7
and level flight 162–6, 220–1
maximum and minimum 220–1, 223
measurement 40–2
and tail planes 154
throttle control 227, 289, 292
and thrust 129, 130
see also stalling
air speed indicator 42–4, 313–14
air to fuel mixture 167
airflow
at hypersonic speeds 373–4
separation from wing surfaces 336–40
shock waves 299, 300–5, 322–8, 342–6
and the speed of sound 28

airflow (continued)
  at subsonic speeds 68–70, 99, 106, 356
  at supersonic speeds 342–57
  visualisation 62–3
airscrew *see* propeller propulsion
altimeters 37–9
altitude
  and air pressure and density 31
  and air speed 166, 223–4
  and the angle of attack 166
  ceiling 224–5
  and jet engine efficiency 227
  performance effects 222–4
  and propeller propulsion efficiency 171–2, 226
  and range flying 171–2, 176
  and stalling speed 315–18
American Federal units 8
ammeter 288
anemometers 62
angle of attack 70, 72, 77, 78
  and air speed 163–6
  and gliding 181–3
  and landing 188–9, 192
  and level flight 163–6
  and supersonic airflow 351–2
  and tail planes 154
anhedral 262, 264
apogee 387
approach to the landing ground 206–12, 294–5
aquaplaning 210
area rule 335–6
artificial stabilisation systems 257
aspect ratio 99, 101
astronautics 377
atmosphere 28–9
  chemical composition 36–7
  compressibility of air 22–3, 53, 298–9
  density of air 29–30, 31
  inertia properties of air 30
  ISA (International Standard Atmosphere) 31–2, 37
  lower atmosphere 377
  mass of air 29–30
  pressure 30–1, 37, 38–9
  re-entry into the atmosphere 406–8
  seeing the air 29
  temperature changes 31–3, 34, 377–80
  up-and-down currents 34–5
  upper atmosphere 377–81
  viscosity properties of air 34
attitude, and landing 192
attraction forces 5, 381
Autogiro 198
balance unit 58–9
balanced controls 274–80
ballistics 377
banking *see* turning and banking
bar 21
Bernoulli's equation 39–40
bi-convex wing 363
Blackbird spy-plane 124, 345
blade angle 132, 134, 136, 143
blown flaps 116, 197
body forces 1, 2, 6
body shapes *see* fuselage
Boeing-Bell V22 Osprey 203
boundary layer 50–2, 94–5, 305, 336–7
  at supersonic speeds 357
brake power 130
brakes
  air brakes 116, 142, 186, 192, 210
  wheel brakes 210–11
British Imperial System 8
buffet boundary 305
bumpy weather 255

Calibrated Air Speed (CAS) 43
camber 93–4, 97, 109
camber flaps 115
canard configuration 152
carburettor icing 289, 295
catapulting 216
ceiling 224–5
centre of drag 148
centre of gravity 25, 86–7, 92, 148, 149, 259, 265
centre of pressure 74, 75–8, 80, 85–7, 91–2, 148
centrifugal force 16
centripetal acceleration 12–16, 233–4
chemical composition of the atmosphere 36–7
chord line 70, 330, 360
Cierva Autogiro 198
circular orbits 389–93
circular velocity 387, 390–1, 393–4
circulation theory 106–8
climbing 147, 217–18, 230
  ceiling 224–5
  rate of climb 220–1, 291
climbing turns 243–4
cockpit 287–9
  see also instrumentation
coefficients of aerofoils 80
compound helicopters 202
compressibility of air 22–3, 53, 298–9
compressive airflow 347–8
computational fluid dynamics (CFD) 64–5

computer assisted landing 214
Concorde 33, 149, 194, 329, 334, 360, 371–2
conservation of momentum principle 20
constant-speed propellers 141
contra-rotating propellers 143, 144, 146
control 259, 273–86
at low speed 280–5
balanced controls 274–80
control column 273–4, 285, 288
directional control 273
engine controls 227, 289, 292
feel of pilots 334
flutter 279
foot pedals 274
longitudinal control 273
manual controls 274
mass balance 279–80
powered controls 274, 285–6, 334
reversal of controls 333–4
roll control 273
rudder 274
servo system 276–7
at supersonic speeds 370–1
throttle control 227, 289, 292
and transonic speeds 333–4
see also ailerons; elevators; flaps
control column 273–4, 285, 288
convergent–divergent nozzle 353–4
cosmic rays 380
critical Mach number 307–9, 325, 330–3
Custer channel wing 197, 198

degrees of stability 258
delta wings 360
density 21–2
of air 29–30, 31
and pressure 22, 33–4
relative density 33
and temperature 33–4
depth of aerofoils 92, 93–4
differential ailerons 283
dihedral angle 260–1, 262–4
directional control 273
directional stability 268–73
dive brakes 116
diving
nose-dives 251–2
pulling-out 235–6
dog-tooth 339
downwash 155–6
drag 1, 46–50, 52–4, 68, 78–80
boundary layer drag 305
centre of drag 148
coefficient 52–4, 80, 91
drag curve 83
form drag 46–50
induced drag 99–106
and level flight 147–50, 156–60, 169–71
lift/drag ratio 83–4, 91, 169–70
minimisation 169–71, 175–6
parasite drag 46
profile drag 46
shock drag 305
skin friction 50–2
and thrust 120
at transonic speeds 309, 311–12
wave drag 305
wing drag 46
droop-snoot 338
Dutch roll 315
dynamic pressure 40, 41, 42, 45
dynamic stability 286

ekranoplanes 205–6
elevators 156, 273, 285
and transonic flight 333
elliptical orbits 394–6
endurance flying
and air speed 221
and gliding 182, 185
and jet propulsion 177
and propeller propulsion 173–4
energy 17–20
and fuel efficiency 167–8
kinetic energy 18
and momentum 20
potential energy 18
energy–work conversion 167–8
engine checks 292
engine controls 227, 289, 292
engine instruments 288
engine power 165–6
and air speed 222
and performance 91, 218–22
throttle control 227, 289, 292
and transonic flight 311–12
see also propulsion systems
engine-assisted approach 211–12
equilibrium forces 2–3
Equivalent Air Speed (EAS) 43
escape from the earth 382–4
escape velocity 383, 393–4
Eurofighter Typhoon 152
Expanding–contracting duct 354–6
expansion wave 350
expansive airflow 340–50
experimental mean pitch of a propeller 137–8

experimentation, and theories 318
external forces 1–2

fan jet engines 129–30
feathered blades 142
feel of pilots 334
fences 338
fin area 267
fineness ratio 48
First Law of Motion 1
flaperons 197
flaps 109–10, 115–16
  and landing 191–2, 196–7, 208–9
  leading-edge flaps 338
  and take-off 216
  and trim 212–14
flat turns 245
flattening out 209
flight levels 39
flight plans 289
fluid pressure 20–1
flutter 279
flying wing 160
foot pedals 274
forces
  attraction forces 5, 381
  body forces 1, 2, 6
  centrifugal force 16
  centripetal force 12–16
  in equilibrium 2–3
  not in equilibrium 3–4
  external forces 1–2
  inertia forces 4, 16
  moment of a force 24–5, 150
  parallelogram of forces 24
  polygon of forces 24
  reactive forces 1–2
  triangle of forces 24
  units of measurement 2, 8
  as a vector quantity 23
form drag 46–50
forward-swept wings 268
Fowler flaps 116, 118
FREDA checks 292
free-flight testing 320–1
friction 50–2
frise 283
fuel checks 290
fuel economy *see* range flying
fuel instruments 288
fuselage
  melting point 368
  slimness of design 308, 330
  and supersonic speeds 366–8

g-forces 233–6
gas law 22–3, 33
gas turbine engines 121
gauge pressure 61
geometric pitch of a propeller 137–8
gliding 180–6
  air brakes 186
  angle of attack 181–3
  efficiency of design 181
  endurance flying 182, 185
  weight effects 184–5
gliding angle 182–3
  flat 185–6
  real and apparent 183
gliding turns 243–4
gravity 5, 381
  and acceleration 6–7
  and escape from the earth 382–4
ground speed 35–6
gyroplane 198
gyroscopic effects 145

Harrier 202–3, 206
height *see* altitude
helicopters 200–2
helix angle 132, 134, 139
high by-pass engines 129–30
high lift aerofoils 111–12, 191–3
high-speed wind tunnels 309, 318–20
hold-off 209
horizontal flight *see* level flight
horizontal velocity
  and landing 186–7
  path of a projectile 384–6
hot wire anemometer 62
hovercraft 203–6
hydraulic analogy 320, 345
hypersonic speeds 124, 372–4

icing problems 118–19
  carburettor icing 289, 295
Indicated Air Speed (IAS) 43, 163, 170
induced drag 99–106
inertia forces 4, 16
inertia and manoeuvrability 255
inertia properties of air 30
instrument error 43
instrumentation 287–8
  air speed indicator 42–4, 313–14
  air speed measurement 40–2
  altimeters 37–9
  engine instruments 288
  fuel instruments 288
  instrument error 43

Mach meters 44, 312–13
navigation systems 44, 289
pitot tube 41
pitot-static tube 41–2
turn and bank indicator 241–2
venturi tube 44–6, 353–4
inverted manoeuvres 252–3
ionosphere 380
ISA (International Standard Atmosphere) 31–2, 37
isothermal processes 22

Jeans, Sir James 391
jet engines 121–5, 129–30, 312
efficiency at altitude 227
and endurance flying 177
and range flying 174–7
see also propulsion systems
jet flaps 116, 197
jet lift aircraft 202–3
jockey weights 59
joystick 274, 288

kinetic energy 18
kinetic heating 368–70
knots 8
Krueger flaps 116

laminar boundary layer 50–2
laminar flow aerofoils 94–5, 324–5
landing 186–214, 294–6
air brakes 192, 210
air speed 187
angle of attack 188–9, 192
approach to the landing ground 206–12, 294–5
attitude 192
computer assisted 214
engine-assisted approach 211–12
and flaps 191–2, 196–7, 208–9
flattening out 209
high lift aerofoils 191–3
hold-off 209
horizontal velocity 186–7
overshooting 208
pulling-up 210
short take-off and landing 198–203
speed of landing 189–90, 191, 194–7
stalling speed 190–1
undershooting 208
vertical take-off and landing 198–203
vertical velocity 186, 188
wind gradient 187–8
wind speed 187
wing loading 193–4, 196–7, 206
laser Doppler anemometer (LDA) 62
lateral axis 230
lateral stability 262–4, 267, 272–3
launching a space-craft 401–4
launching speeds 382–4
law of universal gravitation 381
leading-edge droop 338
leading-edge flaps 338
level flight 147–79
air speed 162–6, 220–1
angle of attack 163–6
drag 147–50, 156–60, 169–71
endurance flying 173–4, 177, 221
equilibrium conditions 148–9
four forces 147–50, 156–60
height effects 166
lift 147–50, 156–60
range flying 167–73, 174–7, 221
rotation prevention 149
tail planes 150–62
and thrust 147–50, 156–60
trimming 150
weight effects 147–50, 156–60, 166–7
lift 67–8, 78–80
coefficient 80–1, 90–1
high-lift devices 111–12, 191–3
and level flight 147–50, 156–60
lift curve 81
lift/drag ratio 83–4, 91, 169–70
and tail planes 151, 154, 156
zero lift line 71, 76–7
see also aerofoils
lift dumpers 116, 212
lifting bodies 374
load factors 236
loads
during a turn 239–41
on tail planes 154, 161–2
on wings 193–4, 196–7, 206
longitudinal axis 230
longitudinal control 273
longitudinal stability 259–61
loops 234, 247–8
inverted loops 253
lower atmosphere 377

Mach Angle 343–6
Mach Cone 345, 358
Mach Line 345, 350
machmeter 44, 312–13
Mach Numbers 305–9, 312–14, 325–7, 330–3
critical Mach Number 307–9, 325, 330–3

magnetos 289, 290
manometers 22, 60–1, 72
manual controls 274
mass 2, 7, 8
  of air 29–30
mass balance control 279–80
materials used in aeroplane construction 368–70
measurement *see* units of measurement
melting points 368
mesosphere 378
meteorites 380
meteorology 34
millibars 21
moment of a force 24–5, 150
moment of inertia 255
momentum 2, 12, 20
  conservation of momentum principle 20
  and thrust 127–9
moon
  going to 396–401
  orbits 405–6
  returning from 406
motion on curved paths 12

NACA 96–8
nautical mile 8
Navier-Stokes equations 64–5
navigation systems 44, 289
neutral lift line 71
neutral stability 257
Newton
  First Law of Motion 1
  Second Law of Motion 3, 12
  Third Law of Motion 4
newton metre 17, 168
newtons (N) 2, 8
nitrogen 36
normal axis 230
nose-dives 251–2
  pulling-out 235–6
NOTAR helicopters 200

oil flow technique 62–3
orbits
  circular 389–93
  elliptical 294–6
  around the moon 405–6
overshooting 208
oxygen 36–7

parallelogram of forces 24
parasite drag 46
Particle Image Velocimetry (PIV) 62
pascal 21
path of a projectile 384–6
patterns and development of shock waves 322–4
performance 218–20
  altitude effects 222–4
  power requirements 219–20
  and weight 225–6
perigee 388
piston engines and altitude 171–2, 226
pitch of a propeller 137–8, 139
  pitch angle 132, 134, 136
  variable pitch 141–2
pitching 230
pitching moment 78–80, 85–7, 260
pitot tube 41
pitot-static tube 41–2
plan area 79
plumb-bob 241
polygon of forces 24
potential energy 18
pound mass 8
pounds force 2
power 17–20
  absorption 142–3
  measurement 17, 130
  and performance 91, 218–20
  and transonic flight 311–12
  see also engine power
powered controls 274, 285–6, 334
pre-flight checks 289–90
pressure
  absolute pressure 61
  atmospheric pressure 30–1, 37, 38–9
  centre of pressure 74, 75–8, 80, 85–7, 91–2, 148
  and density variation 22, 33–4
  distribution on aerofoils 72–4
  distribution of shock waves 324–8
  dynamic pressure 40, 41, 42, 45
  fluid pressure 20–1
  gauge pressure 61
  and lifting surfaces 67–70
  measurement 21, 60–2
  plotting 72
  sea-level atmospheric pressure 38–9
  stagnation pressure 40, 41
  static pressure 40, 41, 45
pressure transducers 61–2
pressurised cabins 33, 36–7
profile drag 46
projectiles 384–9
propeller propulsion 127, 131–46
  advance per revolution 134–7

altitude effects 171–2, 226
blade angle 132, 134, 136, 143
constant-speed propellers 141
contra-rotating propellers 143, 144, 146
efficiency 138–40, 311
endurance flying 173–4
engine checks 292
feathered blades 142
gyroscopic effects 145
helix angle 132, 134, 139
and high-speed flight 129, 311–12
number of blades 142–3
pitch of a propeller 137–8, 139
pitch angle 132, 134, 136
variable pitch 141–2
power absorption 142–3
power curves 218–20
power measurement 130
pusher propeller 131
range flying 167–8
resistance force 131
shape of blades 142–3
slipstream 143–4, 145
solidity of propellers 143
swing on take-off 145–6
thrust force 131, 141
tip speed 140–1
torque forces 131, 145
tractor propeller 131
turboprop engines 122–3, 127, 226
propulsion systems 121–46
gas turbines 121
high by-pass engines 129–30
jet engines 121–5, 129–30, 174–7, 312
and performance 227
ram effect 121, 124
ramjets 123–4, 355
reciprocating engines 168
rocket propulsion 125–7
scramjets 124–5
turbofans 129–30
turbojets 121–2, 227
turboprops 122–3, 127, 226
turboshaft 122–3
see also propeller propulsion
pulling-out of dives 235–6
pulling-up after landing 210
pusher propeller 131

quadrantal system 39

radius of turn 246
RAF (Royal Aircraft Factory) 96
ram effect 121, 124
ramjet engines 123–4, 355
range flying 167–73, 174–7, 221
air to fuel mixture 167
altitude effects 171–2, 176
drag minimisation 169–71, 175–6
energy-work conversion 167–8
jet propulsion 174–7
propeller propulsion 167–8
wind effects 173
rate of climb 220–1, 291
rate of roll 244–5
re-entry into the atmosphere 406–8
reactive forces 1–2
reciprocating engines 168
reference point 86, 92
refuelling in the air 216
relative density 33
resistance *see* air resistance; drag; friction
resistance force 131
resultant moment 150
reversal of controls 333–4
Reynolds Number 320, 412, 431–4
rocket propulsion 125–7
launching a space-craft 401–4
rocket-assisted take-off 216
roll control 273
rolling 250, 315
rotation prevention 149
Royal Aeronautical Society 318
rudder controls 245–6, 274

sailplanes 185
satellites 384–9
scale effects 63, 431–4
schlieren method 302–4
scramjet engines 124–5
sea-level atmospheric pressure 38–9
Second Law of Motion 3, 12
seeding wind tunnels 62
servo system 276–7
shape of propeller blades 142–3
shock drag 305
shock stall 304–5, 315, 316–17
shock waves 299, 300–5, 322–8, 342–6
patterns and development 322–4
pressure distribution 324–8
sonic bangs 328–30
stagnation point 324
shooting stars 380
short take-off and landing 198–203
sideslipping 230, 250–1, 264
skidding 230
skin friction 50–2
slab tailplanes 156, 333

slimness of design 308, 330
slipstream 143–4, 145
slot-cum-aileron control 284
slots 109–14, 191–2, 196–7
slug 8
smoke tunnel 297
solidity of propellers 143
sonic bangs 328–30
space flight 377–410
circular orbits 389–93
elliptical orbits 394–6
escape from the earth 382–4
going to the moon 396–401
launching a spacecraft 401–4
law of universal gravitation 381
moon orbits 405–6
projectiles 384–9
re-entry into the atmosphere 406–8
returning from the moon 406
satellites 384–9
sub-orbital flight 409–10
upper atmosphere 377–81
spars 92
speed
ground speed 35–6
measurement 8, 36, 62
see also air speed; velocity
speed brakes 119
speed of sound 28, 53, 297–309
shock waves 299, 300–5, 322–8, 342–6
temperature variations 307
spins 248–9
inverted spin 253
Spitfire 308, 330
spoilers 116, 284, 334
spring tabs 277
stability 257–73
artificial stabilisation systems 257
and centre of gravity 259, 265
degrees of stability 258
dihedral angle 260–1, 262–4
directional stability 268–73
dynamic stability 286
fin area 267
lateral stability 262–4, 267, 272–3
longitudinal stability 259–61
neutral stability 257
pitching moment 260
power-off/power-on conditions 258
stick-fixed/stick-free conditions 258
at supersonic speeds 370–1
and the tail plane 260–1
and wings 259, 265–6
stagnation point 74, 324
stagnation pressure 40, 41
stalling 82–3, 110, 236, 281, 293
and altitude 315–18
shock stall 304–5, 315, 316–17
speed 190–1, 315–18
warning device 290
starting vortex 106
static pressure 40, 41, 45
static thrust 141
static tube 41–2
stick pusher 285
stick shaker 285
stratosphere 31, 33, 377
streamlining 46–50
subsonic airflow 68–70, 99, 106, 356
suction gauge 288
supercharged engines 37
supersonic speeds 28, 310–11, 342–76
airflow 342–57
body shapes 366–8
boundary layer 357
control 370–1
kinetic heating 368–70
Mach Angle 343–6
Mach Cone 345, 358
Mach Line 345, 350
materials used in aeroplane construction 368–70
shock waves 342–6
stability 370–1
and temperature 368–70
wing shapes 357–66
surface heating 368–70
surface skimming vehicles 203–6
sweepback 267, 285, 330–3, 338, 357–60
swing on take-off 145–6
Système International (SI) 7–8

tabs 277
tachometer 288
tail parachutes 210, 211
tail planes 150–62, 260–1
angle of attack 154
conditions of balance 156–60
downwash 155–6
lift 151, 154, 156
loads on 154, 161–2
slab tailplanes 156, 333
speed effects 154
tail load 161–2
trim tab 151, 276
see also elevators
tail-first aeroplanes 152
tail-less aeroplanes 152

tailerons 197
take-off 147, 215–17, 291
  and flaps 216
  short take-off and landing 198–203
  swing on take-off 145–6
  vertical take-off and landing 198–203
tapered wings 108–9
taxying 291
temperature
  atmospheric changes 31–3, 34, 377–80
  and density 33–4
  lapse rate 31
  melting points 368
  and the speed of sound 307
  and supersonic flight 368–70
  units of measurement 7
terminal velocity 7
test flights 320–1
theories, and experimentation 318
thermosphere 378
thickened trailing edges 338
Third Law of Motion 4
throttle control 227, 289, 292
thrust 1, 120, 131, 141
  and drag 120
  and level flight 147–50, 156–60
  and momentum 127–9
  and speed 129, 130
  static thrust 141
  see also propulsion systems
thrust vectoring aircraft 202–3
tilt-rotor aircraft 203
tip speed 140–1
tip stalling 332
torque forces 131, 145
total resultant force 75
tractor propeller 131
transition point 95
transonic speeds 28, 297–340
  area rule 335–6
  and control 333–4
  drag rise 309, 311–12
  and elevators 333
  Mach numbers 305–9, 312–14, 325–7, 330–3
  pilot's viewpoint 312–15
  power requirements 311–12
  shock stall 304–5, 315, 316–17
  shock waves 299, 300–5, 322–8
  sonic bangs 328–30
  speed of sound 28, 53, 297–309
  vortex generators 336–7
triangle of forces 24
trim 150, 212–14
trim tab 151, 276
tropopause 31, 33
troposphere 31, 377
True Air Speed (TAS) 43
turbofan engines 129–30
turbojet engines 121–2, 227
turboprop engines 122–3, 127, 226
turboshaft engines 122–3
turbulent boundary layer 50–2
turn and bank indicator 241–2
turning and banking 236–46, 272–3, 293–4
  angles 239–42
  climbing turns 243–4
  flat turns 245
  gliding turns 243–4
  holding off bank 243
  loads during a turn 239–41
  plumb-bob 241
  radius of turn 246
  rate of roll 244–5
  rudder controls 245–6
  and wind direction 241

U-tube manometer 22, 60–1, 72
undercarriage 150, 192, 216
  wheel brakes 210–11
undershooting 208
uniform acceleration 9
uniform velocity 9
units of measurement 7–8
  American Federal units 8
  British Imperial System 8
  forces 2, 8
  mass 2, 8
  power 17
  pressure 21
  speed 8, 36
  Système International (SI) 7–8
  temperature 7
  work 17
up-and-down air currents 34–5
upper atmosphere 377–81
upside-down flight 252–3

variable camber 109
variable sweep 361–2
vector quantities 24
velocity 2, 9–10, 78, 382–4
  circular velocity 387, 390–1, 393–4
  escape velocity 383, 393–4
  and the path of flight 384–6
  of the slipstream 143–4
  terminal velocity 7
  uniform velocity 9

velocity squared law 52–3
venturi tube 44–6, 353–4
vertical take-off and landing 198–203
vertical velocity 186, 188
viscosity properties of air 34
vortex generators 336–7
vortices 100, 106

wash-out 283
watt 17
wave drag 305
weather conditions 255
weight 5–6
  and gliding 184–5
  and level flight 147–50, 156–60, 166–7
  mass weight 7
  and performance 225–6
wheel brakes 210–11
wind tunnels 55–9, 431–4
  balance unit 58–9
  errors in the model 63
  flow visualisation 62–3
  high-speed tunnels 309, 318–20
  interference from the walls 63
  jockey weights 59
  oil flow technique 62–3
  pressure measurement 60–2
  scale effects 63, 431–4
  seeding 62
  speed measurements 62
  types of 55–8, 319–20
winds 34–6
  gradient 35, 187–8
  and range flying 173
  and turning and banking 241
  in the upper atmosphere 380
  wind speed 187
wing drag 46
wing loading 193–4, 196–7, 206
wing shapes
  bi-convex 363
  delta 360
  flying wing 160
  forward-swept 268
  leading-edge droop 338
  and stability 259, 265–6
  and supersonic flight 357–66
  sweepback 267, 285, 330–3, 338, 357–60
  tapered 108–9
  thickened trailing edges 338
  variable sweep 361–2
  see also aerofoils
wing-tip vortices 100, 106
work 17–20, 167–8
  units of measurement 17
Wright Flyer 152

Yakovlev Yak 36 203
yawing 230, 273, 315

X15 374

Zap flaps 116
zero lift line 71, 76–7